Piping Design for Process Plants

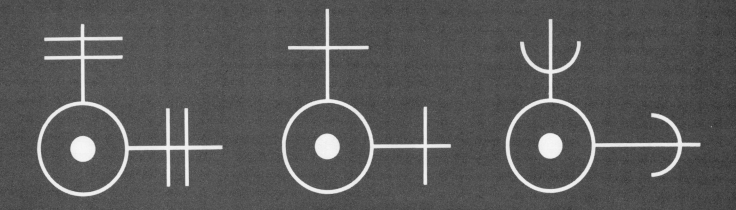

Piping Design for Process Plants

HOWARD F. RASE, *Professor of Chemical Engineering, University of Texas, Austin, Texas*

original illustrations by JAMES R. HOLMES, *Associate Professor of Drawing, University of Texas*

JOHN WILEY & SONS, INC., NEW YORK · LONDON

to my brothers Henry, Louis, and Robert

Preface

Piping constitutes 25 to 35 per cent of the material costs of a process plant; it requires 30 to 40 per cent of the erection labor, and consumes 40 to 48 per cent of the total engineering man-hours. Despite this overwhelming importance there exists no single reference manual devoted to piping design for process plants. Piping designers outnumber all other design specialists; yet, unlike their colleagues engaged in structural or electrical design, they must accumulate information and data from a wide assortment of handbooks, periodicals, and manufacturers' literature. They need a dependable manual written specifically about process plants, containing under one cover all the information required in the routine design of piping. *Piping Design for Process Plants* has been written to fulfill this important need.

All piping designers, though often from widely differing educational backgrounds, have three things in common: (1) A remarkable ability to visualize the most complex three-dimensional systems. (2) A keen insight into problems and a desire for intelligent explanations. (3) A busy schedule that requires the best materials and time-saving methods. Based on these facts the book has been arranged and presented as follows:

1. The reasons for design practices and the background and limitations of design calculations are given.

2. Convenient and clearly presented design charts and tables are included.

3. Examples of design calculations, with every step shown, are grouped conveniently.

4. Pictorial representation is used freely throughout the book in place of wordy explanations. Original illustrations were drawn by one of America's top technical artists.

5. The book is limited to convenient size by avoiding lengthy reprinting of infrequently used portions of codes available in any design office. Complete lists of codes have been included and many frequently used portions have been reproduced in abbreviated easy-to-use form.

6. The complex field of piping stress analysis has been covered in detail in other books. The treatment here is slanted toward minimizing, through proper layout, the number of lines which must be subjected to detailed analysis.

7. A thorough index with ample cross-referencing is provided.

In summary, *Piping Design for Process Plants* was written and arranged so that a single item of information can be found instantly and its proper use immediately comprehended. It is hoped that through use of this book piping designers, and all engineers concerned with piping problems, will find their work load lightened and their effectiveness enhanced.

The author is grateful to Mr. M. H. Barrow, Technical Manager, Soc. Foster Wheeler Italiana (Foster Wheeler Corporation), who called attention to the need for this book and gave competent technical advice, particularly in the preparation of Chapters 6, 7, 8, 9, and 10. Mr. W. H. Hulle (piping design consultant) and Mr. M. H. Barrow afforded me the benefit of their years of engineering design experience by critically reviewing the manuscript.

HOWARD F. RASE

Austin, Texas
August, 1963

Contents

List of figures

List of tables

1

A brief history

Whenever civilizations achieve a stage of development that inspires a desire for daily conveniences, some form of piping is invented to meet these new needs. Thus the use of pipe antedates recorded history. Clay pipe dating from 4000 B.C. was found in the ruins of Babylon and a lead pipe distribution system, complete with bronze plug valves, can be seen in the ruins of ancient Pompeii just as it was the day in 87 B.C. when the little Roman city was covered by volcanic ash. Wooden pipes made from bored logs and staves, as well as hollowed stone pipes, have served in many parts of the world for centuries. Wooden pipe systems were used as late as the early 1900's in the United States and continue to have certain limited applications today.

The use of cast iron for water conduit evidently followed the invention of cast-iron cannons. Such cannons were being made in 1313 in Ghent, and there is evidence that iron pipe was cast in 1455 in Germany. In 1562 cast-iron pipe was laid to supply water for the Rathans fountains. Louis XIV (17 century), in an attempt to copy the fountains of Italian villas, installed cast-iron pipe to bring water to the fountains of Versailles palace in France. This system continues in use today, as do many other early installations of cast-iron pipe both in Europe and the United States. Common use of cast-iron pipe for water conduit began in the first decade of the 19th century. Because of its durability in underground installations it continues to be the piping specified for new water systems.

Iron may have been used for pipe or accessories much earlier than history records, but, probably because of corrosion, evidence of its use in piping has not been found. Iron artifacts, however, which show a knowledge of heat treating and are believed to have been buried more than 50 centuries, have been found in the pyramids.

The development and wide use of the steam engine, beginning in the latter part of the 18th century, created a need for piping materials capable of withstanding higher pressures and temperatures than had previously been encountered. Steel pipe soon became increasingly common and many methods of production were devised such as lap-welding plate formed in a cylindrical shape, a procedure in use today. After World War I as the power and process industries demanded even higher pressure materials, seamless pipe formed from solid billets of steel became common.

Although steel pipe gained increasing acceptance in the late 18th and 19th centuries and was manufactured in large quantities by the early 20th century, valves and fitttings continued to be made from cast iron. Pipe-joining methods at this time were also antique in origin. Bolted and flanged cast-iron connections and threaded connections had been used for several centuries. Flanged-end connections for steel pipe were either forged integrally or the pipe was threaded and a threaded flange attached. Joints in runs of pipe which were not flanged were made by threaded coupling connections. Gasket materials and gasketed joints were not reliable for high pressure and temperature, and many special methods were devised for sealing joints, some of which are still used today. The early Van Stone joint was one of the first such reliable pressure connections.

Although reliable pressure pipe and accessories have been manufactured over thirty years, strong and leak-proof pipe-joining methods other than bolted flanges were unknown until the relatively recent use of welding for joining pipe. The first portable welding method utilized oxygen-acetylene torches which are now used primarily for cutting pipe. Although it was possible to obtain a reasonably reliable welded joint, the spread of the high-temperature zone and atmosphere of the gas flame caused metallurgical problems. Construction of modern piping systems became possible only after portable arc-welding apparatus and the "coated" electrode were developed. Some types of welds, of course, were only possible under shop-controlled conditions. Within the past few years, however, welding methods for all types of materials have been developed to yield reliable welds under any kind of construction conditions. This, of course, simplifies construction and erection problems, thereby reducing initial plant cost.

Modern high-pressure connections for the ranges within the present "American Standard Code for Pressure Piping" were fully accepted only a few decades ago, just prior to the issuance of the first code. The quickening pace of steam-power plant and process plant expansion had required steadily increasing design temperatures and pressures; and standardized piping, valves, and fittings which could withstand these conditions were required. This need was fulfilled by the "ASA Code for Pressure Piping," first issued in 1935 and constantly revised and improved since that date. An important forerunner to this code was a "Code for Safety Rules and Regulations Covering the Installation of High and Low Pressure Steam Lines" issued by the Ohio State Department of Industrial Relations in 1925. Similar codes and standard practices were introduced in Europe. Standardized piping practice within individual countries is now well accepted, a sharp contrast to practices of thirty to forty years ago when each plant operator had his own carefully guarded special designs of flanges and joints. Traces of these unique practices may still be detected in some valve and flange manufacturers' catalogs in which special-drilled flanges or undrilled flanges may be obtained as alternates to the standard flanges.

The designer can now proceed with confidence in the design of a modern process plant system by specifying standardized piping and accessories, modern welding practices, and a wide assortment of valves and fittings for all types of operating conditions. Technical advances and reliable rating procedures have eliminated the dangers of failures in piping systems, and standard dimensions permit freedom in selecting the best and most economical offerings of a number of manufacturers for combining into a single system. A designer freed in this manner from numerous bothersome details is better able to apply his talents to carefully designing piping systems, uniquely adapted to the specific needs of the process.

2

Aids in selecting pipe, valves and fittings

Index for Tables and Figures for Valve and Fitting Selection—page 17

This chapter is largely composed of tables and diagrams assembled and prepared as convenient aids to selecting and applying pipe, valves, and fittings. Tables for determining the proper material of construction for corrosive service have been omitted purposely. Corrosion is a complex phenomenon which cannot be overcome by casual use of tables or handy selector wheels. Comprehensive tabular presentations of corrosion data which can be applied by knowledgeable users have been published, and the reader is referred to these as well as to any competent corrosion engineer.[1, 2, 3]

METALLIC PIPE AND TUBING

Pipe and tubing are both tubular products, but the terms as used commercially have more specific meanings.

Pipe

Tubes produced in accordance with the sizes given in Table 3.2, page 48, and in American Petroleum Institute standards (not shown) are called pipe. The outside diameter of any given nominal size is the same for all weights in that size. Thus the inside diameter varies for pipe of the same nominal size but different wall thickness. Pipe 12 in. and smaller is commonly designated by a nominal diameter which approximates but is not equal to the inside diameter of the Schedule 40 or standard weight size. Pipe 14 in. and larger has outside diameters equal to the pipe size designation.

Pipe thickness is expressed in terms of schedule numbers in accordance with the American Standards Association standard (Table 3.18, page 72). Prior to the introduction of schedule numbers the terms standard, extra strong, and double-extra strong served to designate pipe thickness. Sizes through 10 in. in Schedule 40 are the same as standard weight and sizes through 8 in. in Schedule 80 are the same as extra strong. Double-extra strong has been discontinued in some sizes and Schedule 160 used instead. Currently, in sizes 10 in. and below, it is common to designate pipe according to schedule numbers. In larger sizes actual wall thicknesses are usually given.

Permissible tolerances for pipe apply to thickness only. The usual mill tolerance for steel pipe is -12.5%, which means that the actual thickness may be 12.5% under the standard thicknesses given in Tables 3.2 and 3.18 (see Table 3.25 for other tolerances).

Tubing

All other tubes not produced in standard pipe sizes are called tubes or tubing. Sizes are designated by the outside diameter, and each size is offered in a variety of inside diameters. Tolerances can apply to various dimensions as the use demands.

Manufacture

Manufacturing processes for pipe and tubing are depicted schematically in Fig. 2.1 through 2.6. In addition to the methods shown, large-diameter pipe—that is, 24 to 36 in. outside diameter—is made by forming a circular ring from steel plate and welding the seam by the submerged-arc process. Pipe 6 to 36 in. in diameter

3

is also made by welding a spiral seam produced by forming continuous steel skelp into circular shape.

Common Materials and Sizes

Although approximately 260 different metallic materials are listed in the piping code, only about 40 are readily obtainable. Others require special orders. The most commonly used are shown in the cost comparison Fig. 2.7 on page 13.

Table 2.1 gives a summary of common size practices.

Selecting Pipe

The major variables in pipe selection are temperature, pressure, corrosive influences, and cost. The corrosion problem is complex and cannot be resolved by reference to some handy tables or charts presumably designed for rapid selection of the correct material. The corrosion resistance of materials often varies radically with changes in conditions such as temperature and degree of turbulence, and these conditions can never be anticipated in

the usual tabulation. Competent engineers who specialize in corrosion problems should be consulted for recommendations whenever corrosive fluids are being handled.

After the possible choices that possess the necessary resistance to corrosion are known, the final selection of material can be made on the basis of temperature, pressure, and cost. Figure 2.7 and Table 2.2 are helpful for this purpose. Figure 2.7 shows the relative price for pipe, flanges, and fittings for various piping materials. The ability of the pipe to withstand the anticipated operating pressure and temperature varies with materials and is particularly marked at higher temperatures. This variation is directly related to the allowable stress ("S") specified by the American Standards Association Code (see Table 3.3). Thus a true measure of relative economy is the allowable stress at each temperature divided by the relative cost. The index obtained is essentially the amount of "S" you purchase for a dollar. Other factors, such as corrosion resistance and availability, being equal, one selects the pipe among the choices having the highest "S" per dollar rating. Table 2.2 presents these indecies for a number of common piping materials. Tabulations such as these and Fig. 2.7 should be checked periodically and corrected for significant price changes.

NON-METALLIC PIPE AND TUBING

Tables 2.3, 2.4, and Fig. 2.8 (pages 14 to 16) compare various plastic materials, glass, and graphite with well-known metal piping, and should aid the designer in orienting his thinking when dealing with these newer materials. In addition, a few cautious generalizations may prove helpful.

1. In the low temperature and pressure range where plastic pipe is applicable it has the advantages of light weight, low installation cost, low first cost compared to corrosion-resistant metal alloys, and good resistance to many corrosive chemicals.

2. Major types of plastic pipe are shown in Table 2.3. These plastics have varying degrees of resistance to attack by acids, alkalis, and organic compounds. Fluorocarbon plastics are the most resistant to all types of attack. In general it may be said that plastics supplement metals in the range where metals are most strongly attacked. Dilute acids, for example, do not attack most plastics but strongly attack metals. In contrast, strong acids and alkalis harm plastics but do not affect many metals. Organic compounds such as petroleum products, aromatics, and chlorinated hydrocarbons are also readily handled in metallic piping, but not in all plastics.

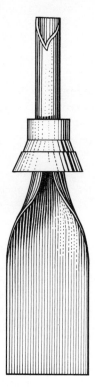

FIG. 2.1. Butt-welded pipe manufacture. Skelp (steel plate) of proper size and with beveled edges is heated to welding temperature (approximately 2600°F) and drawn by tongs through funnel-shaped dies. The dies force the beveled edges to meet squarely and become welded. This operation is followed by a series of rolling operations which compress and elongate the pipe to uniform outside diameter. Continuous mills employ continuous strips of skelp and operate at 300 to 500 ft/min. Range of pipe sizes produced: ½ to 4 in.

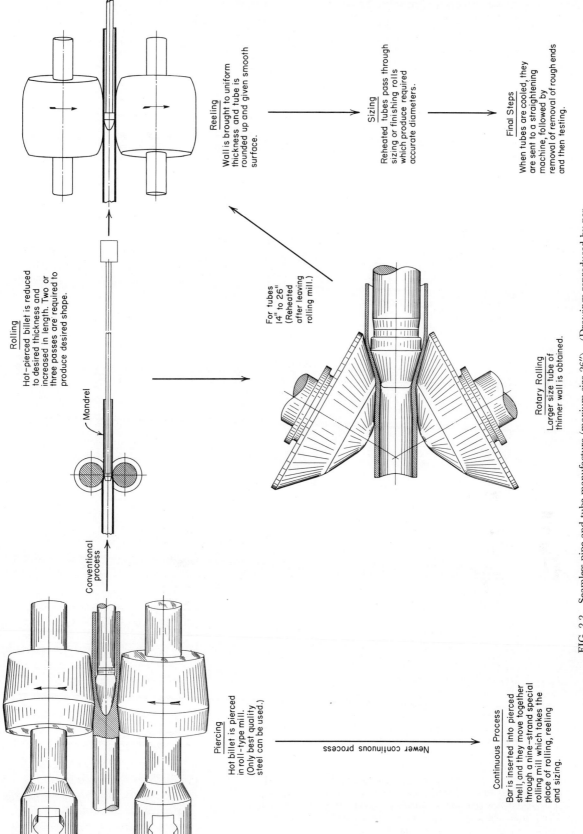

Rolling
Hot-pierced billet is reduced to desired thickness and increased in length. Two or three passes are required to produce desired shape.

Mandrel

Conventional process

Reeling
Wall is brought to uniform thickness and tube is rounded up and given smooth surface.

Sizing
Reheated tubes pass through sizing or finishing rolls which produce required accurate diameters.

Final Steps
When tubes are cooled, they are sent to a straightening machine, followed by removal of removal of rough ends and then testing.

For tubes 14" to 26" (Reheated after leaving rolling mill.)

Rotary Rolling
Larger size tube of thinner wall is obtained.

Piercing
Hot billet is pierced in roll-type mill. (Only best quality steel can be used.)

Continuous Process
Bar is inserted into pierced shell, and they move together through a nine-strand special rolling mill which takes the place of rolling, reeling and sizing.

Newer continuous process

FIG. 2.2. Seamless pipe and tube manufacture (maxium size 26″). (Drawings reproduced by permission of National Tube Division, United States Steel Corporation.)

6

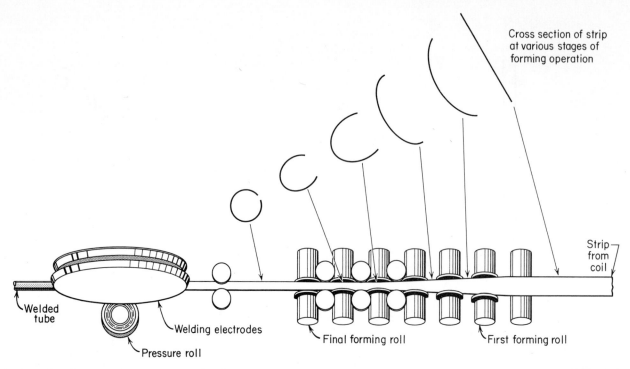

FIG. 2.3. Manufacture of electric resistance-welded pipe and tubing. Used for sizes 4 in. OD and under. Flat steel plate of required width is shaped cold on six to nine pairs of forming rolls. Resistance to flow of electric current across seam heats edges which are forced together, forming weld. Metal flash from the self-welding is removed in subsequent finishing. (Reproduced by permission of National Tube Division, United States Steel Corporation.)

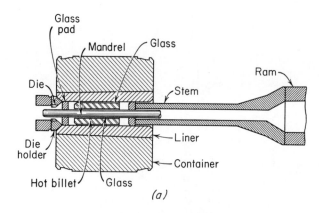

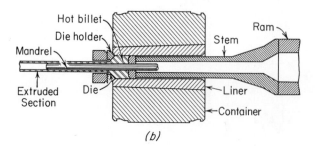

FIG. 2.4. Extrusion process for seamless steel pipe and tubing. This process is particularly adaptable for heavy-wall pipe and hard-to-pierce alloys. (*a*) The hot billet is covered with molten glass for lubrication as it enters extruder. A mandrel, also covered with glass, advances through the billet. (*b*) The extrusion stem forces pierced billet against glass pad and through the die. (Reproduced by permission of National Tube Division, United States Steel Corporation.)

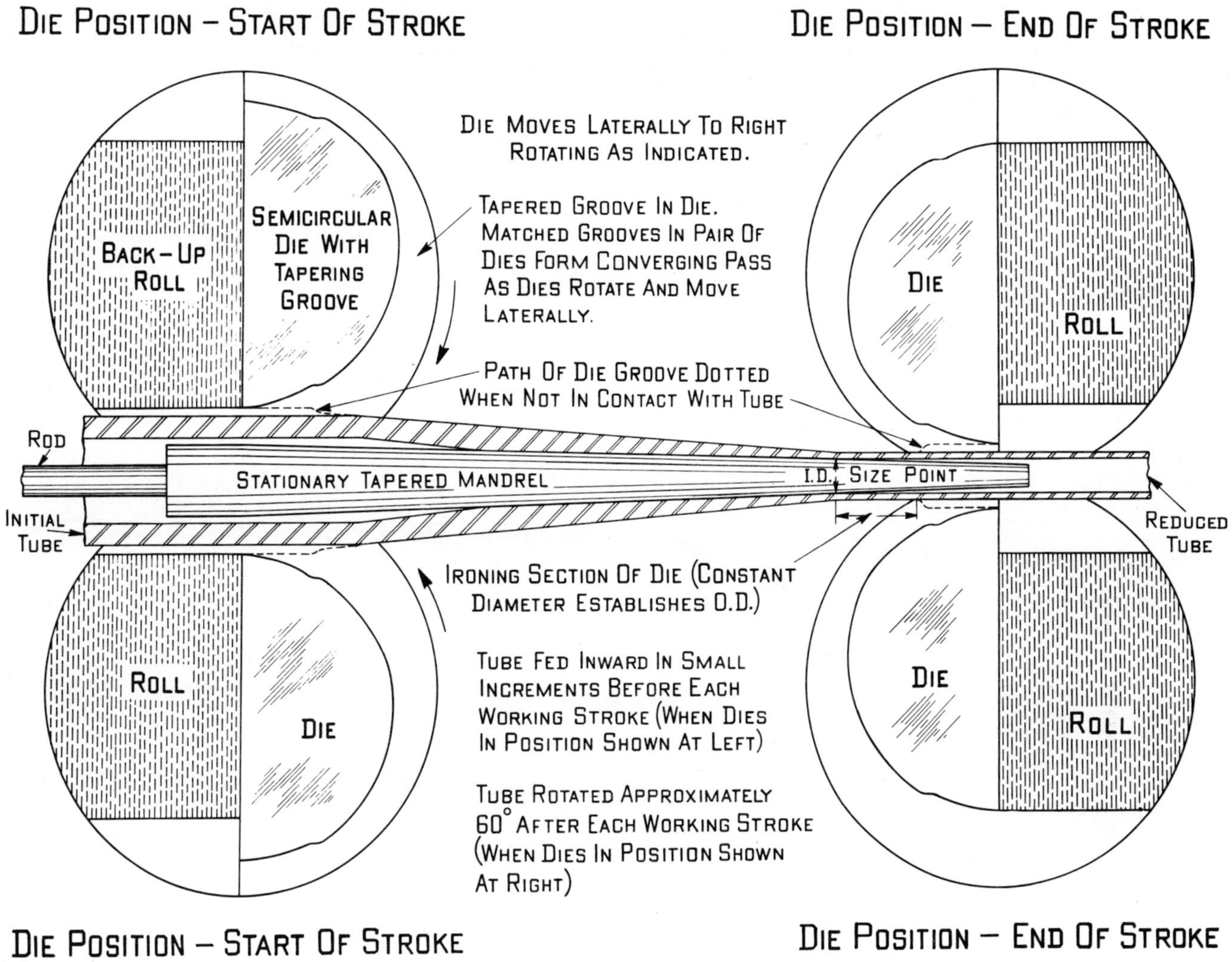

DIE POSITION – START OF STROKE

DIE POSITION – END OF STROKE

DIE MOVES LATERALLY TO RIGHT
ROTATING AS INDICATED.

BACK-UP
ROLL

SEMICIRCULAR
DIE WITH
TAPERING
GROOVE

TAPERED GROOVE IN DIE.
MATCHED GROOVES IN PAIR OF
DIES FORM CONVERGING PASS
AS DIES ROTATE AND MOVE
LATERALLY.

DIE

ROLL

PATH OF DIE GROOVE DOTTED
WHEN NOT IN CONTACT WITH TUBE

ROD

STATIONARY TAPERED MANDREL

I.D. SIZE POINT

INITIAL
TUBE

REDUCED
TUBE

IRONING SECTION OF DIE (CONSTANT
DIAMETER ESTABLISHES O.D.)

ROLL

DIE

TUBE FED INWARD IN SMALL
INCREMENTS BEFORE EACH
WORKING STROKE (WHEN DIES
IN POSITION SHOWN AT LEFT)

DIE

ROLL

TUBE ROTATED APPROXIMATELY
60° AFTER EACH WORKING STROKE
(WHEN DIES IN POSITION SHOWN
AT RIGHT)

DIE POSITION – START OF STROKE

DIE POSITION – END OF STROKE

FIG. 2.5. Tube reducing by Rockrite process. A large reduction is made in a single pass. (Reproduced by permission of National Tube Division, United States Steel Corporation.)

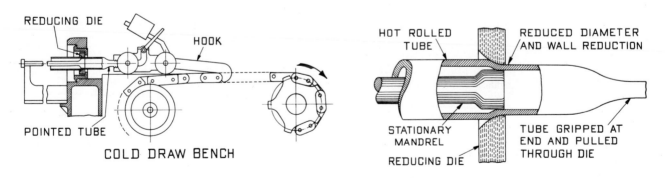

REDUCING DIE

HOOK

POINTED TUBE

COLD DRAW BENCH

HOT ROLLED
TUBE

REDUCED DIAMETER
AND WALL REDUCTION

STATIONARY
MANDREL

REDUCING DIE

TUBE GRIPPED AT
END AND PULLED
THROUGH DIE

COLD DRAWING PROCESS

FIG. 2.6. Cold-drawing process. This process is used for producing pipe and tubing smaller than 1¾₆ in. OD or walls thinner than 0.083 in. and also for sizes requiring close tolerances and superior surface. Tubes are annealed, pickled, washed, and lubricated between each drawing and are final-annealed or heat-treated after the last pass. (Reproduced by permission of National Tube Division, United States Steel Corporation.)

Many plastics also deteriorate gradually when constantly exposed to sunlight. Manufacturers' recommendations and plant experience should be used as guides in selecting plastic piping.

3. Plastics can be used in high-pressure low-temperature service as liners for steel pipe.

4. Synthetic rubbers of various types have good resistance to organics and are frequently used as steel pipe liners or as hose or pipe at low pressures.

5. Glass resists attack of more acids and acidic material than any other pipe. It cannot, however, be used for hydrofluoric acid, fluoride containing phosphoric acid, or alkaline solutions over 100°F. For most other materials glass is useable up to 450°F and has the advantage of easy visibility of flowing contents which may be necessary in some processes. (See Table 3.12 for more data.)

VALVES AND FITTINGS

In compiling the following tables and figures it was assumed that the reader is familiar in a general way with the appearance and function of valves and fittings. There are, however, so many different types of valves and fittings that a reader is often bewildered when attempting to evaluate each type objectively. What features should be compared? What are the advantages and disadvantages? The purpose of the tables and figures given on the remaining pages of this chapter is to aid in answering these questions in an unbiased manner.

Designers should always be receptive to new developments. Large operating companies have planned programs for testing new designs of valves and other piping materials. Such a practice assures that the dependable new developments are used as soon as feasible.

SELECTING FOR ECONOMY

As the process industries inevitably become more competitive, economy in design of a plant is an absolute necessity. Because of the large proportion of plant costs attributable to piping, the piping designer plays a key

TABLE 2.1

PIPE AND TUBING SIZE PRACTICES

PIPE

(See ASA Standard pipe sizes in Table 3.2, page 48.)

	Sizes	Thicknesses
Process Plant Practice with carbon steel	⅜, 1¼, 2½, 3½, and 5″ pipe not commonly used	Schedule 40, 80, 160 standard, extra strong and double extra strong are common thicknesses.
with stainless steel	⅜, 1¼, 2½, 3½, and 5″ pipe not commonly used	Schedule 5, 10, and 40 most common (Schedule 80 also produced).
Manufacturer's Practice	Pipe manufacturers produce and sell pipe in tonnage lots and will make any size within their capabilities. Small lots must be purchased from jobbers, and their stock is most likely to reflect common practice. Jobbers publish stock lists regularly to aid in selecting available pipe.	The following schedule numbers are not currently supplied by the mills except in certain minimum tonnages. Schedule 60 in 12″ and 14″ Schedule 80 in 18″ and 24″ Schedule 100 in 10″, 20″ and 24″ Schedule 120 in 14″, 18″, 20″ and 24″ Schedule 140 in 10″, 12″, 14″, 16″, 18″, 20″ and 24″ Schedule 160 in 14″, 16″, 18″, 20″ and 24″

TUBING

The following tables gives typical tubing sizes and manufacturing capabilities for stainless steel. Large quantities of tubing should be procured directly from mills. Any desired thickness within the limitations of the mill can be produced without penalty. Common outside diameters in inches are:

¼, ⅜, ⁷⁄₁₆, ½, ⅝, ¾, ⅞, 1⅛, 1¼, 1⅜, 1½, 1⅝, 1⅞, 2 (larger sizes are also produced).

Small lots of tubing must be purchased from jobbers whose stocks reflect the tubing sizes and thicknesses commonly used in the area. Request jobber's stock lists.

TABLE 2.1 (*Continued*)
TYPICAL MANUFACTURER'S STANDARDS FOR TUBING
Theoretical weight in pounds per foot—Cold Drawn—round*

Thickness		Outside diameter in inches															
Decimal of inch	B.W.G. or fraction	¼	⅜	⁷⁄₁₆	½	⅝	¾	⅞	1	1⅛	1¼	1⅜	1½	1⅝	1¾	1⅞	2
.028	22	.0664	.1038	.1226	.1441	.1785	.2159	.2533	.2907	.3280	.3654	.4028	.4402	.4776	....	....	....
.035	20	.0804	.1271	.1506	.1738	.2205	.2673	.3140	.3607	.4074	.4542	.5009	.5476	.5943	.6411	.6878	.7345
.049	18	.1052	.1706	.2036	.2360	.3014	.3668	.4323	.4977	.5631	.6285	.6939	.7593	.8248	.8902	.9556	1.021
.058	17	.1189	.1964	.2354	.2738	.3512	.4287	.5061	.5835	.6609	.7384	.8158	.8932	.9707	1.048	1.126	1.203
.065	16	.1284	.2152	.2589	.3020	.3888	.4755	.5623	.6491	.7359	.8226	.9094	.9962	1.083	1.170	1.257	1.343
.072	15	.1369	.2330	.2814	.3291	.4252	.5214	.6175	.7136	.8097	.9058	1.002	1.098	1.194	1.290	1.386	1.483
.083	14	.1480	.2588	.3147	.3696	.4805	.5913	.7021	.8129	.9237	1.034	1.145	1.256	1.367	1.478	1.589	1.699
.095	13	.1573	.2841	.3480	.4109	.5377	.6646	.7914	.9182	1.045	1.172	1.299	1.426	1.552	1.679	1.806	1.933
.109	12	....	.3097	.3830	.4552	.6007	.7462	.8917	1.037	1.183	1.328	1.474	1.619	1.765	1.910	2.056	2.201
.120	11	....	.3268	.4075	.4870	.6472	.8074	.9676	1.128	1.288	1.448	1.608	1.769	1.929	2.089	2.249	2.409
.134	10	....	.3449	.4351	.5238	.7027	.8816	1.060	1.239	1.418	1.597	1.776	1.955	2.134	2.313	2.492	2.670
.156	⁵⁄₃₂	....	.3649	.4698	.5731	.7814	.9897	1.198	1.406	1.614	1.823	2.031	2.239	2.447	2.656	2.864	3.072
.188	³⁄₁₆	....	....	....	.6264	.8774	1.128	1.379	1.630	1.881	2.132	2.383	2.634	2.885	3.136	3.387	3.638
.219	⁷⁄₃₂	....	....	....	....	.9496	1.242	1.534	1.827	2.119	2.411	2.704	2.996	3.289	3.581	3.873	4.166
.250	¼	....	....	....	....	....	1.335	1.669	2.003	2.336	2.670	3.004	3.338	3.671	4.005	4.339	4.673
.281	⁹⁄₃₂	....	....	....	....	....	....	1.783	2.158	2.533	2.908	3.283	3.658	4.033	4.409	4.784	5.159
.313	⁵⁄₁₆	....	....	....	....	....	....	1.879	2.297	2.714	3.132	3.550	3.968	4.386	4.804	5.222	5.639
.344	¹¹⁄₃₂	....	....	....	....	....	....	....	2.410	2.869	3.329	3.788	4.247	4.706	5.166	5.625	6.084
.375	⅜	....	....	....	....	....	....	....	2.503	3.004	3.504	4.005	4.506	5.006	5.507	6.008	6.508
.438	⁷⁄₁₆	....	....	....	....	....	....	....	....	3.798	4.383	4.968	5.553	6.137	6.722	7.307
.500	½	....	....	....	....	....	....	....	....	4.005	4.673	5.340	6.008	6.675	7.343	8.010

Thickness		Outside diameter in inches															
Decimal of inch	B.W.G. or fraction	2¼	2½	2¾	3	3¼	3½	3¾	4	4¼	4½	4¾	5	5½	6	6½	6⅝
.035	20	.8280	.9214	1.015	....	....	....	....	....	....	....	....	....	....	....	....	....
.049	18	1.152	1.283	1.413	1.544	1.675	1.806	1.937	2.068	....	....	....	....	....	....	....	....
.058	17	1.358	1.513	1.668	1.822	1.977	2.132	2.287	2.442	2.597	....	....	....	....	....	....	....
.065	16	1.517	1.690	1.864	2.037	2.211	2.385	2.558	2.732	2.905	3.079	3.252	3.426	....	....	....	....
.072	15	1.675	1.867	2.059	2.252	2.444	2.636	2.828	3.020	3.213	3.405	3.597	3.789	4.174	....	....	....
.083	14	1.921	2.143	2.364	2.586	2.807	3.029	3.251	3.472	3.694	3.915	4.137	4.359	4.802	5.245	....	....
.095	13	2.186	2.440	2.694	2.947	3.201	3.455	3.708	3.962	4.216	4.469	4.723	4.977	5.484	5.991	....	....
.109	12	2.492	2.783	3.074	3.365	3.657	3.948	4.239	4.530	4.821	5.112	5.403	5.694	6.276	6.858	7.440	....
.120	11	2.730	3.050	3.371	3.691	4.011	4.332	4.652	4.973	5.293	5.613	5.934	6.254	6.895	7.536	8.177	8.337
.134	10	3.028	3.386	3.744	4.102	4.459	4.817	5.175	5.533	5.890	6.248	6.606	6.964	7.679	8.395	9.111	9.289
.156	⁵⁄₃₂	3.489	3.905	4.322	4.738	5.155	5.571	5.988	6.404	6.821	7.237	7.654	8.070	8.904	9.737	10.57	10.78
.188	³⁄₁₆	4.140	4.642	5.144	5.646	6.148	6.650	7.152	7.654	8.156	8.658	9.160	9.662	10.67	11.67	12.67	12.92
.219	⁷⁄₃₂	4.750	5.335	5.920	6.505	7.089	7.674	8.259	8.843	9.428	10.01	10.60	11.18	12.35	13.52	14.69	14.98
.250	¼	5.340	6.008	6.675	7.343	8.010	8.678	9.345	10.01	10.68	11.35	12.02	12.68	14.02	15.35	16.69	17.02
.281	⁹⁄₃₂	5.909	6.659	7.410	8.160	8.910	9.660	10.41	11.16	11.91	12.66	13.42	14.16	15.66	17.16	18.66	19.04
.313	⁵⁄₁₆	6.475	7.311	8.147	8.982	9.818	10.65	11.49	12.33	13.16	14.00	14.83	15.67	17.34	19.01	20.68	21.10
.344	¹¹⁄₃₂	7.002	7.921	8.839	9.758	10.68	11.59	12.51	13.43	14.35	15.27	16.19	17.11	18.94	20.78	22.62	23.08
.375	⅜	7.509	8.511	9.512	10.51	11.51	12.52	13.52	14.52	15.52	16.52	17.52	18.52	20.53	22.53	24.53	25.03
.438	⁷⁄₁₆	8.476	9.646	10.82	11.98	13.15	14.32	15.49	16.66	17.83	19.00	20.17	21.34	23.68	26.02	28.36	28.94
.500	½	9.345	10.68	12.02	13.35	14.69	16.02	17.36	18.69	20.03	21.36	22.70	24.03	26.70	29.37	32.04	32.71

* The weights and thicknesses in this table are those adopted by manufacturers of seamless specialities. (Reproduced by permission: National Tube Division, United States Steel Corporation.)

TABLE 2.1 (*Concluded*)
TYPICAL MANUFACTURER'S SIZE LIMITATIONS ON STAINLESS STEEL TUBING*
(Dimensions in Inches)

347			304 and 321			303, 310, 316, 348, 403, 410, 430, 446, 502		
OD	Min. Wall	Max. Wall	OD	Min. Wall	Max. Wall	OD	Min. Wall	Max. Wall
.012 ($\frac{3}{250}$)	.002	.004	.012 ($\frac{3}{250}$)	.002	.004	.012 ($\frac{3}{250}$)	.002	.004
.0312 ($\frac{1}{32}$)	.002	.010	.0312 ($\frac{1}{32}$)	.002	.010	.0312 ($\frac{1}{32}$)	.002	.010
.0625 ($\frac{1}{16}$)	.003	.028	.0625 ($\frac{1}{16}$)	.003	.028	.0625 ($\frac{1}{16}$)	.003	.028
.0937 ($\frac{3}{32}$)	.004	.042	.0937 ($\frac{3}{32}$)	.004	.042	.0937 ($\frac{3}{32}$)	.004	.042
.125 ($\frac{1}{8}$)	.004	.049	.125 ($\frac{1}{8}$)	.004	.049	.125 ($\frac{1}{8}$)	.004	.049
.1875 ($\frac{3}{16}$)	.005	.065	.1875 ($\frac{3}{16}$)	.005	.065	.1875 ($\frac{3}{16}$)	.005	.065
.250 ($\frac{1}{4}$)	.005	.083	.250 ($\frac{1}{4}$)	.005	.083	.250 ($\frac{1}{4}$)	.005	.083
.3125 ($\frac{5}{16}$)	.005	.109	.3125 ($\frac{5}{16}$)	.005	.109	.3125 ($\frac{5}{16}$)	.005	.109
.375 ($\frac{3}{8}$)	.005	.125	.375 ($\frac{3}{8}$)	.005	.125	.375 ($\frac{3}{8}$)	.005	.125
.4375 ($\frac{7}{16}$)	.005	.125	.4375 ($\frac{7}{16}$)	.005	.125	.4375 ($\frac{7}{16}$)	.005	.125
.500 ($\frac{1}{2}$)	.005	.125	.500 ($\frac{1}{2}$)	.005	.125	.500 ($\frac{1}{2}$)	.005	.125
.5625 ($\frac{9}{16}$)	.006	.125	.5625 ($\frac{9}{16}$)	.006	.125	.5625 ($\frac{9}{16}$)	.006	.125
.625 ($\frac{5}{8}$)	.006	.125	.625 ($\frac{5}{8}$)	.006	.125	.625 ($\frac{5}{8}$)	.006	.125
.750 ($\frac{3}{4}$)	.006	.035	.750 ($\frac{3}{4}$)	.006	.035	.750 ($\frac{3}{4}$)	.006	.035
.875 ($\frac{7}{8}$)	.006	.035	.875 ($\frac{7}{8}$)	.006	.035	.875 ($\frac{7}{8}$)	.006	.035
1.00 (1)	.007	.035	1.00 (1)	.007	.035	1.00 (1)	.007	.035
1.125 ($1\frac{1}{8}$)	.008	.035	1.125 ($1\frac{1}{8}$)	.008	.035	1.125 ($1\frac{1}{8}$)	.008	.035
1.250 ($1\frac{1}{4}$)	.010	.035	1.250 ($1\frac{1}{4}$)	.010	.035			
1.375 ($1\frac{3}{8}$)	.010	.035	1.375 ($1\frac{3}{8}$)	.010	.035			
1.500 ($1\frac{1}{2}$)	.010	.025	1.500 ($1\frac{1}{2}$)	.010	.025			
1.625 ($1\frac{5}{8}$)	.010	.025	1.625 ($1\frac{5}{8}$)	.010	.025			
1.750 ($1\frac{3}{4}$)	.010	.025	1.750 ($1\frac{3}{4}$)	.010	.025			
2.000 (2)	.010	.025	2.000 (2)	.010	.025			
2.0625 ($2\frac{1}{16}$)	.010	.025	2.0625 ($2\frac{1}{16}$)	.010	.025			
2.125 ($2\frac{1}{8}$)	.010	.025						
2.250 ($2\frac{1}{4}$)	.010	.025						
2.375 ($2\frac{3}{8}$)	.010	.025						
2.500 ($2\frac{1}{2}$)	.010	.025						

* Reproduced by permission: Superior Tube Co. Cat. Sect. 21.

role in effecting the economies that must be realized. In Chapter 3 numerous tables of standards for valves, pipe, and fittings are presented. However, it is apparent to any piping designer that standards do not cover all situations nor do standards exist for all types of valves, fittings, and pipe. Progressive manufacturers often develop new types of fittings and valves that must be subjected to some years of use before they are accepted and included in standards. Substantial savings can be realized by wisely employing certain of these materials, particularly in non-critical locations. Whenever any manufacturer has developed and exhaustively tested a new fitting, for example, competent engineers within the design organization should assess these and recommend their use in locations that will result in large savings consistent with safety and operability.

As in any field of human affairs, tradition plays an important role in piping design. Many organizations have traditionally specified heavier weight flanges, valves, and fittings than would be necessary for certain applications. These traditions have at times developed because of unfortunate accidents with lighter weight materials that were used early in a company's history. However, in these days of high costs such traditions need to be subjected to thorough analysis. Other factors may now be operable, mitigating the original hazard, or new automatic safety devices may have minimized its potential effects.

TABLE 2.2

ALLOWABLE "S" PER DOLLAR. THE LARGEST NUMBER FOR ANY TEMPERATURE IS THE MOST ECONOMICAL MATERIAL

MATERIALS

TEMPERATURE	A-53-B E.R.W.	A-53-B Seamless	A-106-B Seamless	A 53-B Seamless Galvanized	A-83-A Seamless	B-241 Aluminum	A-209 -T1	A-335 -P2	A-335 P-12	A-335 P-11	A-335 P-3-b	A-335 P-22	A-335 P-21	A-335 P-5	B-42 Copper	A-335 P-16	A-335 P-17
−20 to 100 F	17206	**20000**	19589	17391	13594	1338	6070	5956	6035	5922	5607	5236	5202	5153	1699	4081	3851
200	16396	**19100**	18707	16609	12990	1158	5838	5728	5874	5764	5428	5096	5022	4919	1648	3885	3676
300	15586	**18150**	17777	15783	12319	958	5607	5501	5665	5575	5263	4929	4828	4685	1268	3700	3512
400	14827	**17250**	16895	15000	11725	719	5392	5290	5488	5417	5084	4789	4675	4452	634	3515	3357
500	14068	**16350**	16014	14217	11130		5160	5063	5294	5243	4919	4636	4467	4218		3330	3173
600	13309	**15500**	15181	13478	10493		4929	4836	5117	5069	4740	4482	4287	3985		3145	2999
650	12904	**15000**	14691	13043	10195		4813	4722	5037	4990	4665	4412	4204	3875		3047	2916
700	12348	**14350**	14055	12478	9898		4697	4609	4940	4911	4575	4328	4106	3751		2949	2834
750	11133	**12950**	12684	11261	9091		4581	4495	4844	4833	4486	4273	4023	3641		2851	2742
800	9311	**10800**	10578	9391	7901		4466	4382	4747	4738	4396	4189	3857	3517		2721	2660
850	7439	**8650**	8472	7522	6712		4350	4268	4570	4548	4187	4021	3625	3408		2503	2567
900	5567	**6500**	6366	5652	5525		4135	4057	4216	4138	3738	3658	3330	3160		2068	2465
950								3246	**3540**	3474	2990	3072	2497	2748		1524	2218
1000								2029	2414	**2464**	1854	2178	1942	2006		1088	1746
1050								1298	1609	1737	1256	1620	1526	1429		762	1130
1100								799	901	1263	822	1173	1110	907		544	678
1150										790	523	838	749	605		327	452
1200										379	359	559	416	412		261	308
1250																	
1300																	
1350																	
1400																	
1450																	
1500																	

TABLE 2.2 (Concluded)

ALLOWABLE "S" PER DOLLAR. THE LARGEST NUMBER FOR ANY TEMPERATURE IS THE MOST ECONOMICAL MATERIAL

MATERIALS

Temp.	A-312 Type 304 Welded	A-312 Type 304 Seamless	A-268 Type 410 Seamless	B-165 Monel Seamless	A-268 Type 405 Seamless	A-268 Type 430 Seamless	A-312 Type 347 Welded	A-312 Type 321 Welded	A-312 Type 316 Welded	A-312 Type 347 Seamless	A-312 Type 321 Seamless	B-161 Nickel Seamless	A-312 Type 316 Seamless	A-312 Type 309 Welded	A-312 Type 310 Welded	A-312 Type 309 Seamless	A-312 Type 310 Seamless
−20 to 100 F	2822	3107	2789	2528	2742	2859	2240	2181	2179	2508	2402	1237	2212	1838	1797	2083	1762
200	2496	2759	2662	2425	2588	2716	2240	2181	2179	2508	2402	1237	2212	1838	1797	2083	1762
300	2249	2486	2528	2278	2435	2573	2023	1970	2070	2274	2178	1237	2212	1689	1769	1922	1738
400	2046	2262	2402	2175	2281	2430	1876	1827	2029	2113	2024	1237	2065	1632	1741	1855	1710
500	1869	2071	2275	2161	2135	2287	1806	1758	1989	2033	1947	1237	2029	1620	1691	1844	1663
600	1737	1922	2142	2161	1981	2144	1778	1731	1982	1993	1908	. . . .	2017	1614	1640	1833	1616
650	1675	1856	2075	. . . .	1901	2073	1771	1724	1975	1986	1902	. . . .	2012	1609	1618	1827	1588
700	1623	1790	2015	2161	1820	2001	1764	1718	1968	1980	1896	. . . .	2006	1603	1584	1822	1560
750	1561	1723	1948	. . . .	1725	1930	1750	1704	1954	1966	1883	. . . .	1994	1580	1550	1799	1527
800	1499	1657	1896	2131	1608	1873	1729	1683	1941	1946	1864	. . . .	1976	1534	1500	1744	1475
850	1455	1607	1800	. . . .	1477	1787	1701	1656	1907	1912	1832	. . . .	1947	1453	1427	1655	1400
900	1411	1558	1636	1176	1331	1673	1680	1636	1852	1886	1806	. . . .	1888	1344	1314	1533	1297
950	1367	1508	1309	. . . .	1170	1315	1652	1608	1743	1852	1774	. . . .	1781	1218	1191	1388	1175
1000	1323	1458	952	. . . .	585	929	1610	1568	1621	1805	1729	. . . .	1652	1022	1050	1166	1034
1050	1270	1408	. . . .	. . . .	. . . .	. . . .	1554	1513	1416	1752	1678	. . . .	1439	827	932	944	667
1100	1129	1243	. . . .	. . . .	. . . .	. . . .	1484	1449	1205	1672	1601	. . . .	1227	632	809	722	470
1150	864	953	. . . .	. . . .	. . . .	. . . .	952	927	981	1070	1025	. . . .	1003	488	691	555	338
1200	670	746	. . . .	. . . .	. . . .	. . . .	595	579	790	669	641	. . . .	802	373	573	422	235
1250	485	539	. . . .	. . . .	. . . .	. . . .	427	416	613	481	461	. . . .	625	281	455	322	136
1300	370	406	. . . .	. . . .	. . . .	. . . .	322	314	463	361	346	. . . .	472	224	337	255	70
1350	273	298	. . . .	. . . .	. . . .	. . . .	238	232	347	267	256	. . . .	354	172	225	194	42
1400	212	232	. . . .	. . . .	. . . .	. . . .	182	172	272	207	199	. . . .	277	126	152	144	33
1450	150	166	. . . .	. . . .	. . . .	. . . .	140	136	211	160	154	. . . .	218	86	107	100	23
1500	115	124	. . . .	. . . .	. . . .	. . . .	119	116	177	134	128	. . . .	177	75	73	83	19

MOST ECONOMICAL PIPING MATERIAL

Temp.	Material
−20 to 900 F	A-53—Grade B Seamless
950	A-335—Grade P-12 Seamless
1000	A-335—Grade P-11 Seamless
1050–1150	A-312—Type 347 Seamless
1200–1500	A-312—Type 316 Seamless

(Reproduced by permission: Dan Christopher, *Petroleum Refiner*, **37**, 3, 146 (1958). Copyright by Gulf Publishing Company.)

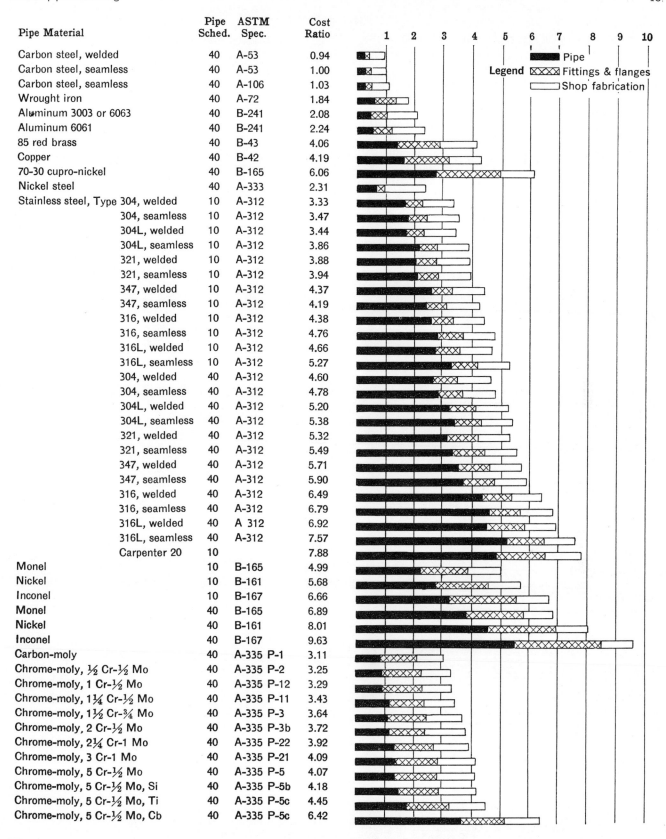

Pipe Material	Pipe Sched.	ASTM Spec.	Cost Ratio
Carbon steel, welded	40	A-53	0.94
Carbon steel, seamless	40	A-53	1.00
Carbon steel, seamless	40	A-106	1.03
Wrought iron	40	A-72	1.84
Aluminum 3003 or 6063	40	B-241	2.08
Aluminum 6061	40	B-241	2.24
85 red brass	40	B-43	4.06
Copper	40	B-42	4.19
70-30 cupro-nickel	40	B-165	6.06
Nickel steel	40	A-333	2.31
Stainless steel, Type 304, welded	10	A-312	3.33
304, seamless	10	A-312	3.47
304L, welded	10	A-312	3.44
304L, seamless	10	A-312	3.86
321, welded	10	A-312	3.88
321, seamless	10	A-312	3.94
347, welded	10	A-312	4.37
347, seamless	10	A-312	4.19
316, welded	10	A-312	4.38
316, seamless	10	A-312	4.76
316L, welded	10	A-312	4.66
316L, seamless	10	A-312	5.27
304, welded	40	A-312	4.60
304, seamless	40	A-312	4.78
304L, welded	40	A-312	5.20
304L, seamless	40	A-312	5.38
321, welded	40	A-312	5.32
321, seamless	40	A-312	5.49
347, welded	40	A-312	5.71
347, seamless	40	A-312	5.90
316, welded	40	A-312	6.49
316, seamless	40	A-312	6.79
316L, welded	40	A 312	6.92
316L, seamless	40	A-312	7.57
Carpenter 20	10		7.88
Monel	10	B-165	4.99
Nickel	10	B-161	5.68
Inconel	10	B-167	6.66
Monel	40	B-165	6.89
Nickel	40	B-161	8.01
Inconel	40	B-167	9.63
Carbon-moly	40	A-335 P-1	3.11
Chrome-moly, ½ Cr-½ Mo	40	A-335 P-2	3.25
Chrome-moly, 1 Cr-½ Mo	40	A-335 P-12	3.29
Chrome-moly, 1¼ Cr-½ Mo	40	A-335 P-11	3.43
Chrome-moly, 1½ Cr-¾ Mo	40	A-335 P-3	3.64
Chrome-moly, 2 Cr-½ Mo	40	A-335 P-3b	3.72
Chrome-moly, 2¼ Cr-1 Mo	40	A-335 P-22	3.92
Chrome-moly, 3 Cr-1 Mo	40	A-335 P-21	4.09
Chrome-moly, 5 Cr-½ Mo	40	A-335 P-5	4.07
Chrome-moly, 5 Cr-½ Mo, Si	40	A-335 P-5b	4.18
Chrome-moly, 5 Cr-½ Mo, Ti	40	A-335 P-5c	4.45
Chrome-moly, 5 Cr-½ Mo, Cb	40	A-335 P-5c	6.42

Legend: ■ Pipe   ⊠ Fittings & flanges   ☐ Shop fabrication

FIG. 2.7. Cost ratios for shop-fabricated metallic pipe. Basis: 100 ft of 3 in. pipe, ten 3-in. weld ells, one 3-in. weld tee, four 3-in. weld neck flanges, and twenty-four 3-in. butt welds. Heat treatment done in accordance with ASA Code. [Reproduced by permission from Otto Mendel, *Chemical Engineering*, **68**, No. 5, 190 (1961).]

TABLE 2.3

COMPARISON OF PHYSICAL AND MECHANICAL PROPERTIES FOR VARIOUS PIPING MATERIALS AT 73°F (23°C)

	Specific Gravity	Flexural Strength, psi	Compressive Strength, psi	Modulus of Elasticity, 100,000 psi	Impact Strength, Izod Notch, ft-lb/in.
Polyethylene (Type I and II)	0.910–0.940	1,700–3,700	. . .	0.2–0.6	Over 15
Polyethylene (Type III)	0.941–0.965	3,700–6,600	. . .	0.6–1.7	1.2–12
Cellulose acetate butyrate (Tenite butyrate)	1.2	5,700–6,200	5,700–6,300	1.2–1.4	1.0–3.0
Acrylonitrile-butadiene-styrene (ABS Type I)	1.04–1.07	7,000–8,000	5,000–7,000	1.8–3.0	4.0–9.0
Acrylonitrile-butadiene-styrene (ABS Type II)	1.06–1.08	12,000	10,000	3.5–4.0	4–6
Vinylidene chloride (saran)	1.7	4,000–7,000	7,500–8,500	0.4–0.8	0.3–1.0
Polyvinyl chloride (PVC Type I)	1.35–1.45^c	12,000–17,000	8,000–11,000	4.0–5.4	0.6–0.9
Polyvinyl chloride (PVC Type II)	1.35–1.45	9,500–11,500	8,000	2.5–3.5	8–18
Phenolic asbestos (Haveg)	1.7	6,500	10,000–14,000	8.7	0.48
Polyester (glass reinforced)	1.5–2.0	Up to 65,000	Up to 40,000	Up to 30	Over 10
Epoxy (glass reinforced)	1.7–2.2	Up to 80,000	Up to 60,000	Up to 50	Over 10
Carbon steel (A53, Grade A)	7.8	. . .	48,000	300	Excellent
Stainless Steel (18 Cr-8 Ni)	8.0	. . .	75,000	280	Excellent
Copper (seamless annealed)	8.9	. . .	30,000	170	Excellent
Aluminum (annealed 3003)	2.7	. . .	14,000	100	Excellent
Borosilicate glass	2.2	1,000	100,000–180,000	98	Poor
Impregnated graphite (Karbate)	1.9	4,700	9,000	23	Poor
Hard rubber (Buna N)	1.2–1.5	11,000	7,000–12,000	3.0	0.3–0.4

Dash indicates property not applicable.   Blank indicates data not available.

Values for plastics are flexural modulus; for other materials, tensile (Young's) modulus is listed.

ASTM test results referring to thermoplastics: Specific gravity, D-792; Flexural strength, D-790; Elasticity modulus, D-638; Impact strength, Izod notch D-256; Deflection Temperature, D-648; Coeff. of thermal expansion, D-696; Flammability, D-635, Compressive strength, D-695.

[a] Straight lengths in 4-in. and 6-in. sizes; smaller sizes are furnished in coils up to 400 ft. long.

[b] Straight lengths in 3 in. to 8 in. sizes; smaller sizes are furnished in coils up to 400 ft. long.

[c] Pipe extruded from Montecatini Vipla resin (Ryertex-Omicron) has a specific gravity of 1.51 and a Deflection Temperature of 171–177°F.

TABLE 2.4

COMPARISON OF ALLOWABLE OPERATING STRESSES, PRESSURES, AND TEMPERATURES (Basis: 2 In. Pipe)

	Pipe Wall Schedule	O.D., in.	I.D., in.	Wall, in.
Polyethylene (Types I and II)	Sch. 40	2.375	2.067	0.154
Polyethylene (Type III)	Sch. 40	2.375	2.067	0.154
Butyrate (Tenite Butyrate)	S.W.P.	2.250	2.000	0.125
Acrylonitrile-butadiene-styrene polymer (ABS Type I)	Sch. 40	2.375	2.067	0.154
Acrylonitrile-butadiene-styrene polymer (ABS Type II)	100# Class	2.250	2.100	0.075
Vinylidene chloride (saran)	Sch. 80	2.375	1.939	0.218
Polyvinyl chloride (PVC Type I)	Sch. 40	2.375	2.067	0.154
Polyvinyl chloride (PVC Type II)	Sch. 40	2.375	2.067	0.154
Phenolic asbestos (Haveg 31)	. . .	3.000	2.000	0.500
Polyester-glass (Spiral-glas)	. . .	2.180	1.930	0.125
Cast epoxy-glass (Fibercast J-700)	. . .	2.375	1.895	0.240
Laminated expoxy-glass (Bondstrand)	. . .	2.375	2.205	0.085
Carbon steel (seamless A53, Gr. A)	Sch. 40	2.375	2.067	0.154
Stainless steel (seamless, Type 304)	Sch. 40	2.375	2.067	0.154
Copper (seamless annealed)	Sch. 40	2.375	2.067	0.154
Aluminum (annealed 3003)	Sch. 40	2.375	2.067	0.154
Borosilicate glass	. . .	2.344	2.000	0.172
Impregnated graphite (Karbate)	. . .	2.750	2.000	0.375
Hard rubber (Buna N)	Sch. 120	2.375	1.875	0.250

[a] Values recommended by the Thermoplastic Pipe Division of the S.P.I.

[b] American code for pressure piping (ASA B. 31)

[c] Calculated from Barlow's Formula (ASA B. 31, Par. 324b) with zero corrosion allowance.

[d] First number is axial tensile strength; second number is hoop strength at burst.

TABLE 2.3 (*Concluded*) [Reproduced by permission: G. Sorell, *Chemical Engineering*, p. 149, March 23, 1959]

Deflection Temperature @ 264 psi Deg. F	Max. Recommended Operating Temp., Deg. F	Coefficient of Thermal Expansion 10^{-5} in./in./F	Thermal Conductivity Btu/sq ft/hr/F/in.	Flammability, in./min	Standard Max. Pipe Length, ft	Standard Pipe Sizes, Diam., in.
105–120	120	6–14	1.8–2.3	0.9–1.1	25[a]	½–6
120–125	150	6–9	2.3–3.2	0.9–1.1	30[b]	½–8
145–155	140	6–10	1.1–2.2	1.3	20	½–8
185–190	170	3.0–6.0	1.0	1.0–1.3	20	½–12
205–225	180	3.8–5.6	1.0–1.8	1.3	20	½–12
130–150	150	7–11	0.6–0.9	Self-extinguishing	10	½–6
160–165(c)	150	2.9–6.7	0.8–1.2	Self-extinguishing	20	¼–12
150–155	140	6–14	1.3	Self-extinguishing	20	¼–12
	265	1.1–1.8	3.0	Nonflammable	10	½–12
	200	1.2–4	1.5	S.-e. to s.b.†	20	1–12
	300	0.7–0.9	0.9–2.5	S.-e. to s.b.†	20	2–8
…	1000	0.67	360	Nonflammable	20	¼–24
…	1500	0.93	110	Nonflammable	20	¼–24
…	400	0.95	2700	Nonflammable	20	⅛–12
…	400	1.3	1320	Nonflammable	20	⅛–10
…	450	0.18	8.0	Nonflammable	10	1–6
…	340	0.24	1020	Nonflammable	9	1–10
275	225	2.6–4.0	1.0	Burns slowly	10	¼–8

† Self-extinguishing to slow-burning.

TABLE 2.4 (*Concluded*) [Reproduced by permission: G. Sorell, *Chemical Engineering*, p. 149, March 23, 1959]

Short-Time Tensile Strength at 73.4F, psi	Allowable Long-Time Hoop Stress at 73.4 F, psi	Allowable Working Pressure, psi		Maximum Recommended Operating Temp., F
		At 73.4 F	At Max Oper. Temp.	
1,400–2,500	385	50	24	120
2,500–5,100	600	75	29	150
5,000–5,500	700[a]	95	50	140
5,000–6,000	1,000[a]	75	40	170
7,500–8,500	1,400	100	25	180
4,000–7,000	800	125	50	150
6,400–9,000	1,200[a]	175	97	150
5,500–6,500	1,000[a]	156	25	140
2,250–4,500		150	110	265
29,000; 58,000[d]		500	500	200
30,000; 25,000[d]		700	350	300
40,000; 8,000[d]		550	275	220
48,000	16,000[b]	1,820[c]	284[c]	1,000
75,000	18,750[b]	2,140[c]	86[c]	1,500
30,000	6,700[b]	765[c]	284[c]	400
14,000	3,350[b]	382[c]	205[c]	400
10,000		50	50	450
2,500		75	75	340
6,500–7,200	1,400	50	50	225

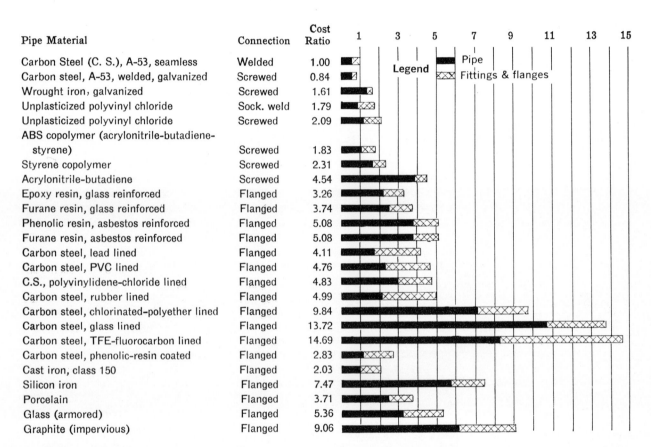

Pipe Material	Connection	Cost Ratio
Carbon Steel (C. S.), A-53, seamless	Welded	1.00
Carbon steel, A-53, welded, galvanized	Screwed	0.84
Wrought iron, galvanized	Screwed	1.61
Unplasticized polyvinyl chloride	Sock. weld	1.79
Unplasticized polyvinyl chloride	Screwed	2.09
ABS copolymer (acrylonitrile-butadiene-styrene)	Screwed	1.83
Styrene copolymer	Screwed	2.31
Acrylonitrile-butadiene	Screwed	4.54
Epoxy resin, glass reinforced	Flanged	3.26
Furane resin, glass reinforced	Flanged	3.74
Phenolic resin, asbestos reinforced	Flanged	5.08
Furane resin, asbestos reinforced	Flanged	5.08
Carbon steel, lead lined	Flanged	4.11
Carbon steel, PVC lined	Flanged	4.76
C.S., polyvinylidene-chloride lined	Flanged	4.83
Carbon steel, rubber lined	Flanged	4.99
Carbon steel, chlorinated-polyether lined	Flanged	9.84
Carbon steel, glass lined	Flanged	13.72
Carbon steel, TFE-fluorocarbon lined	Flanged	14.69
Carbon steel, phenolic-resin coated	Flanged	2.83
Cast iron, class 150	Flanged	2.03
Silicon iron	Flanged	7.47
Porcelain	Flanged	3.71
Glass (armored)	Flanged	5.36
Graphite (impervious)	Flanged	9.06

FIG. 2.8.   Cost ratios for representative non-metallic pipe.   Basis: Flanged—100 ft of 3-in. pipe, ten 3-in. flanged ells, one 3-in. flanged tee, ten pairs of 3-in. flanges.   Screwed or welded—100 ft of 3-in. pipe, ten 3-in. ells, one 3-in. tee, four 3-in. flanges.   Valves are not included. [Reproduced by permission from Otto Mendel, *Chemical Engineering*, **68**, No. 5, 190 (1961).]

INDEX OF TABLES AND FIGURES FOR VALVE AND FITTING SELECTION

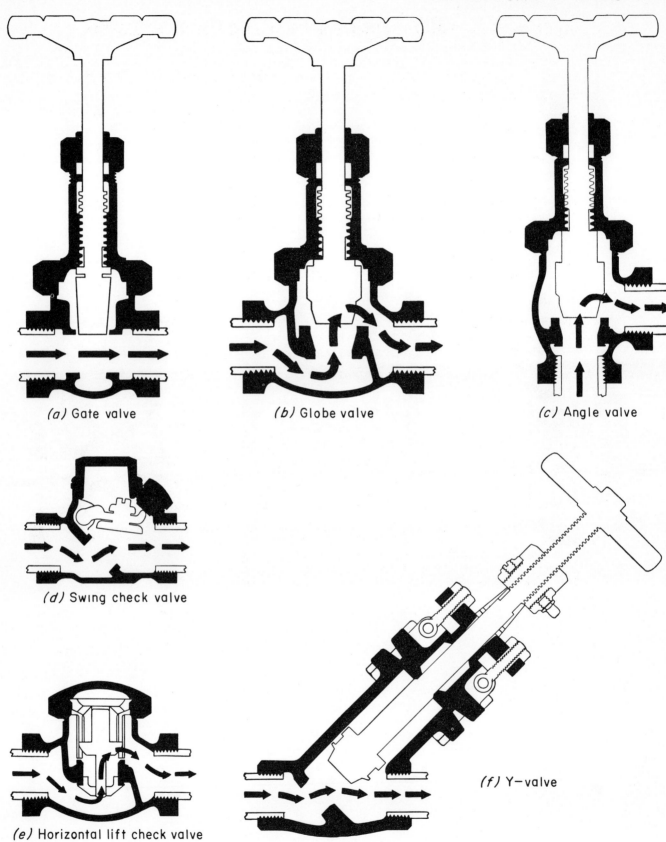

(a) Gate valve

(b) Globe valve

(c) Angle valve

(d) Swing check valve

(e) Horizontal lift check valve

(f) Y-valve

FIG. 2.9.   Valves and valve-flow characteristics.   [Illustrations (a) through (f) reproduced by permission of Wm. Powell Company, Cincinnati, O.]

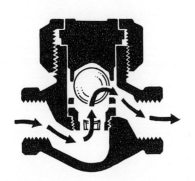

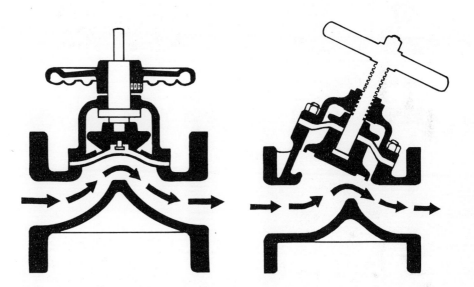

(g) Ball check valve

(h) Two types of diaphragm valves

FIG. 2.9. *(Continued)*.

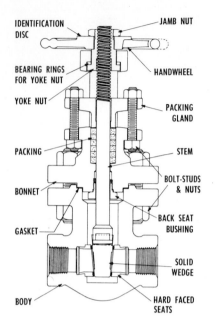

**Forged Steel
Screw End Gate Valve
Bolted Bonnet Type**

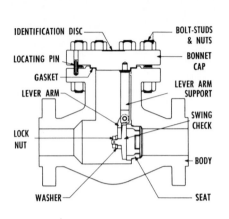

**Forged Steel Flanged
Swing Check Valve
Bolted Bonnet Type**

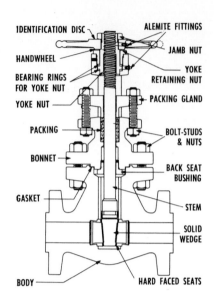

**Forged Steel
Flanged Gate Valve
Bolted Bonnet Type**

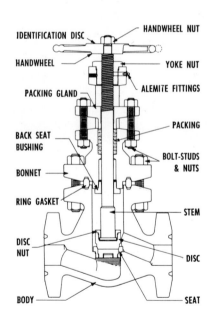

**Forged Steel
Flanged Globe Valve
Bolted Bonnet Type**

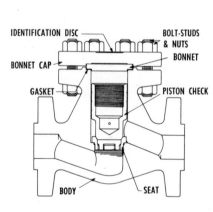

**Forged Steel
Flanged Piston Check Valve
Bolted Bonnet Type**

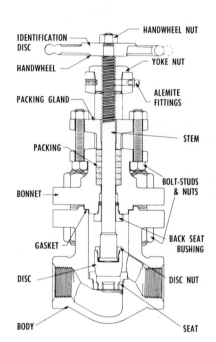

**Forged Steel
Screw End Globe Valve
Bolted Bonnet Type**

FIG. 2.10. Typical nomenclature for gate, globe, angle, and check valves. (Reproduced by permission of Henry Vogt Machine Co., Louisville, Ky.)

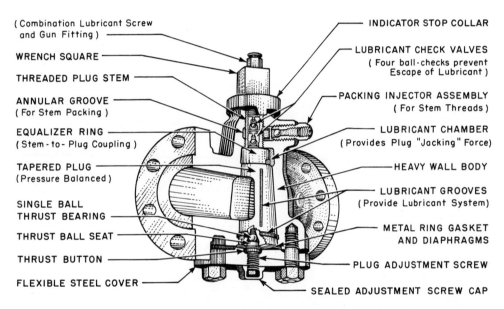

(Combination Lubricant Screw
and Gun Fitting)

WRENCH SQUARE

THREADED PLUG STEM

ANNULAR GROOVE
(For Stem Packing)

EQUALIZER RING
(Stem - to - Plug Coupling)

TAPERED PLUG
(Pressure Balanced)

SINGLE BALL
THRUST BEARING

THRUST BALL SEAT

THRUST BUTTON

FLEXIBLE STEEL COVER

INDICATOR STOP COLLAR

LUBRICANT CHECK VALVES
(Four ball-checks prevent
Escape of Lubricant)

PACKING INJECTOR ASSEMBLY
(For Stem Threads)

LUBRICANT CHAMBER
(Provides Plug "Jacking" Force)

HEAVY WALL BODY

LUBRICANT GROOVES
(Provide Lubricant System)

METAL RING GASKET
AND DIAPHRAGMS

PLUG ADJUSTMENT SCREW

SEALED ADJUSTMENT SCREW CAP

FIG. 2.11.   Typical nomenclature for lubricated plug valve (based on Rockwell-Nordstrom valve).

TABLE 2.5
COMPARISON OF BASIC VALVE TYPES

Valve Type	How it Works	Use	Limitations
Gate	(See Figs. 2.9 and 2.10.) Gatelike disc actuated by a stem; screw and handwheel move at right angles to flow. The disc seats against two faces for shut-off.	On-off service requiring infrequent throttling	Not good for throttling service. Throttling causes wire drawing and erosion (plug-type or slide-type is satisfactory for throttling). Pocket at bottom of valve can fill with foreign material and prevent closing of valve.
Globe	(See Figs. 2.9 and 2.10.) Disc attached to stem seats on circular opening. Fluid changes direction in passing through valve.	Good for throttling service because of increased resistance to flow. Y-valve (Fig. 2.10) produces lower pressure drop and turbulence than standard globe and is preferred for corrosive or erosive service. Many special alloy and plastic valves (e.g. polyvinyl chloride) for corrosive service are Y-valves.	Not recommended for on-off service. Cost and throttling efficiency above 6 in. become unfavorable.
Angle (90°)	Similar to globe valve except inlet and outlet makes 90° angle (Fig. 2.9).	Same as globe. Used for non-critical service in place of globe valve and an elbow.	False economy for industrial uses. Bends in piping systems are subject to strains that should not be placed on valves.
Plug Cock	(See Fig. 2.11.) Tapered plug with hole same shape as interior of valve body, opens or closes valve with minimum effort. ¼ or 1 full-turn required to fully open or close. Three body types are made short, regular and venturi pattern. The short pattern has same face-to-face dimension as gate valves and is preferred for most services. Regular and venturi patterns produce less pressure drop and are used when a minimum ΔP is essential.	General: On-off service—More positive shut-off than gate valve. Also can be used for throttling but characteristics not as satisfactory as globe for such service. For low-pressure drop service. Unexposed seats eliminate corrosion and erosion.	See below.
Lubricated	Screw in top of plug is used to force lubricant into grooves in plug and to bottom chamber. Lubricant reaching bottom chamber forces plug slightly off its seat (Fig. 2.11). Only ¼ turn required to open or close.	"General" uses above and can be used in any service where lubricant does not constitute a disadvantage. Used for critical services which require repacking under pressure.	Lubricant can cause undesirable contamination of high purity products. Lubrication requires extra effort. Lubricant sets maximum service temperature (659 to 1000°F)
Non-lubricated	Cam-crank mechanism lifts plug and turns it without friction between plug and seat. A ¾ turn of handle is required to open or close.	See "General" uses above. Used for services where lubrication is inconvenient or temperatures exceed lubricated plug limits. Excellent for corrosive service requiring special alloys and linings.	Not repackable while under pressure; does not provide as positive a seal as lubricated plug.

TABLE 2.5 (*Continued*)

Valve Type	How it Works	Use	Limitations
Check Valve		General: Prevent back-flow in lines.	
Swing check	Flow keeps swing gate open, while gravity and flow reversal close it. Tilting-type is pivoted at center and insures closing without slamming (Figs. 2.9 and 2.10). Outside levers and weights are used on standard swing checks when greater sensitivity for changes in flow is required.	Where minimum pressure drop is required—Best for liquids and for large line sizes.	Not suitable in line subject to pulsating flow. Some styles operate only in a horizontal position.
Piston check	Flow pattern as in globe valve. Flow forces piston up and reversal and gravity returns it to seat (Figs. 2.9 and 2.10).	Good for vapors, steam, and water. Suitable for pulsating flow.	Many designs are for horizontal service only. Not common in sizes over 6 in. Not recommended for service which deposit solids.
Ball check	A lift-type check consisting of a ball with guides (Fig. 2.9).	Stops flow reversal more rapidly than others. Good for viscous fluids which deposit solid residues that would impair operation of other types. Vertical or horizontal installation is possible.	Not common in sizes over 6 in. Not suitable for lines subject to pulsating flow.
Needle	Similar to globe except disc is pointed at end. Steel valves are often made from barstock and are rugged.	Valves 2 in. and smaller for use on pilot plants and bench-scale equipment and instrument service. Good for manual flow control. Ground-joint union bonnet connection preferable.	Positive shut-off not always possible or desirable. In some designs, seat is scored if shut down tightly.
Automatic Control	(See Fig. 2.16.) Similar in principle to globe valve but precision-built for accurate automatic control. Air pressure actuates diaphragm causing the stem to move, opening or closing valve orifice. Air pressure is controlled by the primary measuring instrument. The valve plug is tapered (parabolic) or has V-ports to give the desired throttling characteristics. Double-port valves give better control range and require only small force to move stem.	Automatic control of flow and pressure in processes.	First cost is high, but in most areas of world resulting labor savings and improved operating results far offset this cost. Often not justified, however, for small-scale production or testing.
Hand Control	Single-port hand control valves with micrometer for setting valve to within 1/100 of a turn are useful in pilot plants and other applications where automatic control is not justified.		
Packless Diaphragm Valve	Diaphragm seals the bonnet, preventing fluid from contacting inner bonnet or stem (Fig. 2.9). Seating member may be a separate disc diaphragm or a solid diaphragm may serve as the closing device.	For corrosive, volatile, and toxic fluid service in which leakage cannot be tolerated. All plastic valves are produced in this design.	Choice of diaphragms limited to rubber-like or plastic materials which cannot withstand temperatures above 400°F or operate effectively at sub-atmospheric temperatures.

TABLE 2.5 (*Concluded*)

Valve Type	How it Works	Use	Limitations
Relief Valve	Valve opens automatically when force on seat exceeds that of spring. Returns to closed position when excess pressure has been relieved (Fig. 2.17).	For protecting equipment and vessels from excessive pressures.	Require periodic inspection to insure operability. Not good for highly corrosive fluids.
Rupture Disc	Thin metal diaphragm ruptures at a predesigned pressure.	For protecting equipment and vessels from excessive pressures when maintenance is difficult and excessive pressure occurrences rare.	Diaphragm must be replaced after each excess pressure occurrence.

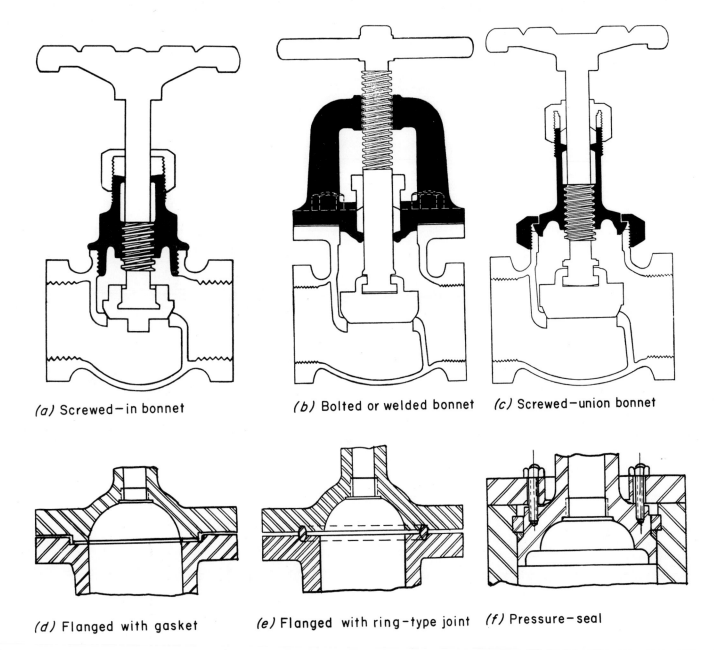

(a) Screwed-in bonnet (b) Bolted or welded bonnet (c) Screwed-union bonnet

(d) Flanged with gasket (e) Flanged with ring-type joint (f) Pressure-seal

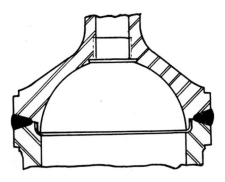

(g) Welded design

FIG. 2.12. Comparison of standard bonnet designs. (a) *Screwed-in bonnet:* Used for small valves in non-critical, low-pressure service. Not recommended for general process use. Satisfactory only when frequent dismantling is not required. (b) *Bolted or welded bonnet:* Most larger-size (1½″ and larger) steel valves used in process plants are of bolted or welded construction. Types shown in (d) through (g) are designed for all pressures and temperatures. (c) *Screwed-union bonnet:* Used for small valves in general process service. Design is such to permit high-pressure service and frequent dismantling. (d) *Flanged with gasket:* Most common design for use up to 925°F. (e) *Flanged with ring-type joint:* Preferred at high temperatures and pressures. (f) *Pressure-seal type:* Designed so that line pressure causes bonnet to seal against triangular-shaped seal ring. Requires adjustment in service due to relaxation of bolting as in flanged designs. Good for severe services requiring valve dismantling. (g) *Welded design:* This design is light in weight and free of bonnet leakage problems. Good for corrosive or other difficult services. Frequent disassembly not practical because of inaccessibility of working parts. [Illustrations (a), (b), (c) reproduced by permission of Wm. Powell Company, Cincinnati, O.; (d), (e), (f), (g) by permission of the Chapman Valve Company, Indian Orchard, Mass.)

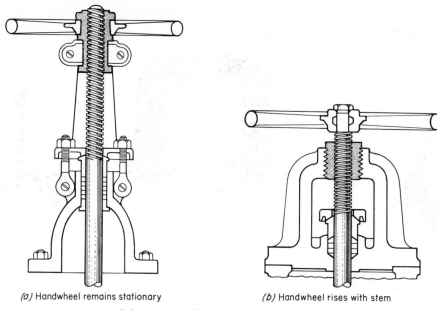

(a) Handwheel remains stationary   (b) Handwheel rises with stem

Rising stem-outside screw and yoke (OS&Y)

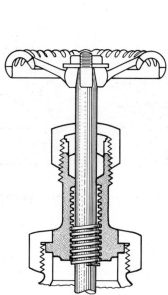

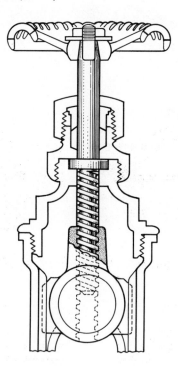

(c) Rising stem Inside screw   (d) Non-rising stem Inside screw

FIG. 2.13. Comparison of stem types for gate, globe, and angle valves. Valves (a) and (b) are recommended for sizes 2½ in. and larger by the ASA "Code for Pressure Piping." They are commonly used for sizes 2 to 3 in. and larger in process plants. Position of stem indicates degree of valve opening. Since threaded portion is never in contact with fluid flowing, it is not subject to corrosion or erosion. Threads are readily lubricated. Used in all sizes for severe services. Rising stem (c) is the most common stem construction in valves 2 in. and smaller. It is used in process plants for fluids which will not damage the thread, such as hydrocarbons, steam, and water. In (d) the stem does not rise; instead the disc rises on the stem. Good when clearance for valve is restricted. Not generally used in process plants. Most manufacturers do not offer this arrangement in steel valves for process service.

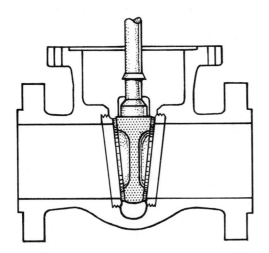

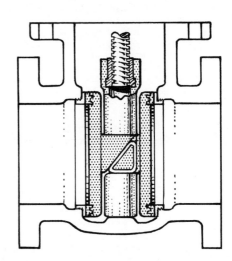

(a) SOLID WEDGE

Characteristics: Simplest and most common type. Good fatigue resistance and no trapping of fluid in bonnet. Can be installed in any position without jamming.

Description: Solid wedge of one-piece construction. The wedge is accurately machined on each face and slotted to receive the "T" head of the stem. Some manufacturers offer either T-slot or hinged connection for the stem.

(b) DOUBLE DISC

Characteristics: Used for moderate temperatures and common only in iron and bronze valves of all sizes.

Description: Discs with parallel faces are forced against the seats by a spreader. A double-wedge disc type (not shown) employs a tapered seat and thus does not require a separate spreader.

FIG. 2.14. Disc variations in gate valves.

Specially designed miscellanous discs are offered by various manufacturers for unusual service conditions. These designs include a plug disc of cone shape for corrosive service and a flexible disc which offers good operability at high pressures and in large valves.

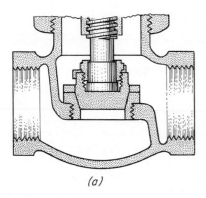
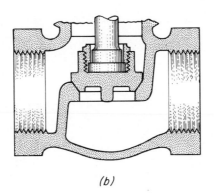
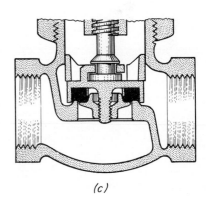

(a)   (b)   (c)

FIG. 2.15. Disc variations in globe valves.

(a) PLUG-TYPE

Wide bearing surface with long taper resists erosion and provides excellent flow characteristics for throttling service.

(b) CONVENTIONAL-TYPE

Taper not as great and seating surface contact less than on plug-type. Not so good for throttling, but provides tighter shut-off because of narrow bearing surface. It is effective even when deposits form on seat.

(c) COMPOSITION-TYPE

Disc is composed of a holder, composition disc, and nut. The discs are made of various compositions of laminated asbestos or resins. Discs are selected for the material flowing through the valve. This valve is used at moderate pressures and temperatures and is constructed of brass or iron. It is not good for throttling.

TABLE 2.6
DESIGN FEATURES OF GATE VALVES COMPARED

The following tabulation of design features is based on the suggestions of O. L. Lewis [*The Petroleum Engineer,* C27 November, 1952]. It is presented here as a guide in comparing gate valves and as a recommendation of the type of analysis that should be made in selecting any type of valve.

Feature	Comments
Stem	T-head method of connecting the stem to the wedge simplest and strongest.
Seats	Screwed-in seat rings are good for moderate service conditions. Welded or rolled rings are best for high pressure, high-temperature service but require well-equipped shop for proper replacement. Integral seats satisfactory for bronze and iron valves in services not requiring replacement.
Wedge	Should be guided through full travel to prevent chatter.
Valve Wheel	Steel preferred—cast iron often fails under abuse of "cheaters," etc.
Stuffing Box	Deep stuffing box required. Cheap valves have shallow stuffing box which receive only a few rings of packing.
Repacking	Can valve be repacked under pressure? This is a desirable feature in many situations.
Number of Turns to Open	Where this is important, it should be checked. There are variations between manufacturers which at times may be significant.

TABLE 2.7
END CONNECTIONS ON VALVES AND FITTINGS IN PROCESS PLANTS

Type	Application
Screwed	Used extensively for pipe 2″ and smaller. Made in iron, steel, brass and various alloys.
Socket Welding	Popular in small sizes 2″ and under for severe services where leakage danger must be eliminated. Self-aligning and easy to install.
Butt-welding	Used almost exclusively for all fittings in process lines 2″ to 3″ and above. Not common for valves.
Flanged	Used as unions and as ends on valves in welded lines. Thus most lines 2″ to 3″ or above in process plants have flanged unions and flanged valves.
Solder Joint	Low temperature joint for copper tubing in plumbing and heating service—Service limited by melting point of tin-lead solder (362–450°F). Manufacturer's recommendations for maximum temperature for valve or fitting should be followed.
Brazing Joint	Joint for brass and copper lines which withstands higher temperature than solder joint—Factory inserted ring of silver brazing alloy makes seal upon heating with oxyacetylene flame. Melting point of silver brazing alloy is 1300°F. Service recommendations of manufacturer for particular valve or fitting should be consulted.
Flared and Compression	For thin-walled tubing, and fittings and valves used with thin-walled tubing. Popular for pilot plant and instrumentation work where disassembly is necessary.
Packed and Poured	For cast iron, water, gas, and sewage piping.

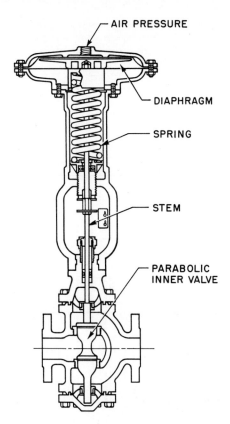

FIG. 2.16. Diaphragm-operated, double-port control valve.

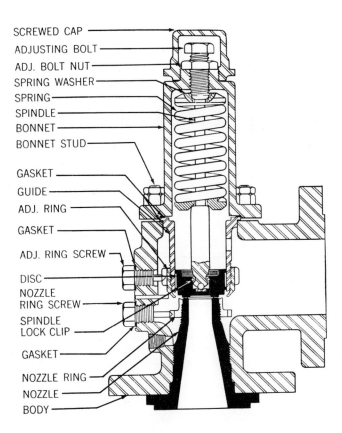

FIG. 2.17. Relief valve. (Reproduced by permission of Crosby Valve and Gage Company.)

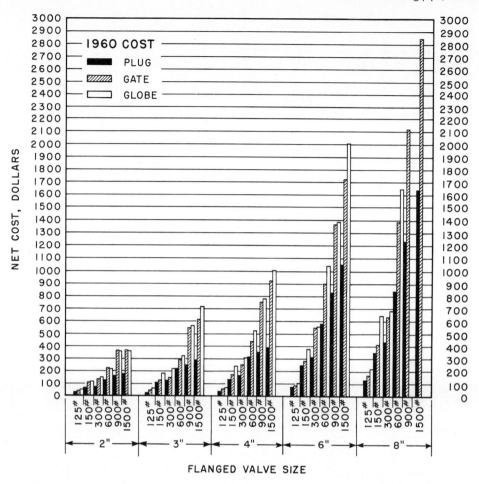

FIG. 2.18. Cost comparison of flanged valves. (A chart similar to this was first introduced by O. L. Lewis in *The Petroleum Engineer,* Nov., 1952.)

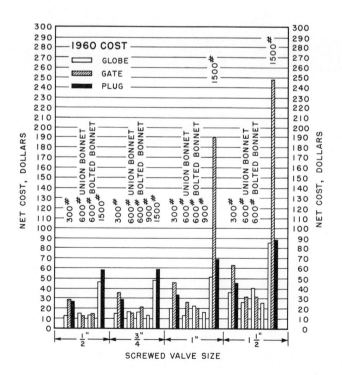

FIG. 2.19. Cost comparison of small steel valves. (A chart similar to this was first introduced by O. L. Lewis in *The Petroleum Engineer,* Nov., 1952.)

TABLE 2.8
FLANGE TYPES

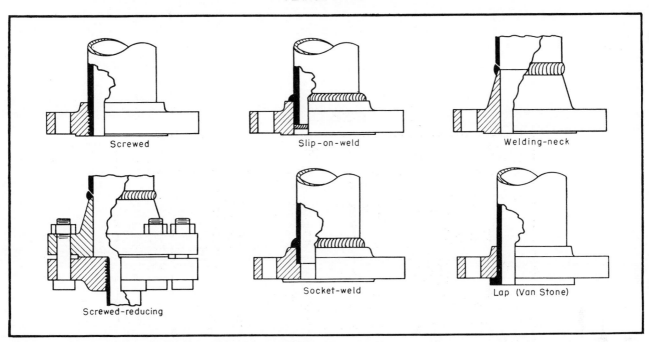

(Drawings reproduced by permission: Theodore R. Olive, *Chemical Engineering*, December, 1953, p. 187.)

Flange Type	Description and Characteristics	Recommended Use
Screwed	Flange screws on to threaded pipe, and no welding required.	High pressure service at moderate temperatures. Not suited for service involving thermal or bending stresses.
Slip-on	Lower cost per flange than welding neck, but installed cost same as welding neck. Less skill required for installing. Calculated strength under internal pressure and life under fatigue is less than welding neck (ASA 16.5 ratings are the same but slip-on not included in standard for 2½ in. and larger 1500-lb flanges and for any size 2500-lb flange). To install, flange is slipped on the pipe and two welds made (see cut) one on the inside and the other on the outside of the pipe.	Moderate service conditions particularly when ease of assembly is a valid consideration.
Welding-neck	Tapered hub terminates at pipe where it is attached by a weld. The long taper makes flange an integral part of pipe and produces a joint which can stand repeated bending.	Severe service (high pressure and/or temperature or low temperature).
Lap Joint	Fit over lap joint stub as illustrated. Fatigue life is ¹⁄₁₀ that of welding neck. Only lap joint stub comes in contact with process fluid.	Services requiring frequent dismantling for inspection and cleaning—For large diameter pipe and other installations for which ability to rotate flange during assembly is an advantage. Avoid using for conditions involving severe bending stresses. Carbon steel flanges can be used with alloy stub ends for certain corrosive services, thus reducing costs.
Socket Welding	Pipe fits into recessed portion of flange. One weld is made at back of flange. Crevice between socket and flange may be subject to excessive corrosion under certain conditions, but internal weld can be made to avoid this difficulty. Initial cost 10% less than slip-on. With internal weld has 50% greater fatigue strength and same static strength as slip-on.	Good for smaller diameter piping where leak-proof fittings are preferred to screwed attachments.

TABLE 2.9
FLANGE FACINGS

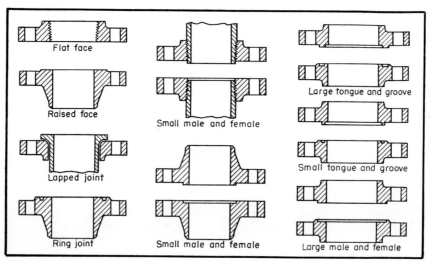

(Drawings reproduced by permission: Theodore R. Olive, *Chemical Engineering*, December, 1953, p. 187.)

Type	Description and Characteristics	Recommendation
Raised	Most common—Both flanges of a pair are identical. Faces are 1/16 in. high for 150- and 300-lb flanges and 1/4 in. high for all others. Gasket is usually less in width than the raised face.	(See Table 2.10 for gasket recommendations.) Preferred for moderate service conditions.
Ring-type	Most expensive but most efficient flange facing. Not easily damaged in assembling.	Preferred for high temperature and pressure service.
Male and Female	Made in large and small designs. Metal gaskets can be used with small design because of high gasket compression. Stocking a problem because both male and female pieces must be stocked.	Used for special services requiring a retained gasket (not common).
Tongue and Groove	Made in large and small designs. Inside diameter does not extend to flange bore thus eliminating contact of gasket by process fluid. Small design gives highest joint efficiency possible with flat gaskets. Both tongue and groove pieces must be stocked.	Used for services requiring a retained gasket and lack of contact with process fluid (not common).
Flat Face	Same as raised face except raised portion removed. Often made by machining face from standard raised-face flanges.	Mating flanges for 125- and 250-lb cast-iron valves and fittings.

TABLE 2.10

COMMON GASKET TYPES

[Gasket drawings reproduced by permission: Tube Turns Division, Chemetron Corporation (Copyright 1958)]

	Description	Application	Recommended Flange Surface Finish	m^*	y^*
Flat Ring Flat Non-Metallic Gasket	Paper, cloth, and rubber.	Up to 250°F	Serrated	0.5–1.75	0–1100
	Compressed asbestos.	Most common flat gasket in process plants use limit 750°F	Serrated	2.75 (for ¹⁄₁₆ in. thk)	3700
	Woven asbestos.	Good for glass-lined flanges and rough surfaces. Use limit 300 to 400°F	...	...	...
Flat Metallic Gasket	Many metals— Selection made to suit service (see Table 2.11).	Satisfactory for maximum temperature metal gasket or flange itself will withstand whichever is lower (see Table 2.11).	Serrated	5.5	18000
Serrated Gasket	Flat metal gaskets with concentric grooves machined into faces.	Requires less bolt load than flat metal gaskets and produces more efficient joint. Replacing flat metal gaskets for many uses.	Very smooth	3.75	7600
Laminated Type Jacketed Type	Metal jacketed asbestos.	For use up to 850°F. Require less bolt load to compress than solid metal gaskets and thus more efficient for high temperature high pressure joints.	Very smooth	3.75	7600
Spiral Wound Type	Interlocked plies of preformed metal, cushioned with asbestos strip spirally wound.		Smooth	2.5	2900
Corrugated Type Asbestos Filled Type	Corrugated metal, asbestos filled.	For use up to 850°F and high pressure. Good for severe service such as hot oil and chemicals.	Very smooth	3.0	4500
Asbestos Inserted Type	Corrugated metal with asbestos laid in corrugations.	For use up to 850°F, but not exceeding 600 psi. Not for hot oil.	Smooth	3.0	4500
Ring Joint Octagonal Type Oval Type	Metal rings commonly furnished in soft iron, low carbon steel, stainless steels, monel, nickel, and copper.	See Table 2.11 for temperature recommendations. Most efficient and costly gasket. Internal pressure expands ring and creates a degree of self-sealing. Preferred for severe service conditions. Octagonal ring most common.	Very smooth	5.5	18000

* These are typical values. Data for metallic gaskets are for soft iron or steel. See Table 3.36 for more complete tabulation.

GASKET SELECTION NOTES FOR TABLE 2.10

To select the proper gasket for a particular service requires careful consideration of the operating temperature and nature of the fluid being contained. If a corrosive fluid is being handled, expert recommendations should be obtained from a corrosion engineer. Jacketed metal gaskets may be used up to 850°F. Above this temperature all metal gaskets must be used and many engineers prefer ring-type joints at elevated temperatures.

If joints are to be disassembled frequently manufacturer's recommendation should be followed for most efficient reuseable gasket.

Other variables being equal the gasket factor m and the design seating stress y are helpful in comparing gaskets. A complete list is given in Table 3.36. These values are related to the bolt load as shown below (ASME Code for Unfired Pressure Vessels 1956). Calculated values of m and y can be used to suggest the best gaskets for a given service. For small low-pressure joints y will govern, but for large flanges and high pressures, m will govern.

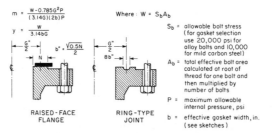

$$m = \frac{W - 0.785G^2P}{(3.14G)(2b)P}$$

$$y = \frac{W}{3.14bG}$$

Where : $W = S_b A_b$

S_b = allowable bolt stress (for gasket selection use 20,000 psi for alloy bolts and 10,000 for mild carbon steel)

A_b = total effective bolt area calculated at root of thread for one bolt and then multiplied by number of bolts

P = maximum allowable internal pressure, psi

b = effective gasket width, in. (see sketches)

RAISED-FACE FLANGE   RING-TYPE JOINT

TABLE 2.11
MAXIMUM SERVICE TEMPERATURES OF GASKET MATERIALS IN OXIDIZING ATMOSPHERE*

Material	Type	Max. Temperature of Continuous Service °F
Tin	...	212
Lead	...	212
Zinc	...	212
Magnesium	...	400
Admiralty Brass	...	500
High Brass	...	500
Copper	...	600
Everdur	...	600
Aluminum	...	800
Stainless Steel	304	800
Stainless Steel	316	800
Rema Iron	...	1000
Armco Iron	...	1000
Low Carbon Steel	...	1000
Silver	...	1200
Gold	...	1200
Chrome Moly Steel	502	1200
Chrome Steel	410	1300
Nickel	...	1400
Monel	...	1500
Stainless Steel	347	1700
Inconel	...	2000
Hastelloy	"A" or "B"	2000
Platinum	...	2300
Tantalum	...	3000

* Temperature listed may be raised or lowered by operating conditions. These values are for general reference only. (Reproduced by permission, "The Gasket," Vol. 1, No. 5–6, Johns Manville Co., New York.)

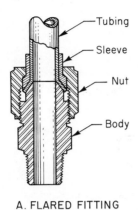

A. FLARED FITTING

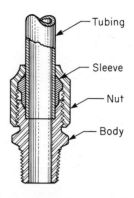

B. FLARELESS FITTING

FIG. 2.20. Small-tubing fittings (⅛ to 2 in.).

A. End of tubing is flared with flaring tool. Nut forces inside of flare against tapered portion of body. Some designs produced without separate sleeve. Flared fittings are usually preferred for thin-wall tubing and soft metals such as brass and copper.

B. No flare required. Sleeve or ferrule grips or bites outer surface of tube, preventing leakage without distorting the inside tube diameter. Preferred for heavy-wall tubing and alloys such as stainless steel. Also used on plastic tubing.

Common materials: Brass, steel, type 316 stainless, and aluminum.

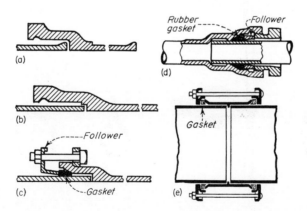

FIG. 2.21. Cast-iron pipe joints. (*a*) and (*b*) Bell and spigot joints: Yarn, braided hemp, or paper calking material is cut slightly longer than the circumference of pipe, wrapped around pipe, and driven into the space between bell and spigot by a special tool. After yarning, the joint is filled with molten lead. Cement has been used in place of lead. Bell and spigot joints are cheapest in first cost but must be assembled in the ditch. (*c*) and (*d*) Mechanical joint: Consists of flanged bell or socket, a rubber gasket fitting into the socket, and a follower ring to compress the gasket. Higher in cost than bell and spigot but cheaper to lay, mechanical joints are quick and easy for even unskilled men to lay since only a rachet wrench is required for assembly. It can be assembled prior to lowering into a ditch and is excellent in services that require deflection, such as underwater lines. (*d*) A mechanical joint for small diameter pipe. (*e*) Coupled joint (Dresser): Used for plain-end cast-iron pipe. The middle ring is forced against gaskets as bolts are evenly tightened. Large sizes must be assembled in ditch. These joints are not commonly used in process plants. Ball and spherical joint (not illustrated): Spigot and bell are spherical permitting 15° deflection. It is good for hilly terrain and underwater installation. (Reproduced by permission from T. Olive, *Chemical Engineering*, p. 187, Dec., 1953.)

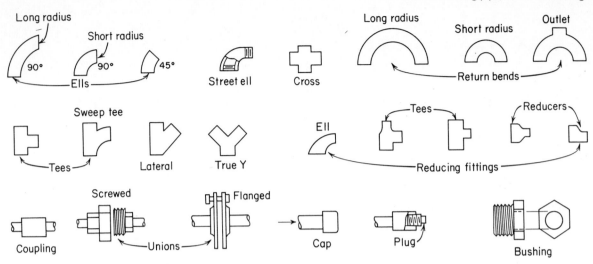

FIG. 2.22. Diagram of fitting nomenclature. (Reproduced by permission from T. Olive, *Chemical Engineering,* p. 187, Dec., 1953.)

FIG. 2.23. Manufactured hangers and supports. (Reproduced by permission of Elcen Metal Products Company.) See Fig. 7.26 for commonly used field-fabricated supports and anchors.

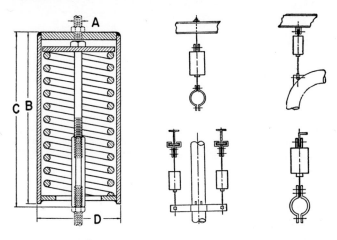

FIG. 2.24. Spring hanger and typical installations. Used for both indoor hot-piping installations and least expensive and most reliable type of flexible pipe support. (Reproduced by permission of Elcen Metal Products Company.)

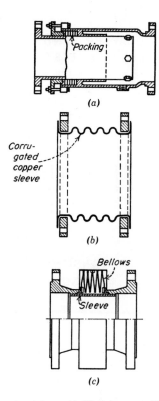

FIG. 2.25. Expansion joints. (a) Slip-joint type: Has stuffing box to prevent leakage and thus requires more maintenance than other types. Design is simplified because length of travel is fixed. Used in power plants. (b) and (c) Packless-bellows type: Requires little maintenance and is usually preferred for process plant use. (b) is low pressure joint and (c) is for high pressures. (Reproduced by permission from T. Olive, *Chemical Engineering*, p. 187, Dec., 1953.)

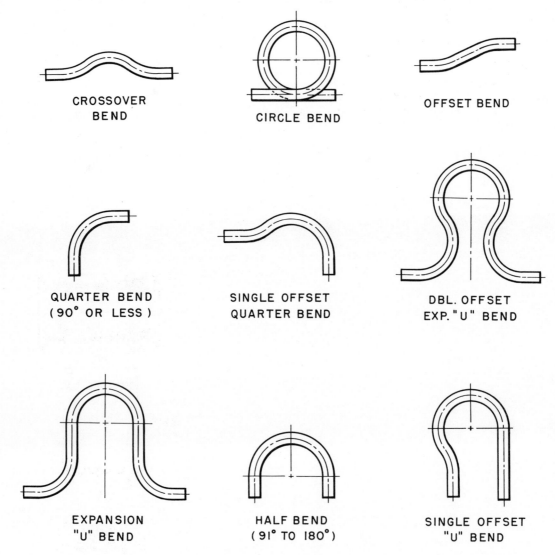

FIG. 2.26. Nomenclature for fabricated pipe bends. (Based on similar presentation by Midwest Piping Division of Crane Company, St. Louis, Mo.)

TABLE 2.12
STEAM TRAP CHARACTERISTICS*

	Operation	Features
1. Inverted bucket	Steam accumulates in inverted bucket causing it to float by increasing the buoyancy of the bucket. Movement of valve lever attached to bucket closes discharge valve. Entrance of more condensate causes bucket to sink closing valve (see cut)	Fast, low steam loss, simple, must be protected from freezing
2. Open bucket	Bucket floats with open end up. Condensate fills bucket causing it to sink and open valve. Steam pressure discharges condensate out discharge pipe in bucket which causes bucket again to float in remaining condensate	Operates well under fluctuating loads
3. Piston operated	Inverted bucket trap which operates a pilot valve which in turn operates a large piston valve	Good for very high capacities (up to 300,000 lb/hr)
4. Thermostatic	Valve attached to bellows containing volatile fluid. When steam contacts bellows, increased vapor pressure of fluid causes expansion of bellows and closing of discharge port	Cannot freeze. Very rapid air venting and high condensate capacity. Not for superheat.
5. Liquid expansion	Similar in principle to No. 4 except expansion of a liquid (oil) is used to move valve	Cannot freeze. Continuous discharge at any desired temperature.
6. Ball float	As condensate collects in trap body it causes a ball float connected to the discharge valve to raise and close the valve	Discharges condensate continuously at steam temperature and without shock.
7. Impulse	One working part, a valve. At or near steam temperature, condensate that has collected in a chamber during discharge flashes and the increase in pressure closes the discharge valve	Cannot freeze. Very small and light weight. High capacity to size ratio.

* See manufacturer's recommendations for selecting and sizing procedures.

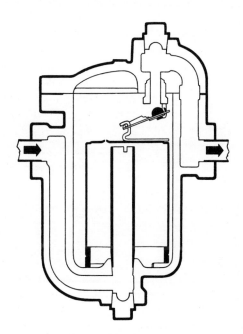

Inverted bucket trap.

(Reproduced by permission: Armstrong Machine Works.)

REFERENCES

1. *Chemical Engineering* **61,** 11, 171–234 (1954). Tabular data on materials of construction with supplemental data presented biennially.

2. *Industrial and Engineering Chemistry* **40,** 1773–1936 (1948). Tabular data on materials of construction with supplemental material presented annually through 1955.

3. Rabald, E., *Corrosion Guide,* Elesvier Publishing Company (1951).

3

Standards for piping

On February 7, 1904, a large portion of the city of Baltimore burned. The fire department, in despair, had sought the aid of nearby Washington, New York, and Philadelphia. Although these cities quickly responded, their generosity had no practical value. The hose couplings of the distant fire departments would not fit the Baltimore hydrants because fire-hose couplings had not been standardized.

The need for standardization is not always as dramatic, but it certainly is just as compelling for the orderly conduct of modern industry.

AMERICAN STANDARDS ASSOCIATION

In 1918 five major engineering societies joined together to form the American Standards Association. This Association consists of more than 100 technical societies and trade associations which legally serve as the owners and operators. Working funds are furnished primarily through the contributions of 2300 company members. To understand the functions of the ASA, it is necessary to turn back the clock to the early 1900's. As the American industrial empire grew, progressive companies realized the value of standardization. Certain design problems which required the application of judgment in their solution recurred frequently, and common agreement among experts in a given company on the solution of such problems saved many design man-hours and costly delays that might have resulted from indecision. In addition, standardization of parts for manufactured equipment not only assured the success of mass production, but also encouraged customer confidence in equipment for which interchangeable parts were readily obtainable.

Often groups of companies within a given industry pooled their talents through trade associations or technical societies and developed industry-wide standards. By 1918 there were several hundred trade associations, engineering societies, and government departments which were originating and issuing standards. Many of these standards overlapped those of other groups, and the duplication caused much confusion. The American Standards Association was founded in 1918 to bring order to this chaotic situation and to serve as the instrument for authorizing national standards and promoting eventual adoption of international standards. It is through this association that interested engineering societies and trade associations can work in creating unified standards on a national level.

ROLE OF FEDERAL GOVERNMENT

The federal government, through its General Service Administration, the Department of Commerce and the military services, issues standards covering thousands of items. In many cases these standards have preceded by a number of years those used by the ASA and trade associations and have encouraged development of industry-wide standards. The present policy of the federal government is to use ASA standards and other recognized standards whenever possible. It remains necessary, however, even in non-military procurement, to prepare government standards for certain new materials for which no ASA standards exist. This proves an excellent service to industry because adoption of industry-wide standardization too early in the development of materials can retard technological advances. The early government standards serve as pilot standards for industry and often as models for the eventual adoption of an industry-wide standard.

The operation of the Department of Commerce in relation to standards is of particular interest to industrial people. Groups of manufacturers, distributors, or users of a specific product may submit to the Commodity Standards Division of the Department of Commerce the necessary data on which a standard may be based. By means of conferences or letter referenda, the Division assists the sponsor group in developing a tentative standard. The Division then submits this tentative standard to other elements of the industry for review and criticism. After final changes and acceptance by the industry, it becomes a voluntary standard for the particular commodity. This procedure is most valuable for new products (e.g., a new plastic pipe) for which not enough technical data have been obtained to establish engineering standards, but for which certain levels of agreement such as dimensions and tolerances can be profitably developed into an acceptable standard.

OTHER STANDARDS ORGANIZATIONS

Standards organizations operate in all parts of the world. Most of these are members of the International Organization for Standardization which promotes interchange of standards and is actively engaged in the development of international standards. Addresses of the member organizations and information concerning standards issued by them can be obtained from the American Standards Association.

The number of standards issued by other countries on metal pipe, valves, and fittings is indicative of the activity in this category. The following are representative:

Canada	16	Great Britian	73
France	184	Sweden	273
Germany	321		

Numbers, of course, do not tell the whole story. Many of these organizations or their predecessors originated and continue to originate standards for important items of commerce. For example, American polyethylene pipe standards are based, in part, on British standards.

Piping designers planning piping in other countries should be familiar with piping materials and standards for those countries. As a service, the American Standards Association maintains a library of foreign standards, and ASA members may borrow standards from this library. Copies of frequently used foreign standards may also be purchased from the ASA.

It is hoped that ultimately international standards will be developed and generally accepted. Some encouraging progress has already been made as can be seen from the following examples of "Recommendations" issued by the International Standards Organization:

CAST-IRON PIPE

ISO/R13	Cast-iron pipes—special castings and cast iron parts for pressure main lines
ISO/R49	Malleable cast-iron pipe fittings screwed in accordance ISO/R7

PIPE THREADS

ISO/R7	Pipe threads for gas list tubes and screwed fittings where pressure-tight joints are made on the threads (⅛ in. to 6 in.)

PIPE LINES

ISO/R51	Pipe lines for the transport of combustible liquids—nominal diameters

STEEL PIPE

ISO/R64	Steel tubes—outside diameters
ISO/R65	Seamless and welded steel tubes suitable for screwing in accordance with ISO/R7
	Draft ISO Recommendation 225—Non-screwed steel tubes for general purposes

ENGINEERING DRAWING—*Methods, Dimensioning, etc.*

ISO Draft Recommendation 140–143

AMERICAN PIPING STANDARDS

The piping standards approved by the American Standards Association comprise the majority of standards in the United States which govern the design of piping systems and the dimensions and ratings of pipe, valves, and fittings. Most of these standards were sponsored by the American Society of Mechanical Engineers, American Water Works Association, and the American Society for Testing Materials (ASTM). A number of other standards which have not yet been approved as American Standards are in wide use. These include standards published by the following organizations:

Organization	Type Standard
American Society for Testing Materials	Materials (some dimensional)
Department of Commerce Commercial Standards	All types of new products
Federal Specifications by General Service Administration	All types
American Water Works Association	Cast-iron pipe and fittings

Organization	Type Standard
American Gas Association	Cast-iron pipe and fittings
Instrument Society of America	Control valves and other instrument data associated with piping
Manufacturers Standardization Society of the Valve and Fitting Industry	Valves and fittings

The most important and frequently used piping standard is the *Code for Pressure Piping* (ASA B31.1). It is a guide for minimum design requirements and, as such, enables a designer to make rapid decisions on recurring design problems with the assurance that such decisions will be accepted, not only by his colleagues but also by the outstanding authorities of the nation. While its use is not mandatory by law, its general acceptance throughout the United States makes it a valuable guide for the designer.

Lists of the most important American standards on piping design, dimensional and mechanical characteristics, and materials are given, beginning on pages 42, 67, and 109, respectively. Each of these lists is accompanied by a series of tables and charts extracted from these standards. These have been designed for rapid location of frequently used data, and should not be considered as substitutes for the complete standards. Those portions of the ASA Codes which are included are designated by the appropriate numbers and were extracted with the permission of the publisher, The American Society of Mechanical Engineers, 345 East 47th St., New York 17, N.Y.

PIPING DESIGN STANDARDS

Title	Designation
Code for Pressure Piping, Sections 1–8:	B31.1
1. Power Piping Systems	
2. Industrial Gas and Air Piping Systems	
3. Refinery and Oil Transportation Systems (issued separately)	B31.3
4. District Heating Piping Systems	B31.4
5. Refrigeration Piping (issued separately)	B31.5
6. Fabrication Details	
7. Materials	
8. Gas Transmission and Distribution Piping Systems (issued separately)	B31.8
Manual for the Computation of Strength and Thickness of Cast-Iron Pipe (AWWA C101 (See also ASA A21, 2, 3, 6, 7, 8, and 9 for Tables of thicknesses for various laying conditions)	A21.1
National Plumbing Code	A40.8
Recommended Practice for Installing Clay Sewer Pipe (Rules for Calculating Supporting Strength)	ASTM C12
Specifications of Crushing Strengths of Clay, Concrete and Asbestos-Cement Pipe	See Materials Standards List

THICKNESS OF PIPE

METAL PIPING AND FITTINGS OTHER THAN CAST IRON

The section of the code applicable to most process plant piping is ASA B31.3, "Petroleum Refinery Piping." For power stations and refrigeration systems reference should be made to Sections 1 and 5, respectively. The following most frequently used data have been adapted from Section 3 and other indicated codes:

Calculation of Minimum Thickness

$$t = M \left(\frac{pD}{2S} + C \right) \qquad (1)^*$$

where t = Wall thickness of pipe or fitting in inches, including 12.5% manufacturing tolerance (see Table 3.1, page 47, for minimum thickness).

p = Internal design service pressure, psig (common practice to use about 10% over maximum anticipated).

D = Outside diameter of pipe in inches (see Table 3.2, page 48, for pipe dimension).

S = Maximum allowable stress, psi, based on maximum anticipated operating temperature (see Tables 3.3 and 3.3A, pages 52 and 54).

M = Manufacturer's tolerance 1.125 for steel pipe psi.

C = Corrosion allowance in inches, plus thread or groove depth in inches (see Table 3.4, page 54, for thread depths). Corrosion allowances must be based on the type of material being handled and are not specified in Section 3. The following allowances are recommended in other sections of the Code and will prove helpful as guides for carbon steel pipe:

Power Piping	
District Heating	0.05 in. for 1 in. and smaller
Refrigeration	0.065 in. above 1 in.
Industrial Gas and Air	0.05 in.

For carbon-steel pipe in process plants operating with moderately corrosive fluids an allowance of 0.125 in. is often used for 2-in. and larger pipe. The allowance for special corrosion resistant alloys is zero if experience indicates that they are essentially corrosion resistant.

* This is a useful formula for estimates. Recent editions of the Code (B31.3—1962) give a more precise formula employing a welded-joint efficiency which should be used for all Code design calculations.

Example Calculation

A 6-in. ASTM 106 Grade A pipe is to operate at 600 psig and 500°F. Select the Schedule Number for this service, using a corrosion allowance of 0.125 in. Welded construction is to be used.

$$S = 13,100 \text{ psi (Table 3.3)}.$$
$$D = 6.625 \text{ in. (Table 3.2)}.$$
$$t = 1.125\left[\frac{(600)(6.625)}{(2)(13100)} + 0.125\right] = 0.311 \text{ in.}$$

From Table 3.2, page 48, the nearest schedule number is Schedule 80. Thus Schedule 80 pipe and Schedule 80 welding fittings should be used.

INSTRUMENT PIPING

The type of tubing selected depends on operating temperature and pressure as shown in Fig. 3.3. Thickness should be checked for pressure and temperature as described above. Minimum sizes are given in Table 3.5. Materials should not be used above temperatures corresponding to listed maximum allowable stresses in Table 3.3. Copper, brass, bronze, and aluminum must not be used in connecting main lines containing flammable fluids.

CAST-IRON PIPE

The use of cast-iron pipe has been limited by the code as shown in Figs. 3.1 and 3.2.

Underground Water Lines, Gas Lines, and Oil Lines.* Tables 3.6, 3.7, 3.8, and 3.9 (pages 55 to 60) give thicknesses for various laying conditions. Table 3.17, page 71, gives standard thickness classes for centrifugally cast pipe. These tables are from designated ASA specifications and were calculated by the methods of ASA A21.1. Thicknesses include foundry allowances, corrosion, and truck load (two trucks passing simultaneously with rear axles both over the pipe at the same time and 9000-lb load on each rear wheel) or water hammer, whichever controls. For conditions and pipe sizes not shown in the tables see ASA A21.1.

Centrifugally cast pipe is stronger than pit cast and is most generally used in process work. Similar data to that presented here can be obtained for pit cast pipe from ASA A21.2 and A21.3.

Example

A 24-in. cast-iron pipe for water at 100 psig is to be placed 5 ft underground in a flat-bottom trench without tamping the backfill. Select the thickness class of cast-iron pipe, centrifugally cast in metal molds.

From cuts underneath tables—Laying condition is A
Table 3.6—Thickness required: 0.73 in.
Table 3.17—Thickness Class: 23

Aboveground Service for Oil, Oil Vapor, and Refinery Gas
Limitations
  150 psi and 300°F with 450°F maximum for condensers and coolers employing cast iron immersed in water.
Method
  Use Equation 1 given for metal piping other than cast iron, substituting for P the term $P + P_n$, where P_n is the water-hammer allowance taken from Table 3.10, page 61. For S, use 4000 psi for pit-cast pipe and 6000 psi for centrifugally cast pipe.

CONCRETE PIPE

Concrete pipe is made both plain and reinforced for surface drains and culverts (not used for plant sewers because acid attacks it). Supporting strengths for plain pipe can be determined from Table 3.11, page 61. (See ASTM C76, C361, and C362 for details on reinforced pipe.)

GLASS PIPE

No official standard pressure-temperature ratings exist. Table 3.12, page 62, gives manufacturers' recommendations for a borosilicate glass pipe.

PLASTIC PIPE

Although some standardization has been achieved by the Society of the Plastics Industry in cooperation with the U.S. Department of Commerce, much remains to be done. Standard ratings for regular polyethylene pipe have been issued (CS–197), but other plastic pipe continues to be rated by the individual manufacturers. Some uniformity is developing and further standardization should be forthcoming. The Commercial Standard for regular polyethylene, together with some typical manufacturers' ratings of other plastic pipe, are included here in Fig. 3.4 and in Tables 3.13 and 3.14 (pages 47 and 63). The American Society for testing materials is considering standards for polyethylene, polyvinyl chloride, and acrylonitrile-butadiene-styrene pipe. These will become ASA standards in the B72 category.

PLASTIC- AND RUBBER-LINED PIPE

Schedule 40 seamless steel pipe with 150-lb ASA flanges is the usual commercial lined pipe. Its pressure-temperature ratings conform to 150-lb ASA flange ratings, except that the temperature is limited by the plastic or rubber lining. For high-temperature service (up to 500°F), fluorocarbon plastics are specified. Manufacturers of lined pipe should be consulted for aid in selecting and applying all plastic-lined pipe.

ASBESTOS-CEMENT PIPE

Pressure Pipe and Water Pipe. Made in 100, 150, and 200 psi classes for operation at these pressures (ASTM C296). Temperature is limited by the type of flange coupling selected and varies from 150 to 200°F. Supporting strengths can be determined from Table 3.11 for pipe in underground service.

Sewer Pipe. Thickness of pipe depends on external loads. Manufacturers designate sewer pipe as 1500, 2400, 3300, 4000, and 5000, which correspond respectively to the value of crushing strength in pounds per linear foot. Federal Specifications SS–P331a employs Class 1 and 2, the crushing strength of which varies with the pipe size and the class.

CLAY SEWER AND DRAINAGE PIPE

Thickness of pipe (standard strength or extra strength) is governed by type and amount of anticipated dead and live loads. Supporting strength for the two classes can be determined from Table 3.11.

---

* Limitations on oil lines are 400 psi and 300°F maximum.

PLUMBING DESIGN

(Excerpts from National Plumbing Code, ASA A40.8)

Minimum Facilities. Table 3.15 lists the recommended minimum number of fixtures for various types of buildings or occupancy.

Sizing Building-Water Distribution and Drainage Systems. Data from the code on line-sizing is given on pages 141 to 146.

Location of Water Supply Wells. In addition to the minimum distances given in Table 3.16, the well site should be at a higher elevation than possible sources of contamination. The top of the well should be at least 2 ft above the highest known water mark and at least 50 ft, measured horizontally, from surface bodies of water. The well should be provided with an outside water-tight casing extending at least 10 ft below and 6 in. above ground surface.

INDEX OF TABLES AND FIGURES FOR PIPE DESIGN

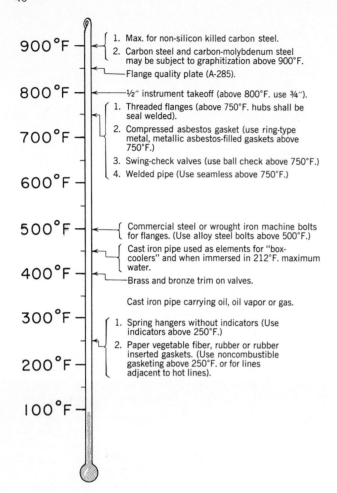

900°F —
1. Max. for non-silicon killed carbon steel.
2. Carbon steel and carbon-molybdenum steel may be subject to graphitization above 900°F.

— Flange quality plate (A-285).

800°F —
½″ instrument takeoff (above 800°F. use ¾″).

1. Threaded flanges (above 750°F. hubs shall be seal welded).
700°F —
2. Compressed asbestos gasket (use ring-type metal, metallic asbestos-filled gaskets above 750°F.)
3. Swing-check valves (use ball check above 750°F.)
600°F —
4. Welded pipe (Use seamless above 750°F.)

500°F —
Commercial steel or wrought iron machine bolts for flanges. (Use alloy steel bolts above 500°F.)

Cast iron pipe used as elements for "box-coolers" and when immersed in 212°F. maximum water.
400°F —
Brass and bronze trim on valves.

Cast iron pipe carrying oil, oil vapor or gas.

300°F —
1. Spring hangers without indicators (Use indicators above 250°F.)

2. Paper vegetable fiber, rubber or rubber inserted gaskets. (Use noncombustible gasketing above 250°F. or for lines adjacent to hot lines).
200°F —

100°F —

FIG. 3.1.   Design temperature limitations.

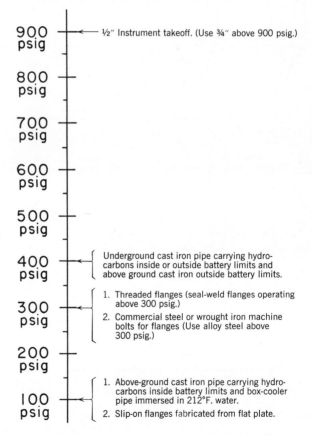

900 psig —
½″ Instrument takeoff. (Use ¾″ above 900 psig.)

800 psig —

700 psig —

600 psig —

500 psig —

400 psig —
Underground cast iron pipe carrying hydro-carbons inside or outside battery limits and above ground cast iron outside battery limits.

300 psig —
1. Threaded flanges (seal-weld flanges operating above 300 psig.)

2. Commercial steel or wrought iron machine bolts for flanges (Use alloy steel above 300 psig.)

200 psig —

100 psig —
1. Above-ground cast iron pipe carrying hydro-carbons inside battery limits and box-cooler pipe immersed in 212°F. water.

2. Slip-on flanges fabricated from flat plate.

FIG. 3.2.   Design pressure limitations.   For other pressure limitations, see listings for valves, fittings, and flanges.

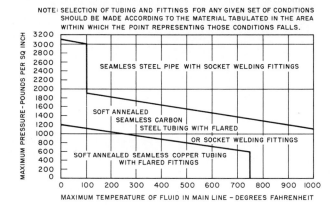

FIG. 3.3. Chart for selection of tubing and fittings for meter and instrument connecting piping (ASA B31.1).

TABLE 3.1
MINIMUM DESIGN THICKNESS FOR CARBON STEEL AND ALLOY PIPE
(Based on Paragraph 324 ASA B31.1)

Nominal Pipe Size, Inches	t − C
½	0.03
¾	0.04
1	0.045
1¼–1½	0.05
2	0.06
2½–3	0.07
4	0.09
6	0.11
8	0.12
10–24	0.13

Note: For mechanical strength it is customary in many plants to use a minimum of Schedule 80 for screwed carbon steel pipe in sizes less than 1½ in. Lighter weight alloy piping, however, is frequently used (see Table 3.2). Lightweight electric resistance-welded carbon steel pipe ranging from ¾ to 12 in. conforming to Schedule 10S is being standardized by the ASTM for use in low pressure services. Savings are expected in fabrication and erection as well as in material cost.

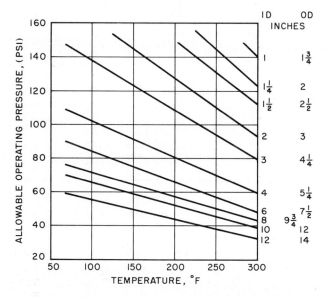

FIG. 3.4. Reinforced plastic pipe. Manufacturer's recommended pressure-temperature ratings for asbestos reinforced phenol-formaldehyde resin and furan resin pipe. (Adapted by permission of Haveg Industries, Inc.)

TABLE 3.2
PROPERTIES OF PIPE

(Reproduced by permission: *Piping Design and Engineering,* copyright 1951, Grinnel Company, Providence, R.I.)

The following formulas were used in the computation of the values shown in the table:

† Weight of pipe per foot (pounds) $= 10.6802t\,(D - t)$
Weight of water per foot (pounds) $= 0.3405d^2$
Square feet outside surface per foot $= 0.2618D$
Square feet inside surface per foot $= 0.2618d$
Inside area (square inches) $= 0.785d^2$
Area of metal (square inches) $= 0.785\,(D^2 - d^2)$
Moment of inertia (inches⁴) $= 0.0491\,(D^4 - d^4)$
$= A_M R_g^2$

Section modulus (inches³) $= \dfrac{0.0982\,(D^4 - d^4)}{D}$

Radius of gyration (inches) $= 0.25\,\sqrt{D^2 + d^2}$

A_M = Area of Metal (square inches)
d = Inside Diameter (inches)
D = Outside Diameter (inches)
R_g = Radius of Gyration (inches)
t = Pipe Wall Thickness (inches)

NOTE: a. A.S.A. B 36.10 Steel Pipe Schedule Numbers.
b. A.S.A. B 36.10 Steel Pipe Nominal Wall Thickness Designations.
c. A.S.A. B 36.19 Stainless Steel Pipe Schedule Numbers.

† The ferritic stainless steels may be about 5% less, and the austenitic stainless steels about 2% greater than the values shown in this table which are based on weights for carbon steel.

Nominal Pipe Size *Outside Diameter,* in.	Schedule Number* a	b	c	Wall Thickness, in.	Inside Diameter, in.	Inside Area, sq in.	Metal Area, sq in.	Sq Ft Outside Surface, per ft	Sq Ft Inside Surface, per ft	Weight per ft, lb	Weight of Water per ft, lb	Moment of Inertia, in.⁴	Section Modulus, in.³	Radius Gyration, in.
⅛ 0.405	—	—	10S	0.049	0.307	0.0740	0.0548	0.106	0.0804	0.186	0.0321	0.00088	0.00437	0.1271
	40	Std	40S	0.068	0.269	0.0568	0.0720	0.106	0.0705	0.245	0.0246	0.00106	0.00525	0.1215
	80	XS	80S	0.095	0.215	0.0364	0.0925	0.106	0.0563	0.315	0.0157	0.00122	0.00600	0.1146
¼ 0.540	—	—	10S	0.065	0.410	0.1320	0.0970	0.141	0.1073	0.330	0.0572	0.00279	0.01032	0.1694
	40	Std	40S	0.088	0.364	0.1041	0.1250	0.141	0.0955	0.425	0.0451	0.00331	0.01230	0.1628
	80	XS	80S	0.119	0.302	0.0716	0.1574	0.141	0.0794	0.535	0.0310	0.00378	0.01395	0.1547
⅜ 0.675	—	—	10S	0.065	0.545	0.2333	0.1246	0.177	0.1427	0.423	0.1011	0.00586	0.01737	0.2169
	40	Std	40S	0.091	0.493	0.1910	0.1670	0.177	0.1295	0.568	0.0827	0.00730	0.02160	0.2090
	80	XS	80S	0.126	0.423	0.1405	0.2173	0.177	0.1106	0.739	0.0609	0.00862	0.02554	0.1991
½ 0.840	—	—	10S	0.083	0.674	0.357	0.1974	0.220	0.1765	0.671	0.1547	0.01431	0.0341	0.2692
	40	Std	40S	0.109	0.622	0.304	0.2503	0.220	0.1628	0.851	0.1316	0.01710	0.0407	0.2613
	80	XS	80S	0.147	0.546	0.2340	0.320	0.220	0.1433	1.088	0.1013	0.02010	0.0478	0.2505
	160	—	—	0.187	0.466	0.1706	0.383	0.220	0.1220	1.304	0.0740	0.02213	0.0527	0.2402
	—	XXS	—	0.294	0.252	0.0499	0.504	0.220	0.0660	1.714	0.0216	0.02425	0.0577	0.2192
¾ 1.050	—	—	5S	0.065	0.920	0.665	0.2011	0.275	0.2409	0.684	0.2882	0.02451	0.0467	0.349
	—	—	10S	0.083	0.884	0.614	0.2521	0.275	0.2314	0.857	0.2661	0.02970	0.0566	0.343
	40	Std	40S	0.113	0.824	0.533	0.333	0.275	0.2157	1.131	0.2301	0.0370	0.0706	0.334
	80	XS	80S	0.154	0.742	0.432	0.435	0.275	0.1943	1.474	0.1875	0.0448	0.0853	0.321
	160	—	—	0.218	0.614	0.2961	0.570	0.275	0.1607	1.937	0.1284	0.0527	0.1004	0.304
	—	XXS	—	0.308	0.434	0.1479	0.718	0.275	0.1137	2.441	0.0641	0.0579	0.1104	0.2840
1 1.315	—	—	5S	0.065	1.185	1.103	0.2553	0.344	0.310	0.868	0.478	0.0500	0.0760	0.443
	—	—	10S	0.109	1.097	0.945	0.413	0.344	0.2872	1.404	0.409	0.0757	0.1151	0.428
	40	Std	40S	0.133	1.049	0.864	0.494	0.344	0.2746	1.679	0.374	0.0874	0.1329	0.421
	80	XS	80S	0.179	0.957	0.719	0.639	0.344	0.2520	2.172	0.311	0.1056	0.1606	0.407
	160	—	—	0.250	0.815	0.522	0.836	0.344	0.2134	2.844	0.2261	0.1252	0.1903	0.387
	—	XXS	—	0.358	0.599	0.2818	1.076	0.344	0.1570	3.659	0.1221	0.1405	0.2137	0.361
1¼ 1.660	—	—	5S	0.065	1.530	1.839	0.326	0.434	0.401	1.107	0.797	0.1038	0.1250	0.564
	—	—	10S	0.109	1.442	1.633	0.531	0.434	0.378	1.805	0.707	0.1605	0.1934	0.550
	40	Std	40S	0.140	1.380	1.496	0.669	0.434	0.361	2.273	0.648	0.1948	0.2346	0.540
	80	XS	80S	0.191	1.278	1.283	0.881	0.434	0.335	2.997	0.555	0.2418	0.2913	0.524
	160	—	—	0.250	1.160	1.057	1.107	0.434	0.304	3.765	0.458	0.2839	0.342	0.506
	—	XXS	—	0.382	0.896	0.631	1.534	0.434	0.2346	5.214	0.2732	0.341	0.411	0.472
1½ 1.900	—	—	5S	0.065	1.770	2.461	0.375	0.497	0.463	1.274	1.067	0.1580	0.1663	0.649
	—	—	10S	0.109	1.682	2.222	0.613	0.497	0.440	2.085	0.962	0.2469	0.2599	0.634

* See note at top of table for definitions of columns a, b, and c.

TABLE 3.2 (*Continued*). PROPERTIES OF PIPE

Nominal Pipe Size *Outside Diameter,* in.	Schedule Number*			Wall Thickness, in.	Inside Diameter, in.	Inside Area, sq in.	Metal Area, sq in.	Sq Ft Outside Surface, per ft	Sq Ft Inside Surface, per ft	Weight per ft, lb	Weight of Water per ft, lb	Moment of Inertia, in.⁴	Section Modulus, in.³	Radius Gyration, in.
	a	b	c											
1½ 1.900	40	Std	40S	0.145	1.610	2.036	0.799	0.497	0.421	2.718	0.882	0.310	0.326	0.623
	80	XS	80S	0.200	1.500	1.767	1.068	0.497	0.393	3.631	0.765	0.391	0.412	0.605
	160	—	—	0.281	1.338	1.406	1.429	0.497	0.350	4.859	0.608	0.483	0.508	0.581
	—	XXS	—	0.400	1.100	0.950	1.885	0.497	0.288	6.408	0.412	0.568	0.598	0.549
2 2.375	—	—	5S	0.065	2.245	3.96	0.472	0.622	0.588	1.604	1.716	0.315	0.2652	0.817
	—	—	10S	0.109	2.157	3.65	0.776	0.622	0.565	2.638	1.582	0.499	0.420	0.802
	40	Std	40S	0.154	2.067	3.36	1.075	0.622	0.541	3.653	1.455	0.666	0.561	0.787
	80	XS	80S	0.218	1.939	2.953	1.477	0.622	0.508	5.022	1.280	0.868	0.731	0.766
	160	—	—	0.343	1.689	2.240	2.190	0.622	0.442	7.444	0.971	1.163	0.979	0.729
	—	XXS	—	0.436	1.503	1.774	2.656	0.622	0.393	9.029	0.769	1.312	1.104	0.703
2½ 2.875	—	—	5S	0.083	2.709	5.76	0.728	0.753	0.709	2.475	2.499	0.710	0.494	0.988
	—	—	10S	0.120	2.635	5.45	1.039	0.753	0.690	3.531	2.361	0.988	0.687	0.975
	40	Std	40S	0.203	2.469	4.79	1.704	0.753	0.646	5.793	2.076	1.530	1.064	0.947
	80	XS	80S	0.276	2.323	4.24	2.254	0.753	0.608	7.661	1.837	1.925	1.339	0.924
	160	—	—	0.375	2.125	3.55	2.945	0.753	0.556	10.01	1.535	2.353	1.637	0.894
	—	XXS	—	0.552	1.771	2.464	4.03	0.753	0.464	13.70	1.067	2.872	1.998	0.844
3 3.500	—	—	5S	0.083	3.334	8.73	0.891	0.916	0.873	3.03	3.78	1.301	0.744	1.208
	—	—	10S	0.120	3.260	8.35	1.274	0.916	0.853	4.33	3.61	1.822	1.041	1.196
	40	Std	40S	0.216	3.068	7.39	2.228	0.916	0.803	7.58	3.20	3.02	1.724	1.164
	80	XS	80S	0.300	2.900	6.61	3.02	0.916	0.759	10.25	2.864	3.90	2.226	1.136
	160	—	—	0.437	2.626	5.42	4.21	0.916	0.687	14.32	2.348	5.03	2.876	1.094
	—	XXS	—	0.600	2.300	4.15	5.47	0.916	0.602	18.58	1.801	5.99	3.43	1.047
3½ 4.000	—	—	5S	0.083	3.834	11.55	1.021	1.047	1.004	3.47	5.01	1.960	0.980	1.385
	—	—	10S	0.120	3.760	11.10	1.463	1.047	0.984	4.97	4.81	2.756	1.378	1.372
	40	Std	40S	0.226	3.548	9.89	2.680	1.047	0.929	9.11	4.28	4.79	2.394	1.337
	80	XS	80S	0.318	3.364	8.89	3.68	1.047	0.881	12.51	3.85	6.28	3.14	1.307
4 4.500	—	—	5S	0.083	4.334	14.75	1.152	1.178	1.135	3.92	6.40	2.811	1.249	1.562
	—	—	10S	0.120	4.260	14.25	1.651	1.178	1.115	5.61	6.17	3.96	1.762	1.549
	40	Std	40S	0.237	4.026	12.73	3.17	1.178	1.054	10.79	5.51	7.23	3.21	1.510
	80	XS	80S	0.337	3.826	11.50	4.41	1.178	1.002	14.98	4.98	9.61	4.27	1.477
	120	—	—	0.437	3.626	10.33	5.58	1.178	0.949	18.96	4.48	11.65	5.18	1.445
	160	—	—	0.531	3.438	9.28	6.62	1.178	0.900	22.51	4.02	13.27	5.90	1.416
	—	XXS	—	0.674	3.152	7.80	8.10	1.178	0.825	27.54	3.38	15.29	6.79	1.374
5 5.563	—	—	5S	0.109	5.345	22.44	1.868	1.456	1.399	6.35	9.73	6.95	2.498	1.929
	—	—	10S	0.134	5.295	22.02	2.285	1.456	1.386	7.77	9.53	8.43	3.03	1.920
	40	Std	40S	0.258	5.047	20.01	4.30	1.456	1.321	14.62	8.66	15.17	5.45	1.878
	80	XS	80S	0.375	4.813	18.19	6.11	1.456	1.260	20.78	7.89	20.68	7.43	1.839
	120	—	—	0.500	4.563	16.35	7.95	1.456	1.195	27.04	7.09	25.74	9.25	1.799
	160	—	—	0.625	4.313	14.61	9.70	1.456	1.129	32.96	6.33	30.0	10.80	1.760
	—	XXS	—	0.750	4.063	12.97	11.34	1.456	1.064	38.55	5.62	33.6	12.10	1.722
6 6.625	—	—	5S	0.109	6.407	32.2	2.231	1.734	1.677	5.37	13.98	11.85	3.58	2.304
	—	—	10S	0.134	6.357	31.7	2.733	1.734	1.664	9.29	13.74	14.40	4.35	2.295
	40	Std	40S	0.280	6.065	28.89	5.58	1.734	1.588	18.97	12.51	28.14	8.50	2.245
	80	XS	80S	0.432	5.761	26.07	8.40	1.734	1.508	28.57	11.29	40.5	12.23	2.195
	120	—	—	0.562	5.501	23.77	10.70	1.734	1.440	36.39	10.30	49.6	14.98	2.153
	160	—	—	0.718	5.189	21.15	13.33	1.734	1.358	45.30	9.16	59.0	17.81	2.104
	—	XXS	—	0.864	4.897	18.83	15.64	1.734	1.282	53.16	8.17	66.3	20.03	2.060
8 8.625	—	—	5S	0.109	8.407	55.5	2.916	2.258	2.201	9.91	24.07	26.45	6.13	3.01
	—	—	10S	0.148	8.329	54.5	3.94	2.258	2.180	13.40	23.59	35.4	8.21	3.00
	20	—	—	0.250	8.125	51.8	6.58	2.258	2.127	22.36	22.48	57.7	13.39	2.962
	30	—	—	0.277	8.071	51.2	7.26	2.258	2.113	24.70	22.18	63.4	14.69	2.953
	40	Std	40S	0.322	7.981	50.0	8.40	2.258	2.089	28.55	21.69	72.5	16.81	2.938
	60	—	—	0.406	7.813	47.9	10.48	2.258	2.045	35.64	20.79	88.8	20.58	2.909
	80	XS	80S	0.500	7.625	45.7	12.76	2.258	1.996	43.39	19.80	105.7	24.52	2.878

* See note on page 48 for definitions of columns a, b, and c.

TABLE 3.2 (*Continued*). PROPERTIES OF PIPE

Nominal Pipe Size *Outside Diameter*, in.	Schedule Number*			Wall Thickness, in.	Inside Diameter, in.	Inside Area, sq in.	Metal Area, sq in.	Sq Ft Outside Surface, per ft	Sq Ft Inside Surface, per ft	Weight per ft, lb	Weight of Water per ft, lb	Moment of Inertia, in.⁴	Section Modulus, in.³	Radius Gyration, in.
	a	b	c											
8 8.625	100	—	—	0.593	7.439	43.5	14.96	2.258	1.948	50.87	18.84	121.4	28.14	2.847
	120	—	—	0.718	7.189	40.6	17.84	2.258	1.882	60.63	17.60	140.6	32.6	2.807
	140	—	—	0.812	7.001	38.5	19.93	2.258	1.833	67.76	16.69	153.8	35.7	2.777
	—	XXS	—	0.875	6.875	37.1	21.30	2.258	1.800	72.42	16.09	162.0	37.6	2.757
	160	—	—	0.906	6.813	36.5	21.97	2.258	1.784	74.69	15.80	165.9	38.5	2.748
10 10.750	—	—	5S	0.134	10.482	86.3	4.52	2.815	2.744	15.15	37.4	63.7	11.85	3.75
	—	—	10S	0.165	10.420	85.3	5.49	2.815	2.728	18.70	36.9	76.9	14.30	3.74
	20	—	—	0.250	10.250	82.5	8.26	2.815	2.683	28.04	35.8	113.7	21.16	3.71
	—	—	—	0.279	10.192	81.6	9.18	2.815	2.668	31.20	35.3	125.9	23.42	3.70
	30	—	—	0.307	10.136	80.7	10.07	2.815	2.654	34.24	35.0	137.5	25.57	3.69
	40	Std	40S	0.365	10.020	78.9	11.91	2.815	2.623	40.48	34.1	160.8	29.90	3.67
	60	XS	80S	0.500	9.750	74.7	16.10	2.815	2.553	54.74	32.3	212.0	39.4	3.63
	80	—	—	0.593	9.564	71.8	18.92	2.815	2.504	64.33	31.1	244.9	45.6	3.60
	100	—	—	0.718	9.314	68.1	22.63	2.815	2.438	76.93	29.5	286.2	53.2	3.56
	120	—	—	0.843	9.064	64.5	26.24	2.815	2.373	89.20	28.0	324	60.3	3.52
	140	—	—	1.000	8.750	60.1	30.6	2.815	2.291	104.13	26.1	368	68.4	3.47
	160	—	—	1.125	8.500	56.7	34.0	2.815	2.225	115.65	24.6	399	74.3	3.43
12 12.750	—	—	5S	0.165	12.420	121.2	6.52	3.34	3.25	19.56	52.5	129.2	20.27	4.45
	—	—	10S	0.180	12.390	120.6	7.11	3.34	3.24	24.20	52.2	140.5	22.03	4.44
	20	—	—	0.250	12.250	117.9	9.84	3.34	3.21	33.38	51.1	191.9	30.1	4.42
	30	—	—	0.330	12.090	114.8	12.88	3.34	3.17	43.77	49.7	248.5	39.0	4.39
	—	Std	40S	0.375	12.000	113.1	14.58	3.34	3.14	49.56	49.0	279.3	43.8	4.38
	40	—	—	0.406	11.938	111.9	15.74	3.34	3.13	53.53	48.5	300	47.1	4.37
	—	XS	80S	0.500	11.750	108.4	19.24	3.34	3.08	65.42	47.0	362	56.7	4.33
	60	—	—	0.562	11.626	106.2	21.52	3.34	3.04	73.16	46.0	401	62.8	4.31
	80	—	—	0.687	11.376	101.6	26.04	3.34	2.978	88.51	44.0	475	74.5	4.27
	100	—	—	0.843	11.064	96.1	31.5	3.34	2.897	107.20	41.6	562	88.1	4.22
	120	—	—	1.000	10.750	90.8	36.9	3.34	2.814	125.49	39.3	642	100.7	4.17
	140	—	—	1.125	10.500	86.6	41.1	3.34	2.749	139.68	37.5	701	109.9	4.13
	160	—	—	1.312	10.126	80.5	47.1	3.34	2.651	160.27	34.9	781	122.6	4.07
14 14.000	10	—	—	0.250	13.500	143.1	10.80	3.67	3.53	36.71	62.1	255.4	36.5	4.86
	20	—	—	0.312	13.376	140.5	13.42	3.67	3.50	45.68	60.9	314	44.9	4.84
	30	Std	—	0.375	13.250	137.9	16.05	3.67	3.47	54.57	59.7	373	53.3	4.82
	40	—	—	0.437	13.126	135.3	18.62	3.67	3.44	63.37	58.7	429	61.2	4.80
	—	XS	—	0.500	13.000	132.7	21.21	3.67	3.40	72.09	57.5	484	69.1	4.78
	—	—	—	0.562	12.876	130.2	23.73	3.67	3.37	80.66	56.5	537	76.7	4.76
	60	—	—	0.593	12.814	129.0	24.98	3.67	3.35	84.91	55.9	562	80.3	4.74
	—	—	—	0.625	12.750	127.7	26.26	3.67	3.34	89.28	55.3	589	84.1	4.73
	—	—	—	0.687	12.626	125.2	28.73	3.67	3.31	97.68	54.3	638	91.2	4.71
	80	—	—	0.750	12.500	122.7	31.2	3.67	3.27	106.13	53.2	687	98.2	4.69
	—	—	—	0.875	12.250	117.9	36.1	3.67	3.21	122.66	51.1	781	111.5	4.65
	100	—	—	0.937	12.126	115.5	38.5	3.67	3.17	130.73	50.0	825	117.8	4.63
	120	—	—	1.093	11.814	109.6	44.3	3.67	3.09	150.67	47.5	930	132.8	4.58
	140	—	—	1.250	11.500	103.9	50.1	3.67	3.01	170.22	45.0	1127	146.8	4.53
	160	—	—	1.406	11.188	98.3	55.6	3.67	2.929	189.12	42.6	1017	159.6	4.48
16 16.000	10	—	—	0.250	15.500	188.7	12.37	4.19	4.06	42.05	81.8	384	48.0	5.57
	20	—	—	0.312	15.376	185.7	15.38	4.19	4.03	52.36	80.5	473	59.2	5.55
	30	Std	—	0.375	15.250	182.6	18.41	4.19	3.99	62.58	79.1	562	70.3	5.53
	—	—	—	0.437	15.126	179.7	21.37	4.19	3.96	72.64	77.9	648	80.9	5.50
	40	XS	—	0.500	15.000	176.7	24.35	4.19	3.93	82.77	76.5	732	91.5	5.48
	—	—	—	0.562	14.876	173.8	27.26	4.19	3.89	92.66	75.4	813	106.6	5.46
	—	—	—	0.625	14.750	170.9	30.2	4.19	3.86	102.63	74.1	894	112.2	5.44
	60	—	—	0.656	14.688	169.4	31.6	4.19	3.85	107.50	73.4	933	116.6	5.43
	—	—	—	0.687	14.626	168.0	33.0	4.19	3.83	112.36	72.7	971	121.4	5.42
	—	—	—	0.750	14.500	165.1	35.9	4.19	3.80	122.15	71.5	1047	130.9	5.40
	80	—	—	0.843	14.314	160.9	40.1	4.19	3.75	136.46	69.7	1157	144.6	5.37

* See note on page 48 for definitions of columns a, b, and c.

TABLE 3.2 (*Concluded*).   PROPERTIES OF PIPE

Nominal Pipe Size Outside Diameter, in.	Schedule Number*			Wall Thickness, in.	Inside Diameter, in.	Inside Area, sq in.	Metal Area, sq in.	Sq Ft Outside Surface, per ft	Sq Ft Inside Surface, per ft	Weight per ft, lb	Weight of Water per ft, lb	Moment of Inertia, in.⁴	Section Modulus, in.³	Radius Gyration, in.
	a	b	c											
16 16.000	—	—	—	0.875	14.250	159.5	41.6	4.19	3.73	141.35	69.1	1193	154.1	5.36
	100	—	—	1.031	13.938	152.6	48.5	4.19	3.65	164.83	66.1	1365	170.6	5.30
	120	—	—	1.218	13.564	144.5	56.6	4.19	3.55	192.29	62.6	1556	194.5	5.24
	140	—	—	1.437	13.126	135.3	65.7	4.19	3.44	223.50	58.6	1760	220.0	5.17
	160	—	—	1.593	12.814	129.0	72.1.	4.19	3.35	245.11	55.9	1894	236.7	5.12
18 18.000	10	—	—	0.250	17.500	240.5	13.94	4.71	4.58	47.39	104.3	549	61.0	6.28
	20	—	—	0.312	17.376	237.1	17.34	4.71	4.55	59.03	102.8	678	75.5	6.25
	—	Std	—	0.375	17.250	233.7	20.76	4.71	4.52	70.59	101.2	807	89.6	6.23
	30	—	—	0.437	17.126	230.4	24.11	4.71	4.48	82.06	99.9	931	103.4	6.21
	—	XS	—	0.500	17.000	227.0	27.49	4.71	4.45	93.45	98.4	1053	117.0	6.19
	40	—	—	0.562	16.876	223.7	30.8	4.71	4.42	104.75	97.0	1172	130.2	6.17
	—	—	—	0.625	16.750	220.5	34.1	4.71	4.39	115.98	95.5	1289	143.3	6.15
	—	—	—	0.687	16.626	217.1	37.4	4.71	4.35	127.03	94.1	1403	156.3	6.13
	60	—	—	0.750	16.500	213.8	40.6	4.71	4.32	138.17	92.7	1515	168.3	6.10
	—	—	—	0.875	16.250	207.4	47.1	4.71	4.25	160.04	89.9	1731	192.8	6.06
	80	—	—	0.937	16.126	204.2	50.2	4.71	4.22	170.75	88.5	1834	203.8	6.04
	100	—	—	1.156	15.688	193.3	61.2	4.71	4.11	207.96	83.7	2180	242.2	5.97
	120	—	—	1.375	15.250	182.6	71.8	4.71	3.99	244.14	79.2	2499	277.6	5.90
	140	—	—	1.562	14.876	173.8	80.7	4.71	3.89	274.23	75.3	2750	306	5.84
	160	—	—	1.781	14.438	163.7	90.7	4.71	3.78	308.51	71.0	3020	336	5.77
20 20.000	10	—	—	0.250	19.500	298.6	15.51	5.24	5.11	52.73	129.5	757	75.7	6.98
	—	—	—	0.312	19.376	294.9	19.30	5.24	5.07	65.40	128.1	935	93.5	6.96
	20	Std	—	0.375	19.250	291.0	23.12	5.24	5.04	78.60	126.0	1114	111.4	6.94
	—	—	—	0.437	19.126	287.3	26.86	5.24	5.01	91.31	124.6	1286	128.6	6.92
	30	XS	—	0.500	19.000	283.5	30.6	5.24	4.97	104.13	122.8	1457	145.7	6.90
	—	—	—	0.562	18.876	279.8	34.3	5.24	4.94	116.67	121.3	1624	162.4	6.88
	40	—	—	0.593	18.814	278.0	36.2	5.24	4.93	122.91	120.4	1704	170.4	6.86
	—	—	—	0.625	18.750	276.1	38.0	5.24	4.91	129.33	119.7	1787	178.7	6.85
	—	—	—	0.687	18.626	272.5	41.7	5.24	4.88	141.71	118.1	1946	194.6	6.83
	—	—	—	0.750	18.500	268.8	45.4	5.24	4.84	154.20	116.5	2105	210.5	6.81
	60	—	—	0.812	18.376	265.2	48.9	5.24	4.81	166.40	115.0	2257	225.7	6.79
	—	—	—	0.875	18.250	261.6	52.6	5.24	4.78	178.73	113.4	2409	240.9	6.77
	80	—	—	1.031	17.938	252.7	61.4	5.24	4.70	208.87	109.4	2772	277.2	6.72
	100	—	—	1.281	17.438	238.8	75.3	5.24	4.57	256.10	103.4	3320	332	6.63
	120	—	—	1.500	17.000	227.0	87.2	5.24	4.45	296.37	98.3	3760	376	6.56
	140	—	—	1.750	16.500	213.8	100.3	5.24	4.32	341.10	92.6	4220	422	6.48
	160	—	—	1.968	16.064	202.7	111.5	5.24	4.21	379.01	87.9	4590	459	6.41
24 24.000	10	—	—	0.250	23.500	434	18.65	6.28	6.15	63.41	188.0	1316	109.6	8.40
	—	—	—	0.312	23.376	430	23.20	6.28	6.12	78.93	186.1	1629	135.8	8.38
	20	Std	—	0.375	23.250	425	27.83	6.28	6.09	94.62	183.8	1943	161.9	8.35
	—	—	—	0.437	23.126	420	32.4	6.28	6.05	109.97	182.1	2246	187.4	8.33
	—	XS	—	0.500	23.000	415	36.9	6.28	6.02	125.49	180.1	2550	212.5	8.31
	30	—	—	0.562	22.876	411	41.4	6.28	5.99	140.80	178.1	2840	237.0	8.29
	—	—·	—	0.625	22.750	406	45.9	6.28	5.96	156.03	176.2	3140	261.4	8.27
	40	—	—	0.687	22.626	402	50.3	6.28	5.92	171.17	174.3	3420	285.2	8.25
	—	—	—	0.750	22.500	398	54.8	6.28	5.89	186.24	172.4	3710	309	8.22
	60	—	—	0.968	22.064	382	70.0	6.28	5.78	238.11	165.8	4650	388	8.15
	80	—	—	1.218	21.564	365	87.2	6.28	5.65	296.36	158.3	5670	473	8.07
	100	—	—	1.531	20.938	344	108.1	6.28	5.48	367.40	149.3	6850	571	7.96
	120	—	—	1.812	20.376	326	126.3	6.28	5.33	429.39	141.4	7830	652	7.87
	140	—	—	2.062	19.876	310	142.1	6.28	5.20	483.13	134.5	8630	719	7.79
	160	—	—	2.343	19.314	293	159.4	6.28	5.06	541.94	127.0	9460	788	7.70
30 30.000	10	—	—	0.312	29.376	678	29.1	7.85	7.69	98.93	293.8	3210	214	10.50
	20	—	—	0.500	29.000	661	46.3	7.85	7.59	157.53	286.3	5040	336	10.43
	30	—	—	0.625	28.750	649	57.6	7.85	7.53	196.08	281.5	6220	415	10.39

* See note on page 48 for definitions of columns a, b, and c.

TABLE 3.3
ALLOWABLE STRESSES FOR PROCESS PIPING, PSI

(Selected values from ASA B31.1 Section 3. Reproduction by permission: Catalog 61, Midwest Piping Division of Crane Co., St. Louis, Mo.)

Material	ASTM or API	Grade	Kind*	−20 to 100	200	300	400	500	600	650	700	750	800	850	900 & above
CARBON STEEL	A-53 or 5L†	—	BW	9000	8600	8200	7800								
		—	LW	11250	10800	10200	9750	9250	8700	8500	8250	7700			
		A	ERW	13600	13000	12300	11750	11100	10500	10200	9900	9100	7900	6700	
		B	ERW	17000	16200	15400	14650	13900	13150	12750	12200	11000	9200	7350	
		A	S	16000	15300	14500	13800	13100	12350	12000	11650	10700	9300	7900	
		B	S	20000	19100	18150	17250	16350	15500	15000	14350	12950	10800	8650	
	A-83	A	S	16000	15300	14500	13800	13100	12350	12000	11650	10700	9300	7900	
	A-106	A	S	16000	15300	14500	13800	13100	12350	12000	11650	10700	9300	7900	
		B	S	20000	19100	18150	17250	16350	15500	15000	14350	12950	10800	8650	
	A-135	A	ERW	13600	13000	12300	11750	11100	10500	10200	9900	9100	7900	6700	
		B	ERW	17000	16200	15400	14650	13900	13150	12750	12200	11000	9200	7350	
	A-155†●	C50	EFW	16650	15900	15200	14450	13650	12900	12500	12100	11150	9600	8050	
		C55	EFW	18350	17500	16700	15850	15000	14200	13750	13250	12050	10200	8350	
		KC60	EFW	20000	19100	18150	17250	16350	15500	15000	14350	12950	10800	8650	
		KC65	EFW	21650	20700	19700	18700	17750	16750	16250	15500	13850	11400	8950	
		KC70	EFW	23350	22250	21250	20150	19100	18050	17500	16600	14750	12000	9250	
	A-333	O	S	18350	17500	16700	15850	15000	14200	13750	13250	12050	10200	8350	
WROUGHT IRON	A-72	—	BW	8000	7650	7250	6900								
		—	LW	10650	10200	9700	9200	8750	8250	8000	7700	7300			
½ Cr-½ Mo / 1 Cr-½ Mo	A-155†	½ CR	EFW	21650	20800	19950	19150	18300	17500	17100	16700	16250	15650	14400	
		1 CR	EFW	20000	19250	18500	17750	17000	16250	15900	15500	15150	14750	14200	
1¼ Cr-½ Mo / 2¼ Cr-1 Mo / 5 Cr-½ Mo		1¼ CR	EFW	20000	19300	18550	17850	17150	16450	16050	15700	15350	15000	14400	
		2¼ CR	EFW	18750	18250	17650	17150	16600	16050	15800	15500	15300	15000	14400	
		5 CR	EFW	18750	17900	17050	16200	15350	14500	14100	13650	13250	12800	12400	
½ Cr-½ Mo / 1 Cr-½ Mo / 1¼ Cr-½ Mo.	A-335	P2	S	18350	17650	16950	16300	15600	14900	14550	14200	13850	13500	13150	
		P12	S	18750	18250	17600	17050	16450	15900	15650	15350	15050	14750	14200	
		P11	S	18750	18250	17650	17150	16600	16050	15800	15550	15300	15000	14400	
2¼ Cr-1 Mo / 3 Cr-1 Mo / 5 Cr-½ Mo		P22	S	18750	18250	17650	17150	16600	16050	15800	15500	15300	15000	14400	
		P21	S	18750	18100	17400	16750	16100	15450	15150	14800	14500	13900	13200	
		P5	S	18750	17900	17050	16200	15350	14500	14100	13650	13250	12800	12400	
5 Cr-½ Mo-Si / 7 Cr-½ Mo / 9 Cr-1 Mo		P5b	S	18750	17900	17050	16200	15350	14500	14100	13650	13250	12800	12400	
		P7	S	18750	17850	17000	16150	15300	14450	14000	13550	13100	12500	11500	
		P9	S	18750	17900	17100	16250	15450	14600	14200	13800	13350	12950	12500	
18 Cr-8 Ni / 16 Cr-13 Ni-2½ Mo	A-312	TP304	S	18750	16650	15000	13650	12500	11600	11200	10800	10400	10000	9700	
		TP316	S	18750	18750	17900	17500	17200	17100	17050	17000	16900	16750	16500	
18 Cr-8 Ni-Ti / 18 Cr-8 Ni-Cb		TP321	S	18750	18750	17000	15800	15200	14900	14850	14800	14700	14550	14300	
		TP347	S	18750	18750	17000	15800	15200	14900	14850	14800	14700	14550	14300	
COPPER	B-42▲	Annealed	S	6000	5900	5000	2500	750							
NICKEL	B-161	Annealed	S	10000	10000	10000	10000	10000	10000						
MONEL	B-165	Annealed	S	17500	16500	15500	14800	14700	14700	14700	14700	14650	14500	12500	
ALUMINUM	B-241■	M1A	S	3600	3000	2500	1900								

(900 & above column: See next page.)

*Abbreviations used:
  BW: Butt-Welded
  LW: Lap-Welded
  S: Seamless
  ERW: Electric-Resistance-Welded
  EFW: Electric-Fusion-Welded
†Stress values given are for Class I pipe.

●Above 875° F, firebox quality plate is recommended.
▲Tensile properties must be verified by mill test.
■The stress values shown apply to pipe 1″ N.P.S. and larger; for pipe under 1″ N.P.S., use the stress values for temper H18 as shown in ASA B31.3.

Values are expressed in psi, and may be interpolated for intermediate temperatures.
Pipe is not to be used at temperatures greater than those for which stress values are indicated.
Graphitization may occur after prolonged exposure of carbon steel to temperatures above 775° F; or carbon molybdenum steel abⱱ 875° F.

TABLE 3.3 (*Concluded*).   ALLOWABLE STRESSES

Material	ASTM or API	Grade	Kind*	850 & below	900	950	1000	1050	1100	1150	1200	1300	1400	1500
					Section 3: PETROLEUM REFINERY PIPING — Temperature (°F)									
CARBON STEEL	A-53 or 5L†	—	BW											
		—	LW											
		A	ERW		5500	3800	2150	1350	850					
		B	ERW		5500	3800	2150	1350	850					
		A	S		6500	4500	2500	1600	1000					
		B	S		6500	4500	2500	1600	1000					
	A-83‡	A	S		6500	4500	2500	1600	1000					
	A-106	A	S		6500	4500	2500	1600	1000					
		B	S		6500	4500	2500	1600	1000					
	A-135	A	ERW		5500	3800	2150	1350	850					
		B			5500	3800	2150	1350	850					
	A-155††‡•	C50	EFW	*(See preceding page.)*	6500	4500	2500	1600	1000					
		C55			6500	4500	2500	1600	1000					
		KC60			6500	4500	2500	1600	1000					
		KC65			6500	4500	2500	1600	1000					
		KC70			6500	4500	2500	1600	1000					
	A-333‡	O	S		6500	4500	2500	1600	1000					
WROUGHT IRON	A-72	—	BW											
			LW											
½ Cr-½ Mo	A-155†	½ CR	EFW		12500	10000	6250	4000	2400					
1 Cr-½ Mo		1 CR			13100	11000	7500	5000	2800	1550	1000			
1¼ Cr-½ Mo		1¼ CR			13100	11000	7800	5500	4000	2500	1200			
2¼ Cr-1 Mo		2¼ CR			13100	11000	7800	5800	4200	3000	2000			
5 Cr-½ Mo		5 CR			11500	10000	7300	5200	3300	2200	1500			
½ Cr-½ Mo	A-335	P2	S		12500	10000	6250	4000	2400					
1 Cr-½ Mo		P12			13100	11000	7500	5000	2800	1550	1000			
1¼ Cr-½ Mo		P11			13100	11000	7800	5500	4000	2500	1200			
2¼ Cr-1 Mo		P22			13100	11000	7800	5800	4200	3000	2000			
3 Cr-1 Mo		P21			12000	9000	7000	5500	4000	2700	1500			
5 Cr-½ Mo		P5			11500	10000	7300	5200	3300	2200	1500			
5 Cr-½ Mo-Si		P5b			10900	9000	5500	3500	2500	1800	1200			
7 Cr-½ Mo		P7			9500	7000	5000	3500	2500	1800	1200			
9 Cr-1 Mo		P9			12000	10800	8500	5500	3300	2200	1500			
18 Cr-8 Ni	A-312	TP304	S		9400	9100	8800	8500	7500	5750	4500	2450	1400	750
16 Cr-13 Ni-2½ Mo		TP316			16000	15100	14000	12200	10400	8500	6800	4000	2350	1500
18 Cr-8 Ni-Ti		TP321			14100	13850	13500	13100	12500	8000	5000	2700	1550	1000
18 Cr-8 Ni-Cb		TP347			14100	13850	13500	13100	12500	8000	5000	2700	1550	1000
COPPER	B-42	Annealed	S											
MONEL	B-165	Annealed	S		8000									

*Abbreviations used:

BW: Butt-Welded
LW: Lap-Welded
S: Seamless
ERW: Electric-Resistance-Welded
EFW: Electric-Fusion-Welded

†Stress values given are for Class I pipe.

‡Above 900° F, silicon killed steel is recommended.

•Above 875° F, firebox quality plate is recommended.

Values are expressed in psi, and may be interpolated for intermediate temperatures.

Pipe is not to be used at temperatures greater than those for which stress values are indicated.

Graphitization may occur after prolonged exposure of carbon steel to temperatures above 775° F; carbon molybdenum steel above 875°F; chromium-molybdenum steel (with chromium under 0.60) above 975° F.

TABLE 3.3A

ALLOWABLE STRESSES FOR LEAD PIPE

[Based on "Lead in Modern Industries" (1952) by permission
Lead Industries Association, New York 17, N.Y.]

Temperature °F	Allowable Stress, psi	
	Chemical Lead	6% Antimonial Lead
70	199	396
100	183	347
120	173	313
140	162	280
160	152	250
180	142	217
200	132	186
220	122	153
240	113	122
260	104	90
280	92	57
300	81	

TABLE 3.4

STANDARD DEPTH OF PIPE THREADS

(ASA B2.1)

Nominal Pipe Size, Inches	Depth of Thread, Inches
⅛	0.02963
¼, ⅜	0.04444
½, ¾	0.05714
1 to 2	0.06957
3 and above	0.10000

TABLE 3.5

MINIMUM SIZE AND THICKNESS OF TUBING

(ASA B31.1 Paragraph 325f)

	ID, Inches	Thickness, Inches
Instrument Connecting Pipe	0.36*	0.049*
Control Piping	0.178	0.028

*Minimum to prevent plugging and give adequate mechanical
strength. If these are not factors, smaller sizes with wall thickness in
proportion may be used.

TABLE 3.6
STANDARD THICKNESSES* OF CAST IRON WATER PIPE
CENTRIFUGALLY CAST IN METAL MOLDS (ASA A21.6)

Laying condition A—Flat-bottom trench, without blocks, untamped backfill
Laying condition B—Flat-bottom trench, without blocks, tamped backfill
Laying condition C—Pipe laid on blocks, untamped backfill
Laying condition D—Pipe laid on blocks, tamped backfill

Size in.	Working Pressure psi.	3½ ft. of Cover				5 ft. of Cover				8 ft. of Cover			
		Laying Condition				Laying Condition				Laying Condition			
		A	B	C	D	A	B	C	D	A	B	C	D
		Thickness—in.											
3	50	0.32	0.32	0.32	0.32	0.32	0.32	0.32	0.32	0.32	0.32	0.38	0.32
	100	0.32	0.32	0.32	0.32	0.32	0.32	0.32	0.32	0.32	0.32	0.38	0.32
	150	0.32	0.32	0.32	0.32	0.32	0.32	0.32	0.32	0.32	0.32	0.38	0.32
	200	0.32	0.32	0.32	0.32	0.32	0.32	0.32	0.32	0.32	0.32	0.38	0.32
	250	0.32	0.32	0.32	0.32	0.32	0.32	0.35	0.32	0.32	0.32	0.38	0.32
	300	0.32	0.32	0.32	0.32	0.32	0.32	0.35	0.32	0.32	0.32	0.38	0.32
	350	0.32	0.32	0.32	0.32	0.32	0.32	0.35	0.32	0.32	0.32	0.38	0.32
4	50	0.35	0.35	0.35	0.35	0.35	0.35	0.35	0.35	0.35	0.35	0.41	0.35
	100	0.35	0.35	0.35	0.35	0.35	0.35	0.35	0.35	0.35	0.35	0.41	0.35
	150	0.35	0.35	0.35	0.35	0.35	0.35	0.38	0.35	0.35	0.35	0.41	0.35
	200	0.35	0.35	0.35	0.35	0.35	0.35	0.38	0.35	0.35	0.35	0.44	0.35
	250	0.35	0.35	0.35	0.35	0.35	0.35	0.38	0.35	0.35	0.35	0.44	0.35
	300	0.35	0.35	0.35	0.35	0.35	0.35	0.38	0.35	0.35	0.35	0.44	0.35
	350	0.35	0.35	0.38	0.35	0.35	0.35	0.38	0.35	0.35	0.35	0.44	0.35
6	50	0.38	0.38	0.41	0.38	0.38	0.38	0.44	0.38	0.38	0.38	0.48	0.38
	100	0.38	0.38	0.41	0.38	0.38	0.38	0.44	0.38	0.38	0.38	0.48	0.38
	150	0.38	0.38	0.41	0.38	0.38	0.38	0.44	0.38	0.38	0.38	0.48	0.38
	200	0.38	0.38	0.41	0.38	0.38	0.38	0.44	0.38	0.38	0.38	0.52	0.38
	250	0.38	0.38	0.44	0.38	0.38	0.38	0.44	0.38	0.38	0.38	0.52	0.38
	300	0.38	0.38	0.44	0.38	0.38	0.38	0.44	0.38	0.38	0.38	0.52	0.38
	350	0.38	0.38	0.44	0.38	0.38	0.38	0.48	0.38	0.38	0.38	0.52	0.38
8	50	0.41	0.41	0.44	0.41	0.41	0.41	0.48	0.41	0.41	0.41	0.52	0.41
	100	0.41	0.41	0.48	0.41	0.41	0.41	0.48	0.41	0.41	0.41	0.56	0.41
	150	0.41	0.41	0.48	0.41	0.41	0.41	0.48	0.41	0.41	0.41	0.56	0.41
	200	0.41	0.41	0.48	0.41	0.41	0.41	0.52	0.41	0.41	0.41	0.56	0.44
	250	0.41	0.41	0.48	0.41	0.41	0.41	0.52	0.41	0.44	0.41	0.56	0.44
	300	0.41	0.41	0.48	0.41	0.41	0.41	0.52	0.41	0.44	0.44	0.60	0.44
	350	0.41	0.41	0.52	0.41	0.44	0.41	0.52	0.44	0.48	0.44	0.60	0.48
10	50	0.44	0.44	0.48	0.44	0.44	0.44	0.52	0.44	0.44	0.44	0.60	0.48
	100	0.44	0.44	0.52	0.44	0.44	0.44	0.52	0.44	0.48	0.44	0.60	0.48
	150	0.44	0.44	0.52	0.44	0.44	0.44	0.56	0.44	0.48	0.44	0.60	0.48
	200	0.44	0.44	0.52	0.44	0.44	0.44	0.56	0.44	0.48	0.48	0.60	0.52
	250	0.44	0.44	0.56	0.44	0.48	0.44	0.56	0.48	0.52	0.48	0.65	0.52
	300	0.48	0.44	0.56	0.48	0.48	0.48	0.56	0.48	0.52	0.52	0.65	0.56
	350	0.48	0.48	0.56	0.48	0.52	0.52	0.60	0.52	0.56	0.52	0.65	0.56
12	50	0.48	0.48	0.52	0.48	0.48	0.48	0.56	0.48	0.52	0.48	0.65	0.52
	100	0.48	0.48	0.56	0.48	0.48	0.48	0.56	0.48	0.52	0.48	0.65	0.52
	150	0.48	0.48	0.56	0.48	0.48	0.48	0.60	0.52	0.52	0.52	0.65	0.56
	200	0.48	0.48	0.56	0.48	0.52	0.52	0.60	0.52	0.56	0.56	0.70	0.60
	250	0.52	0.48	0.60	0.52	0.52	0.52	0.60	0.56	0.60	0.56	0.70	0.60
	300	0.52	0.52	0.60	0.52	0.56	0.52	0.60	0.56	0.60	0.60	0.76	0.65
	350	0.56	0.56	0.60	0.56	0.56	0.56	0.65	0.60	0.60	0.60	0.76	0.65

* Thicknesses include allowances for foundry practice, corrosion, and either water hammer or truck load.

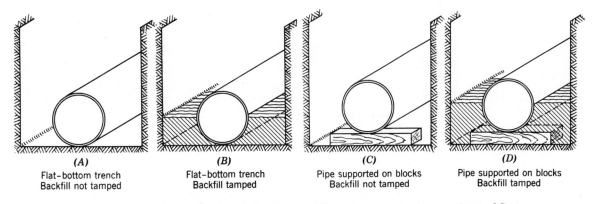

(A)	(B)	(C)	(D)
Flat-bottom trench Backfill not tamped	Flat-bottom trench Backfill tamped	Pipe supported on blocks Backfill not tamped	Pipe supported on blocks Backfill tamped

See Table 3.17 for standard thickness designations. Illustrations reproduced by permission of Cast Iron Pipe Research Association, Chicago, Ill.

TABLE 3.6 *(Concluded)*

STANDARD THICKNESSES* OF CAST IRON WATER PIPE CENTRIFUGALLY CAST IN METAL MOLDS (ASA A21.6)

Laying condition A—Flat-bottom trench, without blocks, untamped backfill
Laying condition B—Flat-bottom trench, without blocks, tamped backfill
Laying condition C—Pipe laid on blocks, untamped backfill
Laying condition D—Pipe laid on blocks, tamped backfill

Size in.	Working Pressure psi.	3½ ft. of Cover Laying Condition				5 ft. of Cover Laying Condition				8 ft. of Cover Laying Condition			
		A	B	C	D	A	B	C	D	A	B	C	D
		Thickness—*in.*											
14	50	0.51	0.48	0.59	0.51	0.51	0.48	0.59	0.55	0.59	0.55	0.69	0.59
	100	0.51	0.48	0.59	0.55	0.55	0.51	0.64	0.55	0.59	0.55	0.69	0.64
	150	0.55	0.51	0.59	0.55	0.55	0.51	0.64	0.59	0.64	0.59	0.75	0.64
	200	0.55	0.51	0.64	0.59	0.55	0.55	0.64	0.59	0.64	0.59	0.75	0.69
	250	0.59	0.55	0.64	0.59	0.59	0.59	0.69	0.59	0.64	0.64	0.75	0.69
	300	0.59	0.59	0.69	0.59	0.64	0.59	0.69	0.64	0.69	0.64	0.81	0.69
	350	0.64	0.64	0.69	0.64	0.64	0.64	0.75	0.69	0.75	0.69	0.81	0.75
16	50	0.54	0.50	0.63	0.58	0.58	0.54	0.63	0.58	0.63	0.58	0.73	0.63
	100	0.54	0.54	0.63	0.58	0.58	0.54	0.68	0.58	0.63	0.58	0.73	0.68
	150	0.58	0.54	0.63	0.58	0.58	0.54	0.68	0.63	0.68	0.63	0.79	0.68
	200	0.58	0.58	0.68	0.63	0.63	0.58	0.68	0.63	0.68	0.63	0.79	0.73
	250	0.63	0.58	0.68	0.63	0.63	0.63	0.73	0.68	0.73	0.68	0.79	0.73
	300	0.63	0.63	0.73	0.68	0.68	0.68	0.73	0.68	0.73	0.73	0.85	0.79
	350	0.68	0.68	0.73	0.68	0.73	0.68	0.79	0.73	0.79	0.73	0.85	0.79
18	50	0.58	0.54	0.63	0.58	0.58	0.54	0.68	0.63	0.68	0.63	0.79	0.68
	100	0.58	0.54	0.68	0.63	0.63	0.58	0.73	0.63	0.68	0.63	0.79	0.73
	150	0.63	0.58	0.68	0.63	0.63	0.58	0.73	0.68	0.73	0.68	0.79	0.73
	200	0.63	0.58	0.73	0.68	0.68	0.63	0.73	0.68	0.73	0.68	0.85	0.79
	250	0.68	0.63	0.73	0.68	0.68	0.68	0.79	0.73	0.79	0.73	0.85	0.79
	300	0.68	0.68	0.79	0.73	0.73	0.73	0.79	0.79	0.79	0.79	0.92	0.85
	350	0.79	0.73	0.79	0.79	0.79	0.79	0.85	0.79	0.85	0.85	0.92	0.85
20	50	0.62	0.57	0.72	0.62	0.67	0.57	0.72	0.67	0.72	0.67	0.78	0.72
	100	0.62	0.57	0.72	0.67	0.67	0.62	0.78	0.67	0.72	0.67	0.84	0.78
	150	0.67	0.62	0.72	0.67	0.67	0.62	0.78	0.72	0.78	0.72	0.84	0.78
	200	0.67	0.62	0.78	0.72	0.72	0.57	0.78	0.72	0.78	0.72	0.91	0.84
	250	0.72	0.67	0.78	0.72	0.78	0.72	0.84	0.78	0.84	0.78	0.91	0.84
	300	0.78	0.72	0.84	0.78	0.78	0.78	0.84	0.84	0.84	0.84	0.98	0.91
	350	0.84	0.78	0.84	0.84	0.84	0.84	0.91	0.84	0.91	0.84	0.98	0.91
24	50	0.68	0.63	0.79	0.68	0.73	0.63	0.79	0.73	0.79	0.73	0.85	0.79
	100	0.73	0.63	0.79	0.73	0.73	0.68	0.85	0.79	0.85	0.73	0.92	0.85
	150	0.73	0.68	0.79	0.79	0.79	0.73	0.85	0.79	0.85	0.79	0.92	0.85
	200	0.79	0.73	0.85	0.79	0.79	0.79	0.92	0.85	0.92	0.85	0.99	0.92
	250	0.79	0.79	0.85	0.85	0.85	0.79	0.92	0.85	0.92	0.85	0.99	0.99
	300	0.85	0.85	0.92	0.85	0.92	0.85	0.99	0.92	0.99	0.92	1.07	0.99
	350	0.92	0.92	0.99	0.92	0.99	0.92	0.99	0.99	1.07	0.99	1.07	1.07

* Thicknesses include allowances for foundry practice, corrosion, and either water hammer or truck load.

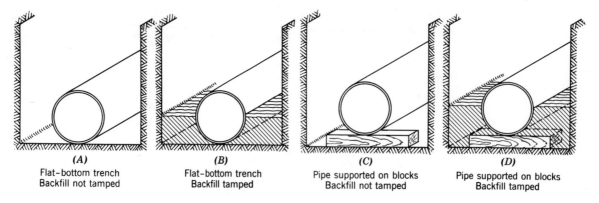

(A)	(B)	(C)	(D)
Flat-bottom trench Backfill not tamped	Flat-bottom trench Backfill tamped	Pipe supported on blocks Backfill not tamped	Pipe supported on blocks Backfill tamped

See Table 3.17 for standard thickness designations. Illustrations reproduced by permission of Cast Iron Pipe Research Association, Chicago, Ill.

TABLE 3.7

STANDARD THICKNESSES* OF CAST IRON WATER PIPE CENTRIFUGALLY CAST IN SAND-LINED MOLDS (ASA A21.8)

Laying condition A—Flat-bottom trench, without blocks, untamped backfill
Laying condition B—Flat-bottom trench, without blocks, tamped backfill
Laying condition C—Pipe laid on blocks, untamped backfill
Laying condition D—Pipe laid on blocks, tamped backfill

Size in.	Working Pressure psi.	3½-ft. Cover Laying Condition				5-ft. Cover Laying Condition				8-ft. Cover Laying Condition			
		A	B	C	D	A	B	C	D	A	B	C	D
						Thickness—in.							
3	50	0.32	0.32	0.32	0.32	0.32	0.32	0.32	0.32	0.32	0.32	0.38	0.32
	100	0.32	0.32	0.32	0.32	0.32	0.32	0.32	0.32	0.32	0.32	0.38	0.32
	150	0.32	0.32	0.32	0.32	0.32	0.32	0.32	0.32	0.32	0.32	0.38	0.32
	200	0.32	0.32	0.32	0.32	0.32	0.32	0.32	0.32	0.32	0.32	0.38	0.32
	250	0.32	0.32	0.32	0.32	0.32	0.32	0.35	0.32	0.32	0.32	0.38	0.32
	300	0.32	0.32	0.32	0.32	0.32	0.32	0.35	0.32	0.32	0.32	0.38	0.32
	350	0.32	0.32	0.32	0.32	0.32	0.32	0.35	0.32	0.32	0.32	0.38	0.32
4	50	0.35	0.35	0.35	0.35	0.35	0.35	0.35	0.35	0.35	0.35	0.41	0.35
	100	0.35	0.35	0.35	0.35	0.35	0.35	0.35	0.35	0.35	0.35	0.41	0.35
	150	0.35	0.35	0.35	0.35	0.35	0.35	0.38	0.35	0.35	0.35	0.41	0.35
	200	0.35	0.35	0.35	0.35	0.35	0.35	0.38	0.35	0.35	0.35	0.44	0.35
	250	0.35	0.35	0.35	0.35	0.35	0.35	0.38	0.35	0.35	0.35	0.44	0.35
	300	0.35	0.35	0.35	0.35	0.35	0.35	0.38	0.35	0.35	0.35	0.44	0.35
	350	0.35	0.35	0.38	0.35	0.35	0.35	0.38	0.35	0.35	0.35	0.44	0.35
6	50	0.38	0.38	0.41	0.38	0.38	0.38	0.44	0.38	0.38	0.38	0.48	0.38
	100	0.38	0.38	0.41	0.38	0.38	0.38	0.44	0.38	0.38	0.38	0.48	0.38
	150	0.38	0.38	0.41	0.38	0.38	0.38	0.44	0.38	0.38	0.38	0.52	0.38
	200	0.38	0.38	0.41	0.38	0.38	0.38	0.44	0.38	0.38	0.38	0.52	0.38
	250	0.38	0.38	0.44	0.38	0.38	0.38	0.44	0.38	0.38	0.38	0.52	0.38
	300	0.38	0.38	0.44	0.38	0.38	0.38	0.44	0.38	0.38	0.38	0.52	0.38
	350	0.38	0.38	0.44	0.38	0.38	0.38	0.48	0.38	0.38	0.38	0.52	0.38
8	50	0.41	0.41	0.44	0.41	0.41	0.41	0.48	0.41	0.41	0.41	0.52	0.41
	100	0.41	0.41	0.48	0.41	0.41	0.41	0.48	0.41	0.41	0.41	0.56	0.41
	150	0.41	0.41	0.48	0.41	0.41	0.41	0.48	0.41	0.41	0.41	0.56	0.41
	200	0.41	0.41	0.48	0.41	0.41	0.41	0.52	0.41	0.41	0.41	0.56	0.44
	250	0.41	0.41	0.48	0.41	0.41	0.41	0.52	0.41	0.44	0.41	0.56	0.44
	300	0.41	0.41	0.48	0.41	0.41	0.41	0.52	0.41	0.44	0.44	0.60	0.44
	350	0.41	0.41	0.52	0.41	0.44	0.41	0.52	0.44	0.48	0.44	0.60	0.48
10	50	0.44	0.44	0.48	0.44	0.44	0.44	0.52	0.44	0.44	0.44	0.60	0.48
	100	0.44	0.44	0.52	0.44	0.44	0.44	0.52	0.44	0.48	0.44	0.60	0.48
	150	0.44	0.44	0.52	0.44	0.44	0.44	0.56	0.44	0.48	0.48	0.60	0.52
	200	0.44	0.44	0.52	0.44	0.44	0.44	0.56	0.44	0.52	0.48	0.65	0.52
	250	0.44	0.44	0.56	0.44	0.48	0.44	0.56	0.48	0.52	0.52	0.65	0.56
	300	0.48	0.44	0.56	0.48	0.48	0.48	0.56	0.48	0.56	0.52	0.65	0.56
	350	0.48	0.48	0.56	0.48	0.52	0.52	0.60	0.52	0.56	0.56	0.65	0.56
12	50	0.48	0.48	0.52	0.48	0.48	0.48	0.56	0.48	0.52	0.48	0.65	0.52
	100	0.48	0.48	0.56	0.48	0.48	0.48	0.56	0.48	0.52	0.48	0.65	0.52
	150	0.48	0.48	0.56	0.48	0.48	0.48	0.56	0.48	0.52	0.52	0.65	0.56
	200	0.48	0.48	0.56	0.48	0.48	0.48	0.60	0.52	0.56	0.52	0.65	0.56
	250	0.52	0.48	0.60	0.52	0.52	0.52	0.60	0.52	0.56	0.56	0.70	0.60
	300	0.52	0.52	0.60	0.52	0.56	0.52	0.60	0.56	0.60	0.56	0.70	0.60
	350	0.56	0.56	0.60	0.56	0.56	0.56	0.65	0.60	0.60	0.60	0.76	0.65
14	50	0.51	0.48	0.59	0.51	0.51	0.48	0.59	0.55	0.59	0.55	0.69	0.59
	100	0.51	0.48	0.59	0.55	0.55	0.51	0.64	0.55	0.59	0.55	0.69	0.64
	150	0.55	0.51	0.59	0.55	0.55	0.51	0.64	0.59	0.64	0.59	0.75	0.69
	200	0.55	0.51	0.64	0.59	0.55	0.55	0.64	0.59	0.64	0.64	0.75	0.69
	250	0.59	0.55	0.64	0.59	0.59	0.59	0.69	0.64	0.69	0.64	0.81	0.69
	300	0.59	0.59	0.69	0.59	0.64	0.59	0.69	0.64	0.75	0.69	0.81	0.75
	350	0.64	0.64	0.69	0.64	0.64	0.64	0.75	0.69	0.75	0.69	0.81	0.75

* Thicknesses include allowances for foundry practice, corrosion, and either water hammer or truck load.

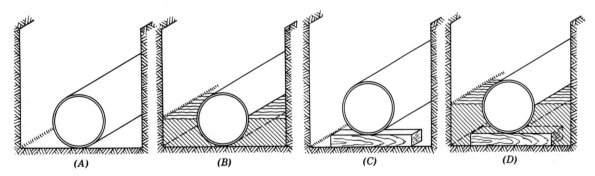

(A)   (B)   (C)   (D)

See Table 3.17 for standard thickness designations. Illustrations reproduced by permission of Cast Iron Pipe Research Association, Chicago, Ill.

TABLE 3.7 (*Concluded*)
STANDARD THICKNESSES* OF CAST IRON WATER PIPE
CENTRIFUGALLY CAST IN SAND LINED MOLDS (ASA A21.8)

Laying condition A—Flat-bottom trench, without blocks, untamped backfill
Laying condition B—Flat-bottom trench, without blocks, tamped backfill
Laying condition C—Pipe laid on blocks, untamped backfill
Laying condition D—Pipe laid on blocks, tamped backfill

Size in.	Working Pressure psi.	3½-ft. Cover				5-ft. Cover				8-ft. Cover			
		A	B	C	D	A	B	C	D	A	B	C	D
		Thickness—in.											
16	50	0.54	0.50	0.63	0.58	0.58	0.54	0.63	0.58	0.63	0.58	0.73	0.63
	100	0.54	0.54	0.63	0.58	0.58	0.54	0.68	0.58	0.63	0.58	0.73	0.68
	150	0.58	0.54	0.63	0.58	0.58	0.54	0.68	0.63	0.68	0.63	0.79	0.68
	200	0.58	0.58	0.68	0.63	0.63	0.58	0.68	0.63	0.68	0.63	0.79	0.73
	250	0.63	0.58	0.68	0.63	0.63	0.63	0.73	0.68	0.73	0.68	0.79	0.73
	300	0.63	0.63	0.73	0.68	0.68	0.68	0.73	0.68	0.73	0.73	0.85	0.79
	350	0.68	0.68	0.73	0.68	0.73	0.68	0.79	0.73	0.79	0.73	0.85	0.79
18	50	0.58	0.54	0.63	0.58	0.58	0.54	0.68	0.63	0.68	0.63	0.79	0.68
	100	0.58	0.54	0.68	0.63	0.63	0.58	0.73	0.63	0.68	0.63	0.79	0.73
	150	0.63	0.58	0.68	0.63	0.63	0.58	0.73	0.68	0.73	0.68	0.79	0.73
	200	0.63	0.58	0.73	0.68	0.68	0.63	0.73	0.68	0.73	0.68	0.85	0.79
	250	0.68	0.63	0.73	0.68	0.68	0.68	0.79	0.73	0.79	0.73	0.85	0.79
	300	0.68	0.68	0.79	0.73	0.73	0.73	0.79	0.79	0.79	0.79	0.92	0.85
	350	0.79	0.73	0.79	0.79	0.79	0.79	0.85	0.79	0.85	0.85	0.92	0.85
20	50	0.62	0.57	0.72	0.62	0.67	0.57	0.72	0.67	0.72	0.67	0.78	0.72
	100	0.62	0.57	0.72	0.67	0.67	0.62	0.78	0.67	0.72	0.67	0.84	0.78
	150	0.67	0.62	0.72	0.67	0.67	0.62	0.78	0.72	0.78	0.72	0.84	0.78
	200	0.67	0.62	0.78	0.72	0.72	0.67	0.78	0.72	0.78	0.72	0.91	0.84
	250	0.72	0.67	0.78	0.72	0.78	0.72	0.84	0.78	0.84	0.78	0.91	0.84
	300	0.78	0.72	0.84	0.78	0.78	0.78	0.84	0.84	0.84	0.84	0.98	0.91
	350	0.84	0.78	0.84	0.84	0.84	0.84	0.91	0.84	0.91	0.84	0.98	0.91
24	50	0.68	0.63	0.79	0.68	0.73	0.63	0.79	0.73	0.79	0.73	0.85	0.79
	100	0.73	0.63	0.79	0.73	0.73	0.68	0.85	0.79	0.85	0.73	0.92	0.85
	150	0.73	0.68	0.79	0.79	0.79	0.73	0.85	0.79	0.85	0.79	0.92	0.85
	200	0.79	0.73	0.85	0.79	0.79	0.79	0.92	0.85	0.92	0.85	0.99	0.92
	250	0.79	0.79	0.85	0.85	0.85	0.79	0.92	0.85	0.92	0.85	0.99	0.99
	300	0.85	0.85	0.92	0.85	0.92	0.85	0.99	0.92	0.99	0.92	1.07	0.99
	350	0.92	0.92	0.99	0.92	0.99	0.92	0.99	0.99	1.07	0.99	1.07	1.07
30	50	0.85	0.73	0.85	0.85	0.85	0.79	0.92	0.85	0.92	0.85	0.99	0.92
	100	0.85	0.79	0.92	0.85	0.92	0.79	0.99	0.92	0.99	0.85	1.07	0.99
	150	0.92	0.79	0.92	0.92	0.92	0.85	0.99	0.92	0.99	0.92	1.07	0.99
	200	0.92	0.85	0.99	0.92	0.99	0.92	1.07	0.99	1.07	0.99	1.16	1.07
	250	0.99	0.92	1.07	0.99	1.07	0.99	1.07	1.07	1.16	1.07	1.16	1.16
36	50	0.94	0.81	1.02	0.94	1.02	0.87	1.10	0.94	1.10	0.94	1.19	1.02
	100	0.94	0.87	1.02	0.94	1.02	0.87	1.10	1.02	1.10	1.02	1.19	1.10
	150	1.02	0.87	1.10	1.02	1.10	0.94	1.10	1.02	1.19	1.02	1.19	1.19
	200	1.10	0.94	1.10	1.02	1.10	1.02	1.19	1.10	1.19	1.10	1.29	1.19
	250	1.10	1.02	1.19	1.10	1.19	1.10	1.29	1.19	1.29	1.19	1.39	1.29
42	50	1.05	0.90	1.13	1.05	1.13	0.97	1.13	1.05	1.22	1.05	1.32	1.13
	100	1.05	0.97	1.13	1.05	1.13	0.97	1.22	1.13	1.22	1.13	1.32	1.22
	150	1.13	0.97	1.22	1.13	1.22	1.05	1.22	1.13	1.32	1.13	1.32	1.32
	200	1.22	1.05	1.22	1.13	1.22	1.13	1.32	1.22	1.32	1.22	1.43	1.32
	250	1.32	1.13	1.32	1.22	1.32	1.22	1.43	1.32	1.43	1.32	1.54	1.43
48	50	1.14	0.98	1.23	1.14	1.23	1.06	1.33	1.14	1.33	1.14	1.44	1.33
	100	1.23	1.06	1.23	1.14	1.23	1.06	1.33	1.23	1.33	1.23	1.44	1.33
	150	1.23	1.14	1.33	1.23	1.33	1.14	1.44	1.33	1.44	1.33	1.56	1.44
	200	1.33	1.14	1.44	1.33	1.44	1.23	1.44	1.33	1.56	1.33	1.56	1.44
	250	1.44	1.33	1.44	1.33	1.44	1.33	1.56	1.44	1.56	1.44	1.68	1.56

* Thicknesses include allowances for foundry practice, corrosion, and either water hammer or truck load.

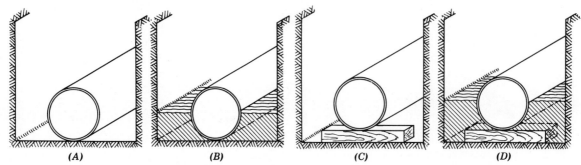

(A)   *(B)*   *(C)*   *(D)*

See Table 3.17 for standard thickness designations. Illustrations reproduced by permission of Cast Iron Pipe Research Association, Chicago, Ill.

TABLE 3.8
STANDARD THICKNESSES FOR CAST IRON GAS PIPE CENTRIFUGALLY CAST IN METAL MOLDS (ASA A21.7)

Thickness in inches. Working pressure in pounds per square inch.
Thicknesses include allowances for foundry practice and corrosion.

Laying condition A—Flat-bottom trench, without blocks, untamped backfill
Laying condition B—Flat-bottom trench, without blocks, tamped backfill
Laying condition C—Pipe laid on blocks, untamped backfill
Laying condition D—Pipe laid on blocks, tamped backfill

Size Inches	Working Pressure	3½ FEET OF COVER				5 FEET OF COVER				8 FEET OF COVER			
		A	B	C	D	A	B	C	D	A	B	C	D
4	10	¹.35	.35	.35	.35	.35	.35	.35	.35	.35	.35	.41	.35
		².38	.38	.38	.38	.38	.38	.38	.38	.38	.38	.41	.38
	50	¹.35	.35	.35	.35	.35	.35	.35	.35	.35	.35	.41	.35
		².38	.38	.38	.38	.38	.38	.38	.38	.38	.38	.41	.38
	100	¹.35	.35	.35	.35	.35	.35	.35	.35	.35	.35	.41	.35
		².38	.38	.38	.38	.38	.38	.38	.38	.38	.38	.41	.38
	150	¹.35	.35	.35	.35	.35	.35	.38	.35	.35	.35	.41	.35
		².38	.38	.38	.38	.38	.38	.38	.38	.38	.38	.41	.38
6	10	¹.38	.38	.41	.38	.38	.38	.41	.38	.38	.38	.48	.38
		².41	.41	.41	.41	.41	.41	.41	.41	.41	.41	.48	.41
	50	¹.38	.38	.41	.38	.38	.38	.41	.38	.38	.38	.48	.38
		².41	.41	.41	.41	.41	.41	.41	.41	.41	.41	.48	.41
	100	¹.38	.38	.41	.38	.38	.38	.44	.38	.38	.38	.48	.38
		².41	.41	.41	.41	.41	.41	.44	.41	.41	.41	.48	.41
	150	¹.38	.38	.41	.38	.38	.38	.44	.38	.38	.38	.48	.38
		².41	.41	.41	.41	.41	.41	.44	.41	.41	.41	.48	.41
8	10	.41	.41	.44	.41	.41	.41	.48	.41	.41	.41	.52	.41
	50	.41	.41	.44	.41	.41	.41	.48	.41	.41	.41	.52	.41
	100	.41	.41	.48	.41	.41	.41	.48	.41	.41	.41	.56	.41
	150	.41	.41	.48	.41	.41	.41	.48	.41	.41	.41	.56	.41
10	10	.44	.44	.48	.44	.44	.44	.52	.44	.44	.44	.60	.44
	50	.44	.44	.48	.44	.44	.44	.52	.44	.44	.44	.60	.44
	100	.44	.44	.52	.44	.44	.44	.52	.44	.44	.44	.60	.48
	150	.44	.44	.52	.44	.44	.44	.56	.44	.48	.44	.60	.48
12	10	.48	.48	.52	.48	.48	.48	.56	.48	.48	.48	.60	.52
	50	.48	.48	.52	.48	.48	.48	.56	.48	.48	.48	.60	.52
	100	.48	.48	.56	.48	.48	.48	.56	.48	.52	.48	.65	.52
	150	.48	.48	.56	.48	.48	.48	.56	.48	.52	.48	.65	.52
16	10	.54	.50	.58	.54	.54	.50	.63	.58	.58	.54	.73	.63
	50	.54	.50	.63	.54	.54	.50	.63	.58	.63	.58	.73	.63
	100	.54	.54	.63	.58	.58	.54	.68	.58	.63	.58	.73	.68
20	10	.62	.57	.67	.62	.62	.57	.72	.67	.67	.62	.78	.72
	50	.62	.57	.72	.62	.67	.57	.72	.67	.72	.62	.78	.72
	100	.62	.57	.72	.67	.67	.62	.78	.67	.72	.67	.84	.78
24	10	.63	.63	.73	.68	.73	.63	.79	.73	.79	.68	.85	.79
	50	.68	.63	.79	.68	.73	.63	.79	.73	.79	.73	.85	.79
	100	.73	.63	.79	.73	.73	.68	.85	.79	.79	.73	.92	.85

¹Class 22 thickness.
²Class 23 thickness offers increased factor of safety and is recommended for use in areas of dense population and heavy traffic.

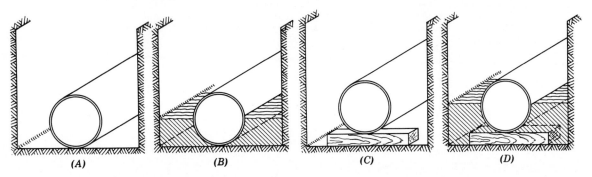

(A) *(B)* *(C)* *(D)*

See Table 3.17 for standard thickness designations. Illustrations reproduced by permission of Cast Iron Pipe Research Association, Chicago, Ill.

TABLE 3.9
STANDARD THICKNESSES OF CAST IRON GAS PIPE CENTRIFUGALLY CAST IN SAND-LINED MOLDS (ASA A21.9)

Thickness in inches.   Working pressure in pounds per square inch.
Thicknesses include allowances for foundry practice and corrosion.

Laying condition A—Flat-bottom trench, without blocks, untamped backfill
Laying condition B—Flat-bottom trench, without blocks, tamped backfill
Laying condition C—Pipe laid on blocks, untamped backfill
Laying condition D—Pipe laid on blocks, tamped backfill

Size Inches	Working Pressure	3½ FEET OF COVER				5 FEET OF COVER				8 FEET OF COVER			
		A	B	C	D	A	B	C	D	A	B	C	D
4	10	[1].35	.35	.35	.35	.35	.35	.35	.35	.35	.35	.41	.35
		[2].38	.38	.38	.38	.38	.38	.38	.38	.38	.38	.41	.38
	50	[1].35	.35	.35	.35	.35	.35	.35	.35	.35	.35	.41	.35
		[2].38	.38	.38	.38	.38	.38	.38	.38	.38	.38	.41	.38
	100	[1].35	.35	.35	.35	.35	.35	.35	.35	.35	.35	.41	.35
		[2].38	.38	.38	.38	.38	.38	.38	.38	.38	.38	.41	.38
	150	[1].35	.35	.35	.35	.35	.35	.38	.35	.35	.35	.41	.35
		[2].38	.38	.38	.38	.38	.38	.38	.38	.38	.38	.41	.38
6	10	[1].38	.38	.41	.38	.38	.38	.41	.38	.38	.38	.48	.38
		[2].41	.41	.41	.41	.41	.41	.41	.41	.41	.41	.48	.41
	50	[1].38	.38	.41	.38	.38	.38	.41	.38	.38	.38	.48	.38
		[2].41	.41	.41	.41	.41	.41	.41	.41	.41	.41	.48	.41
	100	[1].38	.38	.41	.38	.38	.38	.44	.38	.38	.38	.48	.38
		[2].41	.41	.41	.41	.41	.41	.44	.41	.41	.41	.48	.41
	150	[1].38	.38	.41	.38	.38	.38	.44	.38	.38	.38	.48	.38
		[2].41	.41	.41	.41	.41	.41	.44	.41	.41	.41	.48	.41
8	10	.41	.41	.44	.41	.41	.41	.48	.41	.41	.41	.52	.41
	50	.41	.41	.44	.41	.41	.41	.48	.41	.41	.41	.52	.41
	100	.41	.41	.48	.41	.41	.41	.48	.41	.41	.41	.56	.41
	150	.41	.41	.48	.41	.41	.41	.48	.41	.41	.41	.56	.41
10	10	.44	.44	.48	.44	.44	.44	.52	.44	.44	.44	.60	.44
	50	.44	.44	.48	.44	.44	.44	.52	.44	.44	.44	.60	.44
	100	.44	.44	.52	.44	.44	.44	.52	.44	.44	.44	.60	.48
	150	.44	.44	.52	.44	.44	.44	.56	.44	.48	.44	.60	.48
12	10	.48	.48	.52	.48	.48	.48	.56	.48	.48	.48	.60	.52
	50	.48	.48	.52	.48	.48	.48	.56	.48	.48	.48	.60	.52
	100	.48	.48	.56	.48	.48	.48	.56	.48	.52	.48	.65	.52
	150	.48	.48	.56	.48	.48	.48	.56	.48	.52	.48	.65	.52
16	10	.54	.50	.58	.54	.54	.50	.63	.58	.58	.54	.73	.63
	50	.54	.50	.63	.54	.54	.50	.63	.58	.63	.58	.73	.63
	100	.54	.54	.63	.58	.58	.54	.68	.58	.63	.58	.73	.68
20	10	.62	.57	.67	.62	.62	.57	.72	.67	.67	.62	.78	.72
	50	.62	.57	.72	.62	.67	.57	.72	.67	.72	.62	.78	.72
	100	.62	.57	.72	.67	.67	.62	.78	.67	.72	.67	.84	.78
24	10	.68	.63	.73	.68	.73	.63	.79	.73	.79	.68	.85	.79
	50	.68	.63	.79	.68	.73	.63	.79	.73	.79	.73	.85	.79
	100	.73	.63	.79	.73	.73	.68	.85	.79	.79	.73	.92	.85
30	10	.79	.73	.85	.79	.85	.73	.92	.85	.92	.79	.99	.92
	50	.85	.73	.85	.85	.85	.79	.92	.85	.92	.85	.99	.92
36	10	.87	.81	.94	.87	.94	.81	1.02	.94	1.02	.87	1.10	1.02
	50	.94	.81	1.02	.94	1.02	.87	1.10	.94	1.10	.94	1.19	1.02
42	10	1.05	.90	1.05	.97	1.05	.90	1.13	1.05	1.13	.97	1.22	1.13
	50	1.05	.90	1.13	1.05	1.13	.97	1.13	1.05	1.22	1.05	1.32	1.13
48	10	1.14	.98	1.14	1.06	1.14	.98	1.23	1.14	1.33	1.06	1.33	1.23
	50	1.14	.98	1.23	1.14	1.23	1.06	1.33	1.14	1.33	1.14	1.44	1.33

[1]Class 22 Thickness.
[2]Class 23 Thickness offers increased factor of safety and is recommended for use in areas of dense population and heavy traffic.

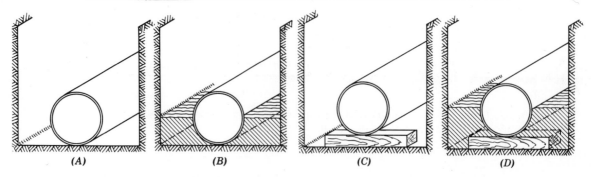

(A)   (B)   (C)   (D)

See Table 3.17 for standard thickness designations.   Illustrations reproduced by permission of Cast Iron Pipe Research Association, Chicago, Ill.

TABLE 3.10

WATER HAMMER ALLOWANCE, P_w FOR

ABOVE-GROUND CAST-IRON PIPE

(ASA B31.1)

Pipe Diameter, inches Sizes, Inclusive	Water Hammer Allowance psi
4 to 10	120
12 to 14	110
16 to 18	100
20	90
24	85
30	80
36	75
42 to 60	70

TABLE 3.11

CRUSHING STRENGTH* OF ASBESTOS-CEMENT CONCRETE AND CLAY PIPE

Based on Designated ASTM Specifications

Using Three-Point Bearing Method

(Reproduced by permission: American Society for Testing Materials)

Values are in lb per lineal foot

Nominal Size, Inches	Asbestos-Cement Pressure Pipe ASTM C296			Concrete Sewer Pipe ASTM C14†		Clay Sewer Pipe ASTM C13, C200	
	Class 100	Class 150	Class 200	Standard Strength	Extra Strength	Standard Strength	Extra Strength
4	4100	5400	8700	1000	2000	1000	2000
6	3900	5400	9000	1100	2000	1100	2000
8	3700	5500	9300	1300	2000	1300	2000
10	3700	7000	11000	1400	2000	1400	2000
12	4000	7600	11800	1500	2250	1500	2250
14	4400	8600	13500	. . .	. . .	. . .	. . .
15	. . .	. . .	. . .	1750	2750	1750	2750
16	4800	9200	15400	. . .	. . .	. . .	. . .
18	5200	10100	17400	2000	3300	2000	3300
20	5600	10900	19400	. . .	. . .	. . .	. . .
21	. . .	. . .	. . .	2200	3850	2200	3850
24	6300	12700	22600	2400	4000	2400	4400
27	. . .	. . .	. . .	. . .	. . .	2750	4700
30	7500	15900	28400	. . .	. . .	3200	5000
33	. . .	. . .	. . .	. . .	. . .	3500	5500
36	8800	19600	33800	. . .	. . .	3900	6000

* Actual supporting strength varies with method of bedding pipe. For pipe having the bottom quadrant (90°) of the barrel uniformly supported on the bottom of the trench, the design supporting strength can be assumed as 1.5 times the listed crushing strengths. See ASTM C12 for details on Clay pipe and various types of bedding.

† For reinforced-concrete culvert and sewer pipe see ASTM C76 and C361.

TABLE 3.12

BOROSILICATE GLASS PIPE
MANUFACTURERS RECOMMENDED OPERATING
TEMPERATURES AND PRESSURES

(Reproduced by permission: Corning Glass Works)

Pipe Size, ID, Inches	Wall Thickness, Inches	Max. Recommended† Working Pressure, psi	Max. Sudden Temperature Differential, °F	Max. Operating Temperatures,* °F
1	0.156	50	200	450
1½	0.172	50	200	450
2	0.172	50	200	450
3	0.203	50	200	450
4	0.265	35	175	450
6	0.328	20	160	450

* Glass must be insulated or protected so that *Maximum Sudden Temperature Differential Figures* are not exceeded.

† Under certain circumstances, the recommended working pressure of 1-in. pipe can be increased to 125 psi; 1½ in. to 80 psi; and 2 in. to 70 psi. Consult manufacturer.

TABLE 3.13
POLYETHYLENE FLEXIBLE PIPE
PRESSURE-TEMPERATURE RATINGS
(Based on Commercial Standard CS–197 except as noted)

Nominal Pipe Size, Inches*	Recommended Working Pressure, PSI at 73.4°F(23°C)			
	Regular Grade			High-Molecular Weight
	Series 1 (Sched. 40)	Series 2	Series 3	Series 1 (Sched. 40)†
½	100	75	100	150
¾	83	75	100	130
1	78	75	100	120
1¼	65	75	100	100
1½	59	75	100	90
2	50	75	100	75
2½	55	. . .	. . .	. . .
3	48	. . .	. . .	. . .
4	40	. . .	. . .	. . .
6	30	. . .	. . .	. . .

CORRECTION FACTOR FOR OTHER TEMPERATURES
Working pressure = (Working pressure at 73.4°F)(Factor)

Temperature °F	Factor for Regular Grade‡	Factor for High Molecular Weight Grade†
60	1.10	1.125
65	1.05	1.06
80	0.95	0.95
90	0.80	0.85
100	0.70	0.75
110	0.60	0.68
120	0.50 (Maximum use temperature	0.59
130	. . .	0.52
140	. . .	0.45
150	. . .	0.38 (Maximum use temperature)

* See Table 3.25 for dimensional standards.
† Not a standard but typical manufacturer's recommendations.
‡ Inferred from CS–197.

TABLE 3.14
POLYVINYL CHLORIDE PIPE—RIGID
MANUFACTURER'S RECOMMENDED PRESSURE-TEMPERATURE RATINGS
(Reproduced by permission: Carlon Sales Corporation, Aurora, Ohio)
Maximum Working Pressures at 73.4°F(23°C)

Nominal Size Inches*	Schedule 40		Schedule 80		Schedule 120	
	PVC-1 Normal Impact	PVC-2 High Impact	PVC-1 Normal Impact	PVC-2 High Impact	PVC-1 Normal Impact	PVC-2 High Impact
¼	456	391	617	528	. . .	. . .
⅜	425	324	600	448	. . .	. . .
½	410	311	575	420	566	486
¾	335	258	470	352	454	388
1	310	243	435	326	426	365
1¼	255	202	360	276	362	311
1½	230	182	325	252	331	284
2	195	156	280	220	294	252
2½	198	170	. . .	. . .	. . .	. . .
3	185	148	260	206	280	240
4	155	126	225	180	272	233
5	130	111	. . .	. . .	. . .	. . .
6	125	101	195	157	237	203
8	101	86	. . .	. . .	. . .	. . .
10	94	81	. . .	. . .	. . .	. . .
12	89	76	. . .	. . .	. . .	. . .

CORRECTION FACTOR FOR OTHER TEMPERATURES (PLAIN-END PIPE)
Working Pressure = (Working pressure at 73.4°F)(Factor)

Temperature °F	Factor Normal Impact	Factor High Impact
50	1.15	1.19
60	1.08	1.10
70	1.01	1.02
80	0.93	0.93
90	0.88	0.88
100	0.84	0.82
110	0.75	0.71
120	0.70	0.46
130	0.65	0.27
140	0.58	. . .
150	0.52	. . .

* See Table 3.24 for dimensional standards. Sizes 8 in. and larger are not standardized.

TABLE 3.15

MINIMUM PLUMBING FACILITIES[1] (ASA A40.8)

(See Notes on Next Page)

Type of Building or Occupancy[2]	Water Closets		Urinals	Lavatories		Bathtubs or Showers	Drinking Fountains[3]
Dwelling or apt. houses[4]	1 for each dwelling or apartment unit		. . . .	1 for each apt. or dwlg. unit		1 for each apt. or dwlg. unit	. . . .
Schools[5]	Male	Female					
Elementary	1 per 100	1 per 35	1 per 30 males	1 per 60 persons			1 per 75 persons
Secondary	1 per 100	1 per 45	1 per 30 males	1 per 100 persons			1 per 75 persons
Office or Public buildings	No. of Persons	No. of. Fixtures	Whenever urinals are provided for men, one water closet less than the number specified may be provided for each urinal installed[2] except that the number of water closets in such cases shall not be reduced to less than $\frac{2}{3}$ of the minimum specified.	No. of Persons	No. of Fixtures		1 for each 75 persons
	1–15	1		1–15	1		
	16–35	2		16–35	2		
	36–55	3		36–60	3		
	56–80	4		61–90	4		
	81–110	5		91–125	5		
	111–150	6		1 fixture for each 45 additional persons			
	1 fixture for each 40 additional persons						
Manufacturing, Warehouses, workshops, loft buildings, foundries and similar establishments[6]	No. of Persons	No. of Fixtures	Same substitution as above	1–100 persons, 1 fixture for each 10 persons Over 100, 1 for each 15 persons[7,8]		1 shower for each 15 persons exposed to excessive heat or to skin contamination with poisonous, infectious, or irritating material	1 for each 75 persons
	1–9	1					
	10–24	2					
	25–49	3					
	50–74	4					
	75–100	5					
	1 fixture for each additional 30 employees						
Dormitories[9]	Male: 1 for each 10 persons Female: 1 for each 8 persons Over 10 persons, add 1 fixture for each 25 additional males and 1 for each 20 additional females		1 for each 25 men Over 150 persons, add 1 fixture for each additional 50 men	1 for each 12 persons. (Separate dental lavatories should be provided in community toilet rooms. Ration of dental lavatories for each 50 persons is recommended.) Add 1 lavatory for each 20 males, 1 for each 15 females.		1 for each 8 persons. In the case of women's dormitories, additional bathtubs should be installed at the ratio of 1 for each 30 females. Over 150 persons, add 1 fixture for each 20 persons	1 for each 75 persons
Theatres, auditoriums	No. of Persons	No. of Fixtures	No. of Persons	No. of Fixtures	No. of Persons	No. of Fixtures		1 for each 100 persons
		Male Female	(Male)		1–200	1		
	1–100	1 1	1–200	1	201–400	2		
	101–200	2 2	201–400	2	401–750	3		
	201–400	3 3	401–600	3	Over 750, 1 for each additional 500 persons			
	Over 400, add 1 fixture for each additional 500 males and 1 for each 300 females		Over 600; 1 for each additional 300 males					

NOTES FOR TABLE 3.15

[1] The figures shown are based upon one fixture being the minimum required for the number of persons indicated or any fraction thereof.

[2] Building category not shown on this table. Will be considered separately by the Administrative Authority.

[3] Drinking fountains shall not be installed in toilet rooms.

[4] Laundry trays—one single compartment tray for each dwelling unit or 2 compartment trays for each 10 apartments. Kitchen sinks—1 for each dwelling or apartment unit.

[5] This schedule has been adopted (1945) by the National Council on Schoolhouse Construction.

[6] As required by the American Standard Safety Code for Industrial Sanitation in Manufacturing Establishments (ASA Z4.1—1935).

[7] Where there is exposure to skin contamination with poisonous, infectious, or irritating materials, provide 1 lavatory for each 5 persons.

[8] 24 lineal inches of wash sink or 18 inches of a circular basin, when provided with water outlets for such space, shall be considered equivalent to 1 lavatory.

[9] Laundry trays, 1 for each 50 persons. Slop sinks, 1 for each 100 persons.

General. In applying this schedule of facilities, consideration must be given to the accessibility of the fixtures. Conformity purely on a numerical basis may not result in an installation suited to the need of the individual establishment. For example, schools should be provided with toilet facilities on each floor having classrooms.

Temporary workingmen facilities:

1 water closet and 1 urinal for each 30 workmen

24-in. urinal trough = 1 urinal.   48-in. urinal trough = 2 urinals.
36-in. urinal trough = 2 urinals.   60-in. urinal trough = 3 urinals.
72-in. urinal trough = 4 urinals.

TABLE 3.16
WATER SUPPLY WELLS
MINIMUM DISTANCES FROM SOURCE OF
CONTAMINATION (ASA A40.8)

	Distance (feet)*
Sewer	50
Septic tanks	50
Subsurface pits	50
Subsurface disposal fields	100
Seepage pits	100
Cesspools	150

*These distances constitute minimum separation and should be increased in areas of creviced rock or limestone, or where the direction of movement of the ground water is from sources of contamination toward the well.

DIMENSIONAL AND MECHANICAL
STANDARDS FOR PIPE, VALVES, AND FITTINGS

Listed here are commonly used standards for pipe, valves, and fittings. These standards give dimensions, tolerances, and pressure-temperature ratings. For convenience certain frequently used data have been assembled in tables and charts beginning on pages 70 and 91.

PIPE AND TUBING — Designations*

FERROUS

Cast Iron

	Designations*
Centrifugally cast pipe in metal molds, for water or other liquids (AWWA C106)	A21.6‡
For gas	A21.7
Centrifugally cast pipe in sand-lined molds for water or other liquids (AWWA C108)	A21.8‡
For gas	A21.9
Pit-cast pipe, for water or other liquids (AWWA C102)	A21.2
For gas	A21.3
Soil pipe	A40.1
Threaded pipe for drainage, vent, and waste services	A40.5

Steel

Line pipe (Pipe line)	API 5L
Line pipe, high-test	API 5LX
Seamless and welded steel and welded wrought-iron pipe, standard sizes and weight	B36.10
Stainless steel pipe	B36.19

NON-FERROUS

Aluminum alloy pipe	ASTM B241
Copper and red-brass pipe and copper tube, standard sizes	ASTM B251
Lead pipe	Fed. Spec. WW–P–325
Nickel and nickel-alloy pipe (made in same schedules as stainless steel), consult manufacturer for common sizes	Refer to B36.19

MISCELLANEOUS NON-METALLIC PIPE

Asbestos-Cement

Pressure pipe (nominal ID only together with crushing loads and flexural strength)	ASTM C296 (see also AWWA C400 and Fed. Spec. SS–P–351a)
Sewer pipe	AWWA C400 (Water only) & Fed. Spec. SS–P–331a

Bituminized Fiber

	Designations*
Sewer pipe and fittings	Fed. Spec. SS–P–356

Clay Pipe

Sewer pipe, standard strength	ASTM C13, C261
Sewer pipe, extra strength	ASTM C200, C278
Perforated pipe, standard strength	ASTM C211

Concrete

Drain tile	ASTM C412
Pressure pipe, low-head	ASTM C361
Reinforced culvert	ASTM C76
Sewer pipe	ASTM C14
Sewer pipe, internal pressure	ASTM C362

Plastic

IPS size for extruded ABS (acrylonitrile-butadiene-styrene)	ASTM D1527
Polyethylene pipe	Commercial Standard CS–197
Polyvinyl chloride pipe, dimensions and tolerances for	Commercial Standard CS–207
Solvent-welded plastic pipe sizes—cellulose-acetate-butyrate pipe and	Commercial Standard CS–206
ABS (acrylonitrile-butadiene-styrene) pipe	ASTM D1503 ASTM D1528

FABRICATED PIPE SPECIFICATIONS

Backing rings, end preparation, and internal machining for	PFI–ES1
Bosses (metal), built-up	PFI–ES6
Branch connections (90° unreinforced for seamless steel pipe), pressure ratings	PFI–TB2

* Designations refer to American Standards Association numbers unless otherwise noted as follows:

API = American Petroleum Industry
ASTM = American Society for Testing Materials
ASRE = American Society of Refrigeration Engineers
AWWA = American Water Works Association
Commercial Standard = Commercial Standard, U.S. Dept. of Commerce
ISA = Instrument Society of America
Fed. Spec. = Federal Specification
MSS = Manufacturer's Standard of the Value and Fitting Industry
PFI = Pipe Fabrication Institute
† Federal Specification WWP–421 was in general use prior to issuance of ASA 21.6 and 21.8 and continues to be used by some.

PIPE AND TUBING (*Continued*)	Designations*
Cleaning, shop testing, and inspection, classification of	PFI–ESB
Cleaning fabricated piping	PFI–ES5
Dimensioning welded assemblies, method for	PFI–ES2
Hydrostatic shop testing of fabricated piping	PFI–ES4
Linear tolerances, bending radii, minimum tangents	PFI–ES3
Nozzles (welded), minimum length and spacing	PFI–ES7
Preheat and postheat welding practices for low chromium-molybdenum steel pipe	PFI–ES(M)8
Preheat and postheat welding practices for medium chromium-molybdenum steel pipe	PFI–ES(M)12
Shielded metal-arc welding-dissimilar ferritic steels	PFI–ES(M)9
Stress relieving welded attachments	PFI–ES(M)10

VALVES

API flanged steel gate and plug valves for drilling and production service	API 6–C
API flanged steel OS & Y wedge gate valves and plug valves for refinery service	API 600
API pipeline valves	API 6–D
Bronze gate valves, 100 lb	MSS–SP–38
Bronze gate valves, 125 lb	MSS–SP–37
Cast-iron pipeline valves	MSS–SP–52
Control valves, dimensions	ISA–RP4.1
Corrosion resistant valves, cast flanged 150 lb	MSS–SP–42
Face-to-face dimensions of ferrous flanged and welding-end valves	B16.10
Gate valves for water works service	AWWA C500
Refrigerant expansion valves, method of rating and testing (ASRE 17–R)	B60.1

FITTINGS

FLANGES AND FLANGED FITTINGS

Brass or Bronze

Flanges (150 and 300 lb)	B16.24

Cast-Iron Flanges and Flanged Fittings

25 lb, maximum WSP	B16b.2
125 lb	B16.1
250 lb	B16.b
300 lb, for refrigerant piping (300 psi at 300°F to −100°F)	B16.16
800-lb hydraulic pressure at ordinary temperature	B16.B1

Steel

Assembly of steel raised-faced flanges to cast iron, brass, bronze or stainless steel flanges	MSS–SP–46

FITTINGS (*Continued*)	Designations*
Branch connections, unreinforced 90°, ratings for	PFI–TB2
Corrosion resistant flanges and flanged fittings, cast, 150 lb	MSS–SP–33
Pipeline flanges	MSS–SP–44 API 6–B
Ring-joint flanges	B16.5
Steel pipe flanges and flanged fittings (general dimensional and pressure-temperature specs.)	B16.5

GASKETS AND BOLTING FOR FLANGES

Bolting

Bolt and stud lengths, number required	B16.5
Screw threads for high-strength bolting	B1.4
Square and hexagonal bolts and nuts	B18.2

Gasketing

Non-metallic gaskets	B16.21
Raised-face flange gaskets, limiting dimensions of which meet requirements of B16.5	MSS–SP–47
Ring-joint gaskets and grooves for steel flanges	B16.20

FLARE-TYPE FITTINGS

Refrigeration type (SAE SP–95)	B70.1
Brass fittings for flared copper tubes (175 psig cold water pressure)	A40.2

JOINTS AND FITTINGS FOR CAST-IRON PIPE

Bell and Spigot Fittings

AWWA fittings (Long body pattern)	AWWA C100
Bell, sockets, and spigot beads, standard dimensions for pit cast and centrifugally cast pipe for both water and gas	A21.2, A21.3, A21.6, A21.7, A21.8, A21.9
Short-body fittings, 3- to 12-in. for 250 psi water pressure	A21.10

Mechnical Joint

For cast-iron pressure pipe and fittings (AWWA C111)	A21.11

SCREWED FITTINGS

Brass or bronze

125 lb	B16.15
250 lb	B16.17

Cast iron

125 and 250 lb	B16.4
Drainage	B16.12
Ferrous plugs, bushings, and locknuts with pipe threads (for use with 125-lb cast-iron and 150 lb malleable-iron screwed fittings)	B16.14
Fire-hose couplings	B26

FITTINGS (*Continued*)	Designations*
Forged steel	
2000, 3000, and 6000 lb	MSS–SP–49
Plugs and bushings	MSS–SP–50
Malleable iron	
150 lb	B16.3
300 lb	B16.19

SOLDER-JOINT FITTINGS

Cast brass	B16.18
Cast brass for drainage	B16.23
Wrought copper and bronze	B16.22

WELDING FITTINGS

Butt welding ends	B16.25
Stainless steel butt-welding	MSS–SP–43
Steel butt-welding	B16.9
Steel butt-welding (26 in. and larger)	MSS–SP–48
Steel socket-welding	B16.11

THREADS

Hose-coupling threads	B33.1
Pipe threads	B2.1
Screw threads for high-strength bolting	B1.4
Threads in valves, fittings, and flanges	API 6–A

THREADS (*Continued*)	Designations*
Unified and American threads for screws, bolts, nuts, and other threaded parts	B1.1

MISCELLANEOUS

DRAFTING PRACTICES AND SYMBOLS

Abbreviations for use on drawings	Z32.13
Drawings and drafting room practice	Z14.1
Graphical symbols for:	
Process flow diagrams	Y32.11
Plumbing	Z32.4
Welding	Z32.2.1
Pipe fittings, valves and piping	Z32.2.3
Heating, ventilating, and air conditioning	Z32.2.4

OTHER USEFUL DATA

Drain and bypass connection standard	MSS–SP–45
Identification of piping systems	A13.1
Finishes for contact faces for connecting end flanges of ferrous valves and fittings	MSS–SP–6
Marking system for valves, fittings, flanges and unions	MSS–SP–25
Spot-facing	MSS–SP–9

INDEX OF SELECTED TABLES OF STANDARD DIMENSIONS
FOR PIPE, VALVES, FLANGES, AND FITTINGS

* See also Table 2.1, page 8, and Table 3.2, page 48.
† See also Table 2.1, page 8, Table 3.5, page 54, and Table A.3, page 236.

TABLE 3.17
CAST-IRON PIPE
STANDARD THICKNESS CLASSES
(Based on ASA A21 Series)

Nominal Pipe Size in.	Outside* Diam. in.		Pipe Wall Thickness (in.) for Standard Thickness Class No.:									
			21	22	23	24	25	26	27	28	29	30
3	3.96	—		0.32	0.35	0.38	0.41	0.44	0.48	0.52	0.56	0.60
4	4.80	5.00		0.35	0.38	0.41	0.44	0.48	0.52	0.56	0.60	0.65
6	6.90	7.10		0.38	0.41	0.44	0.48	0.52	0.56	0.60	0.65	0.70
8	9.05	9.30		0.41	0.44	0.48	0.52	0.56	0.60	0.65	0.70	0.76
10	11.10	11.40		0.44	0.48	0.52	0.56	0.60	0.65	0.70	0.76	0.82
12	13.20	13.50		0.48	0.52	0.56	0.60	0.65	0.70	0.76	0.82	0.89
14	15.30	15.65	0.48	0.51	0.55	0.59	0.64	0.69	0.75	0.81	0.87	0.94
16	17.40	17.80	0.50	0.54	0.58	0.63	0.68	0.73	0.79	0.85	0.92	0.99
18	19.50	19.92	0.54	0.58	0.63	0.68	0.73	0.79	0.85	0.92	0.99	1.07
20	21.60	22.06	0.57	0.62	0.67	0.72	0.78	0.84	0.91	0.98	1.06	1.14
24	25.80	26.32	0.63	0.68	0.73	0.79	0.85	0.92	0.99	1.07	1.16	1.25
30	32.00		0.73	0.79	0.85	0.92	0.99	1.07	1.16	1.25	1.35	1.46
36	38.30		0.81	0.87	0.94	1.02	1.10	1.19	1.29	1.39	1.50	1.62
42	44.50		0.90	0.97	1.05	1.13	1.22	1.32	1.43	1.54	1.66	1.79
48	50.80		0.98	1.06	1.14	1.23	1.33	1.44	1.56	1.68	1.81	1.95

* Two diameters indicate pipe is made in both sizes shown. See ASA A21 series for details.

TABLE 3.18
DIMENSIONS OF WELDED AND SEAMLESS STEEL PIPE (ASA B36.10)
(Listed by Schedule Numbers)

Nominal Pipe Size	Outside Diameter	Nominal Wall Thickness										
		Sched. 10	Sched. 20	Sched. 30	Sched. 40	Sched. 60	Sched. 80	Sched. 100	Sched. 120	Sched. 140	Sched. 160
⅛	0.405	....	....	....	**0.068**	....	**0.095**	....	....	....	....
¼	0.540	....	....	....	**0.088**	....	**0.119**	....	....	....	....
⅜	0.675	....	....	....	**0.091**	....	**0.126**	....	....	....	....
½	0.840	....	....	....	**0.109**	....	**0.147**	....	....	....	0.187
¾	1.050	....	....	....	**0.113**	....	**0.154**	....	....	....	0.218
1	1.315	....	....	....	**0.133**	....	**0.179**	....	....	....	0.250
1¼	1.660	....	....	....	**0.140**	....	**0.191**	....	....	....	0.250
1½	1.900	....	....	....	**0.145**	....	**0.200**	....	....	....	0.281
2	2.375	....	....	....	**0.154**	....	**0.218**	....	....	....	0.343
2½	2.875	....	....	....	**0.203**	....	**0.276**	....	....	....	0.375
3	3.500	....	....	....	**0.216**	....	**0.300**	....	....	....	0.438
3½	4.000	....	....	....	**0.226**	....	**0.318**	....	....	....	....
4	4.500	....	....	....	**0.237**	....	**0.337**	....	0.438	....	0.531
5	5.563	....	....	....	**0.258**	....	**0.375**	....	0.500	....	0.625
6	6.625	....	....	....	**0.280**	....	**0.432**	....	0.562	....	0.718
8	8.625	....	0.250	0.277	**0.322**	0.406	**0.500**	0.593	0.718	0.812	0.906
10	10.750	....	0.250	0.307	**0.365**	**0.500**	0.593	0.718	0.843	1.000	1.125
12	12.750	....	0.250	0.330	0.406	0.562	0.687	0.843	1.000	1.125	1.312
14	14.000	0.250	0.312	0.375	0.438	0.593	0.750	0.937	1.093	1.250	1.406
16	16.000	0.250	0.312	0.375	0.500	0.656	0.843	1.031	1.218	1.438	1.593
18	18.000	0.250	0.312	0.438	0.562	0.750	0.937	1.156	1.375	1.562	1.781
20	20.000	0.250	0.375	0.500	0.593	0.812	1.031	1.281	1.500	1.750	1.968
24	24.000	0.250	0.375	0.562	0.687	0.968	1.218	1.531	1.312	2.062	2.343
30	30.000	0.312	0.500	0.625	....	....	....	....	....	....	....

All dimensions given in inches.

The decimal thicknesses listed for the respective pipe sizes represent their nominal or average wall dimensions. For tolerances on wall thicknesses, see appropriate material specifications.

Thicknesses shown in bold face type for Schedule 40 are identical with thicknesses shown in bold face type for Standard Wall pipe in Table 3.19.

Those in bold face type in Schedules 60 and 80 are identical with thicknesses in bold face type for Extra Strong Wall pipe in Table 3.19.

Some of the larger, heavier wall sections are beyond the capabilities of seamless mill production and must be obtained from turned-and-bored billets or other sources.

TABLE 3.19

DIMENSIONS OF WELDED AND SEAMLESS STEEL PIPE (ASA B36.10)

(Listed as Standard Wall, Extra Strong Wall and Double Extra Strong Wall)

Nominal Pipe Size	Outside Diameter	Nominal Wall Thickness		
		Standard Wall	Extra Strong Wall	Double Extra Strong Wall
⅛	0.405	**0.068**	**0.095**	. . . .
¼	0.540	**0.088**	**0.095**	. . . .
⅜	0.675	**0.091**	**0.126**	. . . .
½	0.840	**0.109**	**0.147**	0.294
¾	1.050	**0.113**	**0.154**	0.308
1	1.315	**0.133**	**0.179**	0.358
1¼	1.660	**0.140**	**0.191**	0.382
1½	1.900	**0.145**	**0.200**	0.400
2	2.375	**0.154**	**0.218**	0.436
2½	2.875	**0.203**	**0.276**	0.552
3	3.500	**0.216**	**0.300**	0.600
3½	4.000	**0.226**	**0.318**	. . . .
4	4.500	**0.237**	**0.337**	0.674
5	5.563	**0.258**	**0.375**	0.750
6	6.625	**0.280**	**0.432**	0.864
8	8.625	**0.322**	**0.500**	0.875
10	10.750	**0.365**	**0.500**	. . . .
12	12.750	0.375	0.500	. . . .
14	14.000	0.375	0.500	. . . .
16	16.000	0.375	0.500	. . . .
18	18.000	0.375	0.500	. . . .
20	20.000	0.375	0.500	. . . .
24	24.000	0.375	0.500	. . . .

All dimensions given in inches.

The decimal thicknesses listed for the respective pipe sizes represent their nominal or average wall dimensions. For tolerances on wall thicknesses, see appropriate material specifications.

Thicknesses shown in bold face type for Standard Wall are identical with corresponding thicknesses shown in bold face type for Schedule 40 in Table 3.18. Those shown in bold face type for Extra Strong Wall are identical with corresponding thicknesses shown in bold face type in Schedules 60 and 80 in Table 3.18.

Double Extra Strong Wall has no corresponding schedule numbers.

TABLE 3.20

DIMENSIONS OF WELDED AND SEAMLESS STAINLESS STEEL PIPE

Nominal Pipe Size	Outside Diameter	Nominal Wall Thickness			
		Schedule 5S†	Schedule 10S†	Schedule 40S	Schedule 80S
⅛	0.405	. . .	0.049	0.068	0.095
¼	0.540	. . .	0.065	0.088	0.119
⅜	0.675	. . .	0.065	0.091	0.126
½	0.840	0.065	0.083	0.109	0.147
¾	1.050	0.065	0.083	0.113	0.154
1	1.315	0.065	0.109	0.133	0.179
1¼	1.660	0.065	0.109	0.140	0.191
1½	1.900	0.065	0.109	0.145	0.200
2	2.375	0.065	0.109	0.154	0.218
2½	2.875	0.083	0.120	0.203	0.276
3	3.500	0.083	0.120	0.216	0.300
3½	4.000	0.083	0.120	0.226	0.318
4	4.500	0.083	0.120	0.237	0.337
5	5.563	0.109	0.134	0.258	0.375
6	6.625	0.109	0.134	0.280	0.432
8	8.625	0.109	0.148	0.322	0.500
10	10.750	0.134	0.165	0.365	0.500*
12	12.750	0.156	0.180	0.375*	0.500*

All dimensions are given in inches.

The decimal thicknesses listed for the respective pipe sizes represent their normal or average wall dimensions.

Tolerances: +12.5%.

* These do not conform to ASA B36.10 Schedule Numbers, but correspond to standard weight (0.375) and extra strong (0.500).

† Schedule 5S and 10S wall thicknesses do not permit threading in accordance with ASA B2.1.

TABLE 3.21
ALUMINUM AND NICKEL PIPE ASA SCHEDULE NUMBERS

Material	Thickness Schedule Numbers	Specification
Aluminum	ASA Schedules 5S, 10S, 40S, and 80S (see Table 3.19 for thicknesses). Also Schedule 30 for Sizes 8 in. and 10 in., Schedule 60 for 10 in. and 12 in., and 0.279-in. wall for 10 in. and 0.330-in. wall for 12 in.	ASTM B241
Nickel and Nickel-Copper	ASA Schedule 5S and 10S up through 6 in. and 40S and 80S up through 8 in. (see Table 3.19 for thicknesses).	Standard Practice (See ASTM B161 and B165)

TABLE 3.22
PREFERRED SIZES OF ROUND SEAMLESS COPPER AND COPPER-ALLOY TUBE
APPLICABLE TO ASTM DESIGNATIONS B75 AND B135 (ASTM B251)

It is recommended that wherever possible, material purchased to these specifications be ordered to the diameters and wall thicknesses indicated below.

Preferred Sizes of Round Seamless Copper and Copper-alloy Tube[a,b]

[× indicates the preferred sizes]

Outside Diameter, in.	Wall thickness, in.																		
	0.010	0.013	0.016	0.020	0.025	0.032	0.040	0.049	0.065	0.083	0.100	⅛	5/32	3/16	¼	5/16	⅜	½	⅝
⅛	×	×	×	×	×	×	×	…	…	…	…	…	…	…	…	…	…	…	…
3/16	×	×	×	×	×	×	×	×	…	…	…	…	…	…	…	…	…	…	…
¼	…	…	×	×	×	×	×	×	×	×	…	…	…	…	…	…	…	…	…
5/16	…	…	×	×	×	×	×	×	×	×	×	…	…	…	…	…	…	…	…
⅜	…	…	×	×	×	×	×	×	×	×	×	×	…	…	…	…	…	…	…
½	…	…	×	×	×	×	×	×	×	×	×	×	×	…	…	…	…	…	…
⅝	…	…	×	×	×	×	×	×	×	×	×	×	×	…	…	…	…	…	…
¾	…	…	×	×	×	×	×	×	×	×	×	×	×	×	…	…	…	…	…
⅞	…	…	×	×	×	×	×	×	×	×	×	×	×	×	…	…	…	…	…
1	…	…	…	×	×	×	×	×	×	×	×	×	×	×	×	…	…	…	…
1¼	…	…	…	×	×	×	×	×	×	×	×	×	×	×	×	×	…	…	…
1½	…	…	…	×	×	×	×	×	×	×	×	×	×	×	×	×	×	…	…
1¾	…	…	…	×	×	×	×	×	×	×	×	×	×	×	×	×	×	…	…
2	…	…	…	…	×	×	×	×	×	×	×	×	×	×	×	×	×	×	…
2¼	…	…	…	…	×	×	×	×	×	×	×	×	×	×	×	×	×	×	…
2½	…	…	…	…	×	×	×	×	×	×	×	×	×	×	×	×	×	×	×
2¾	…	…	…	…	×	×	×	×	×	×	×	×	×	×	×	×	×	×	×
3	…	…	…	…	…	×	×	×	×	×	×	×	×	×	×	×	×	×	×
3½	…	…	…	…	…	×	×	×	×	×	×	×	×	×	×	×	×	×	×
4	…	…	…	…	…	…	×	×	×	×	×	×	×	×	×	×	×	×	×
4½	…	…	…	…	…	…	×	×	×	×	×	×	×	×	×	×	×	×	×
5	…	…	…	…	…	…	…	×	×	×	×	×	×	×	×	×	×	×	×
5½	…	…	…	…	…	…	…	×	×	×	×	×	×	×	×	×	×	×	×
6	…	…	…	…	…	…	…	…	×	×	×	×	×	×	×	×	×	×	×
7	…	…	…	…	…	…	…	…	×	×	×	×	×	×	×	×	×	×	×
8	…	…	…	…	…	…	…	…	…	×	×	×	×	×	×	×	×	×	×
9	…	…	…	…	…	…	…	…	…	×	×	×	×	×	×	×	×	×	×
10	…	…	…	…	…	…	…	…	…	…	×	×	×	×	×	×	×	×	×
11	…	…	…	…	…	…	…	…	…	…	×	×	×	×	×	×	×	×	×
12	…	…	…	…	…	…	…	…	…	…	…	×	×	×	×	×	×	×	×

[a] In conformance with the Simplified Practice Recommendations R 235–48 for Copper and Copper-Alloy Round Seamless Tube issued by the U.S. Department of Commerce.

[b] This tube is not necessarily available in all alloys in the full range of sizes shown.

TABLE 3.23
DIMENSIONS OF COPPER AND BRASS PIPE
(Based on ASTM B251 by permission: American Society for Testing and Materials, Philadelphia, Pa.)

Copper Water Tube (ASTM B88)					Copper and Red Brass Pipe (ASTM B42 and B43)					
						Nominal Dimensions, in.				
Standard Water Tube Size, in.	Actual Outside Diameter, in.	Nominal Wall Thickness, in.			Nominal Pipe Size in.	Regular Weight			Extra Strong	
		Type K*	Type L*	Type M*		Outside Diameter	Inside Diameter	Wall Thickness	Inside Diameter	Wall Thickness
¼	0.375....	0.035	0.030	. . .	⅛	0.405	0.281	0.062	0.205	0.100
⅜	0.500....	0.049	0.035	. . .	¼	0.540	0.376	0.082	0.294	0.123
½	0.625....	0.049	0.040	. . .	⅜	0.675	0.495	0.090	0.421	0.127
⅝	0.750....	0.049	0.042	. . .	½	0.840	0.626	0.107	0.542	0.149
¾	0.875....	0.065	0.045	. . .	¾	1.050	0.822	0.144	0.736	0.157
1	1.125....	0.065	0.050	. . .	1	1.315	1.063	0.126	0.951	0.182
1¼	1.375....	0.065	0.055	0.042	1¼	1.660	1.368	0.146	1.272	0.194
1½	1.625....	0.072	0.060	0.049	1½	1.900	1.600	0.150	1.494	0.203
2	2.125....	0.083	0.070	0.058	2	2.375	2.063	0.156	1.933	0.221
2½	2.625....	0.095	0.080	0.065	2½	2.875	2.501	0.187	2.315	0.280
3	3.125....	0.109	0.090	0.072	3	3.500	3.062	0.219	2.892	0.304
3½	3.625....	0.120	0.100	0.083	3½	4.000	3.500	0.250	3.358	0.321
4	4.125....	0.134	0.110	0.095	4	4.500	4.000	0.250	3.818	0.341
5	5.125....	0.160	0.125	0.109	5	5.562	5.062	0.250	4.812	0.375
6	6.125....	0.192	0.140	0.122	6	6.625	6.125	0.250	5.751	0.437
8	8.125....	0.271	0.200	0.170	8	8.625	8.001	0.312	7.625	0.500
10	10.125....	0.338	0.250	0.212	10	10.750	10.020	0.365	9.750	0.500
12	12.125....	0.405	0.280	0.254	12	12.750	12.000	0.375		

* Recommendations:

Type K: General Plumbing and heating systems and underground service, for severe conditions

Type L: For interior use in general plumbing and heating
Type M: Non-pressure applications (drain, vents, etc.)

TABLE 3.24

DIMENSIONS OF PLASTIC PIPE*

Note: At this time many systems of dimensions are in use. The following are typical.
Those which have been adopted as standards are so indicated.

Designation IPS System		NORMAL PIPE SIZE, INCHES								
		⅛	¼	⅜	½	¾	1	1¼	1½	2
Schedule 10	OD Thickness	Same as stainless steel Schedule 10S (see Table 3.20)								
Schedule 40	OD Thickness	Same as steel Schedule 40 (see Table 3.19)								
Schedule 80	OD Thickness	Same as steel Schedule 80 (see Table 3.19)								
Schedule 120	OD	Same as steel pipe OD (see Table 3.19)								
	Thickness	...	...	...	0.170	0.170	0.200	0.215	0.225	0.250
SWP System (solvent welded pipe)										
	OD	...	...	...	0.600	0.855	1.140	1.420	1.730	2.250
	Thickness	...	...	...	0.050	0.053	0.070	0.085	0.115	0.125
Polyethylene System										
Series 1 (Sched. 40)	OD Thickness	Same as Schedule 40 steel ½″ through 6″ (No 3½ or 5″)								
Series 2 (75-lb class)	OD	...	...	...	0.782	1.024	1.299	1.710	2.000	2.567
	Thickness	...	...	...	0.080	0.100	0.125	0.165	0.195	0.250
Series 3 (100-lb class)	OD	...	...	...	0.840	1.110	1.409	1.860	2.170	2.777
	Thickness	...	...	...	0.109	0.143	0.180	0.240	0.280	0.355

* See Tables 3.13 and 3.14 for pressure-temperature ratings.

TABLE 3.24 (*Concluded*)

2½	3	3½	4	5	6	Use and Standards†
						PVC in schedules 10, 40, 80 and 120—Commercial Standard CS–207.
						ABS in Schedules 40 and 80—ASTM D1527.
0.300	0.350	0.350	0.438	0.500	0.562	Polyethylene pipe series 1 is same as schedule 40 with 3½ and 5″ sizes omitted—Commercial Standard CS–197. Other plastic pipe is also sold in IPS, but no other standards exist. Manufacturers should be consulted.
2.570	3.250	...	4.100	...	6.220	ABS Pipe ASTM D2503
0.125	0.125	...	0.150	...	0.230	CAB Pipe ASTM D1528 and Commercial Standard CS–206.
						Commercial Standard CS–197
...	...	...				ID same as Schedule 40, CS–197
...	...	...				
...	...	...				ID same as Schedule 40, CS–197
...	...	...				

† ABS = acrylonitrile-butadiene-styrene pipe
CAB = cellulose acetate-butyrate pipe
PVC = Polyvinyl chloride pipe

TABLE 3.25
APPROXIMATE THICKNESS TOLERANCES FOR PIPE*

Material	Tolerance, Per cent
Steel	−12.5
Nickel	±10 to 12.5
Copper	±5 to 8
Aluminum	±12.5 for Schedule 5S and 10S
	Others: −12.5
Plastic, IPS size	+10% (Schedule 40, 80, and 120) and +15% (Schedule 10)
Plastic, SWP size	±5%
Plastic, Polyethylene	±4.5 to 5%

* Consult applicable ASTM and Commercial Standard Specifications for details.

TABLE 3.26
CAST STEEL BUTT-WELDING AND FLANGED GATE AND GLOBE VALVE DIMENSIONS
(Based on ASA B16.10)

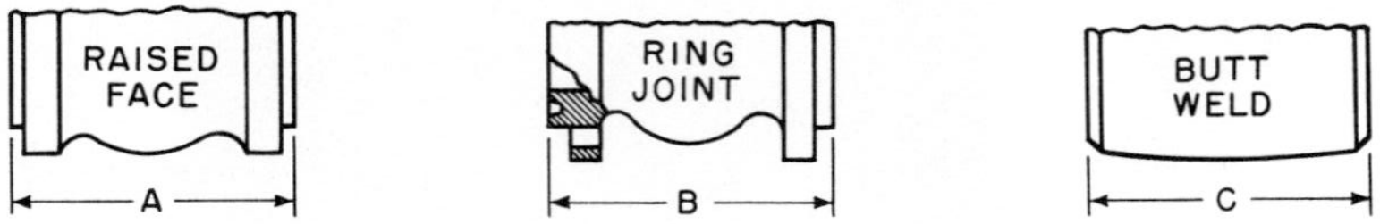
RAISED FACE
A
RING JOINT
B
BUTT WELD
C

Nominal Valve Size	ASA RATING: 150 Gate A	150 Gate B	150 Gate C	150 Globe A&C	150 Globe B	300 Gate A&C	300 Gate B	300 Globe A&C	300 Globe B
¼	4	. . .	4	4	. . .	. . .	. . .	. . .	. . .
½	4¼	4 11/16	4¼	4¼	4 11/16	. . .	. . .	6	6 7/16
¾	4⅝	5⅛	4⅝	4⅝	5⅛	. . .	. . .	7	7½
1	5	5½	5	5	5½	. . .	. . .	8	8½
1½	6½	7	6½	6½	7	7½	8	9	9½
2	7	7½	9½	8	8½	8½	9⅛	10½	11⅛
3	8	8½	11⅛	9½	10	11⅛	11¾	12½	13⅛
4	9	9½	12	11½	12	12	12⅝	14	14⅝
6	10½	11	15⅞	16	16½	15⅞	16½	17½	18⅛
8	11½	12	16½	19½	20	16½	17⅛	22	22⅝
10	13	13½	18	24½	25	18	18⅝	24½	25⅛
12	14	14½	19¾	27½	28	19¾	20⅜	28	28⅝
14	15	15½	22½	31	31½	30	30⅝	. . .	. . .
16	16	16½	24	36	36½	33	33⅝	. . .	. . .
18	17	17½	26	. . .	. . .	36	36⅝	. . .	. . .
20	18	18½	28	. . .	. . .	39	39¾	. . .	. . .
24	20	20½	32	. . .	. . .	45	45⅞	. . .	. . .

Nominal Valve Size	400 Gate A&C	400 Gate B	400 Globe A&C	400 Globe B	600 Gate A&C	600 Gate B	600 Globe A&C	600 Globe B
¼	. . .	. . .	. . .	. . .	. . .	. . .	. . .	. . .
½	6½*	6 7/16*			6½*	6 7/16*		
¾	7½*	7½*			7½*	7½*		
1	8½	8½			8½	8½		
1½	9½	9½			9½	9½		
2	11½	11⅝			11½	11⅝		
3	14	14⅛			14	14⅛		
4	16	16⅛			17	17⅛		
6	19½	19⅝	Same as Gate	Same as Gate	22	22⅛	Same as Gate	Same as Gate
8	23½	23⅝			26	26⅛		
10	26½	26⅝			31	31⅛		
12	30	30⅛			33	33⅛		
14	32½	32⅝	. . .	. . .	35	35⅛	. . .	. . .
16	35½	35⅝	. . .	. . .	39	39⅛	. . .	. . .
18	38½	38⅝	. . .	. . .	43	43⅛	. . .	. . .
20	41½	41¾	. . .	. . .	47	47¼	. . .	. . .
24	48½	48⅞	. . .	. . .	55	55⅜	. . .	. . .

Nominal Valve Size	900 Gate A&C	900 Gate B	900 Globe A&C	900 Globe B	1500 Gate A&C	1500 Gate B	1500 Globe A&C	1500 Globe B	2500 Gate A&C	2500 Gate B	2500 Globe A&C	2500 Globe B	Nominal Valve Size
¼	. . .	. . .	. . .	. . .	. . .	. . .	. . .	. . .	. . .	. . .	. . .	. . .	¼
½	. . .	. . .	. . .	. . .	. . .	. . .	. . .	. . .	10⅜*	10⅜*			½
¾	. . .	. . .	9	9	. . .	. . .	9	9	10¾*	10¾*			¾
1	10*	10*			10*	10*			12⅛*	12⅛*			1
1½	12*	12*			12*	12*			15⅛*	15¼*			1½
2	14½	14⅝			14½	14⅝			17¾	17⅞			2
3	15	15⅛			18½	18⅝			22¾	23			3
4	18	18⅛			21½	21⅝			26½	26⅞			4
6	24	24⅛	Same as Gate	Same as Gate	27¾	28	Same as Gate	Same as Gate	36	36½	Same as Gate	Same as Gate	6
8	29	29⅛			32¾	33⅛			40¼	40⅞			8
10	33	33⅛			39	39⅜			50	50⅞			10
12	38	38⅛			44½	45⅛			56	56⅞			12
14	40½	40⅞			49½	50¼			. . .	. . .	. . .	. . .	14
16	44½	44⅞	. . .	. . .	54½	55⅜	. . .	. . .	. . .	. . .	. . .	. . .	16
18	48	48½	. . .	. . .	60½	61⅜	. . .	. . .	. . .	. . .	. . .	. . .	18
20	52	52½	. . .	. . .	65½	66⅜	. . .	. . .	. . .	. . .	. . .	. . .	20
24	61	61¾	. . .	. . .	76½	77⅝	. . .	. . .	. . .	. . .	. . .	. . .	24

Notes. 1. All dimensions in inches.
2. For larger sizes and special designs see ASA B16.10.
3. Dimensions marked * are for solid wedge gate valves only. All other gate valve dimensions are for both solid wedge and double disc.
4. See manufacturer for height when open and for hand wheel diameter. These vary with manufacturer.
5. See Table 3.28 for dimension between connecting RTJ flanges.

TABLE 3.27

DIMENSIONS OF OTHER FLANGED VALVES (BASED ON ASA B16.10)

(See Table 3.26 for dimension references)

Check Valves

Lift Check — Same as globe valve dimensions A & B in Table 3.26

Swing Check — Same as globe valve dimensions A & B in Table 3.26 with following exceptions

Nominal Valve Size, Inches	150 lb A	150 lb B	Nominal Valve Size, Inches	300 lb A	300 lb B
¼	Not listed		½	Not listed	
6	14	14½	¾	Not listed	
16	Not listed		1	8½	9
			1½	9½	10

Control Valves (Conforms to Instrument Society of America Standard ISA RP4.1)

Nominal Valve Size, Inches	ASA Rating					
	150		300		600	
	A	B	A	B	A	B
½	. . .	. . .	7½	7 15/16	8	7 15/16
¾	. . .	. . .	7⅝	8⅛	8⅛	8⅛
1	7¼	7¾	7¾	8¼	8¼	8¼
1½	8¾	9¼	9¼	9¾	9⅞	9⅞
2	10	10½	10½	11⅛	11¼	11⅜
3	11¾	12¼	12½	13⅛	13¼	13⅜
4	13⅞	14⅜	14½	15⅛	15½	15⅝
6	17¾	18¼	18⅝	19¼	20	20⅛
8	21⅜	21⅞	22⅜	23	24	24⅛

Plug Valves: Plug valves with same dimensions as shown in Table 3.26 for gate valves can be specified as follows:

150 lb	Short pattern, 1 in. to 12 in. (Exception: For short pattern 1-in. size A = 5½ B = 6)
300 lb	Short pattern, 1½ in. to 12 in. Venturi pattern, 6 in. to 24 in.
400 lb	Regular pattern, 1 in. to 12 in. Venturi pattern, 6 in. to 24 in.
600 lb	Regular pattern, 1 in. to 10 in. Venturi pattern, 6 in. to 24 in.
900 lb	Regular pattern, 1 in. to 10 in. Venturi pattern, 6 in. to 16 in.
1500 lb	Regular pattern, 1 in. to 12 in. Venturi pattern, 6 in. to 12 in.
2500 lb	Regular pattern, 2 in. to 12 in.

TABLE 3.28

APPROXIMATE DISTANCE BETWEEN CONNECTING FLANGES HAVING OCTAGONAL OR OVAL RING GASKETS WHEN RINGS ARE COMPRESSED (ASA B16.10)

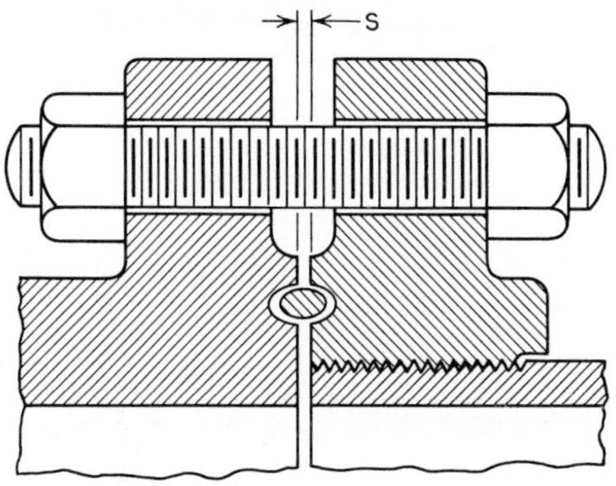

Distance Between Flanges *(S)*

Nominal Valve Size	150 Lb	300 Lb	400 Lb	600 Lb	900 Lb	1500 Lb	2500 Lb
½	⅛	⅛	⅛	⅛	. . .	. . .	5/32
¾	5/32	5/32	5/32	5/32	5/32	5/32	5/32
1	5/32	5/32	5/32	5/32	5/32	5/32	5/32
1¼	5/32	5/32	5/32	5/32	5/32	5/32	⅛
1½	5/32	5/32	5/32	5/32	5/32	5/32	⅛
2	5/32	7/32	3/16	3/16	⅛	⅛	⅛
2½	5/32	7/32	3/16	3/16	⅛	⅛	⅛
3	5/32	7/32	3/16	3/16	5/32	⅛	⅛
4	5/32	7/32	7/32	3/16	5/32	⅛	5/32
5	5/32	7/32	7/32	3/16	5/32	⅛	5/32
6	5/32	7/32	7/32	3/16	5/32	⅛	5/32
8	5/32	7/32	7/32	3/16	5/32	5/32	3/16
10	5/32	7/32	7/32	3/16	5/32	5/32	¼
12	5/32	7/32	7/32	3/16	5/32	3/16	5/16
14	⅛	7/32	7/32	3/16	5/32	7/32	. . .
16	⅛	7/32	7/32	3/16	5/32	5/16	. . .
18	⅛	7/32	7/32	3/16	3/16	5/16	. . .
20	⅛	7/32	7/32	3/16	3/16	⅜	. . .
22	. . .	¼	¼	7/32	. . .	. . .	. . .
24	⅛	¼	¼	7/32	7/32	7/16	. . .
26	. . .	¼	. . .	7/32	. . .	. . .	. . .
28	. . .	¼	. . .	7/32	. . .	. . .	. . .
30	. . .	¼	. . .	7/32	. . .	. . .	. . .

TABLE 3.29A

DIMENSIONS[2][3][4] OF STEEL FLANGES—LENGTH, DIAMETER AND THICKNESS

(Based on ASA 16.5 for sizes up to 24″ and MSS-SP-44 for larger sizes[5])

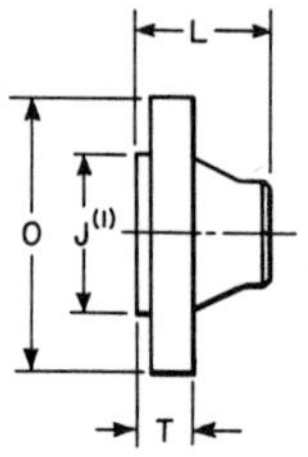

WELDING NECK FLANGES

Welding Neck Flanges

Nominal Pipe Size	Raised-Face L (includes raised face)							Ring-Type Joint L						
	150	300	400	600	900	1500	2500	150	300	400	600	900	1500	2500
3/4	2 1/16	2 1/4	2 1/2	2 1/2	3	3	3 3/8	...	2 1/2	2 1/2	2 1/2	...	3	3 3/8
1	2 3/16	2 7/16	2 11/16	2 11/16	3 1/8	3 1/8	3 3/4	2 7/16	2 11/16	2 11/16	2 11/16	...	3 1/8	3 3/4
1 1/2	2 7/16	2 11/16	3	3	3 1/2	3 1/2	4 5/8	2 11/16	2 15/16	3	3	...	4	4 11/16
2	2 1/2	2 3/4	3 1/8	3 1/8	4 1/4	4 1/4	5 1/4	2 3/4	3 1/16	3 3/16	3 3/16	...	4 5/16	5 5/16
3	2 3/4	3 1/8	3 1/2	3 1/2	4 1/4	4 7/8	6 7/8	3	3 7/16	3 9/16	3 9/16	4 5/16	4 15/16	7
4	3	3 3/8	3 3/4	4 1/4	4 3/4	5 1/8	7 3/4	3 1/4	3 11/16	3 13/16	4 5/16	4 13/16	5 3/16	7 15/16
6	3 1/2	3 7/8	4 5/16	4 7/8	5 3/4	7	11	3 3/4	4 3/16	4 3/8	4 15/16	5 13/16	7 1/8	11 1/4
8	4	4 3/8	4 7/8	5 1/2	6 5/8	8 5/8	12 3/4	4 1/4	4 11/16	4 15/16	5 9/16	6 11/16	8 13/16	13 1/16
10	4	4 5/8	5 1/8	6 1/4	7 1/2	10 1/4	16 3/4	4 1/4	4 15/16	5 3/16	6 5/16	7 9/16	10 7/16	17 3/16
12	4 1/2	5 1/8	5 5/8	6 3/8	8 1/8	11 3/8	18 1/2	4 3/4	5 7/16	5 11/16	6 7/16	8 3/16	11 11/16	18 15/16
14	5	5 5/8	6 1/8	6 3/4	8 5/8	12	...	5 1/4	5 15/16	6 3/16	6 13/16	8 13/16	12 3/8	...
16	5	5 3/4	6 1/4	7 1/4	8 3/4	12 1/2	...	5 1/4	6 1/16	6 5/16	7 5/16	8 15/16	12 15/16	...
18	5 1/2	6 1/4	6 3/4	7 1/2	9 1/4	13 1/8	...	5 3/4	6 9/16	6 13/16	7 9/16	9 1/2	13 9/16	...
20	5 11/16	6 3/8	6 7/8	7 3/4	10	14 1/4	...	5 5/16	6 3/4	7	7 7/8	10 1/4	14 11/16	...
24	6	6 5/8	7 1/8	8 1/4	11 3/4	16 1/4	...	6 1/4	7 1/16	7 5/16	8 7/16	12 1/8	16 13/16	...
26	5	7 1/4	7 7/8	9	11 1/2	...	...	...	7 3/4	8 1/8	9 1/4	11 15/16	...	...
28	5 1/16	7 3/4	8 3/8	9 1/2	12	...	...	...	8 1/4	8 5/8	9 3/4	12 7/16	...	...
30	5 1/8	8 1/4	8 7/8	10	12 1/2	...	...	...	8 3/4	9 1/8	10 1/4	12 15/16	...	...
32	5 1/4	8 3/4	9 3/8	10 1/2	13 1/4	...	...	...	9 5/16	9 11/16	10 13/16	13 11/16	...	...
34	5 5/16	9 1/8	9 3/4	10 7/8	14	...	...	...	9 11/16	10 1/16	11 3/16	14 9/16	...	...
36	5 3/8	9 1/2	10 1/8	11 3/8	14 1/2	...	...	...	10 1/16	10 7/16	11 11/16	15 1/16	...	...
42	5 5/8	...	...	...	...	...	...	...	...	...	...	...	...	...

[1] See Table 3.30 for dimension J for all raised-face flanges only.
[2] See ASA 16.5 and MSS-SP-44 for other dimensions and MSS-SP-44 for sizes larger than 24″.
[3] See Table 3.28 for distances between connecting RTJ flanges.

TABLE 3.29A (*Concluded*)

DIMENSIONS[2][3][4] OF STEEL FLANGES—LENGTH, DIAMETER AND THICKNESS

(Based on ASA 16.5 for sizes up to 24″ and MSS–SP–44 for larger sizes[5])

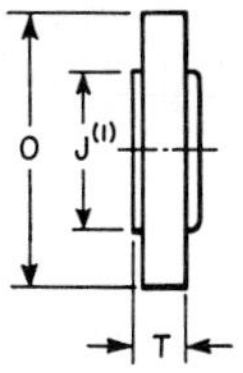

SLIP-ON AND SCREWED FLANGES

Welding Neck, Slip-On, and Screwed Flanges

Raised-Face T							Ring-Type Joint T							Raised-Face and Ring-Type Joint O							Nominal Pipe Size
150	300	400	600	900	1500	2500	150	300	400	600	900	1500	2500	150	300	400	600	900	1500	2500	
½	⅝	⅞	⅞	1¼	1¼	1½	...	⅞	⅞	⅞	...	1¼	1½	3⅞	4⅝	4⅝	4⅝	5⅛	5⅛	5½	¾
9/16	11/16	15/16	15/16	1⅜	1⅜	1⅝	13/16	15/16	15/16	15/16	...	1⅜	1⅝	4¼	4⅞	4⅞	4⅞	5⅞	5⅞	6¼	1
11/16	13/16	1⅛	1⅛	1½	1½	2	15/16	1 1/16	1⅛	1⅛	...	1½	2 1/16	5	6⅛	6⅛	6⅛	7	7	8	1½
¾	⅞	1¼	1¼	1¾	1¾	2¼	1	1 3/16	1 5/16	1 5/16	...	1 13/16	2 5/16	6	6½	6½	6½	8½	8½	9¼	2
15/16	1⅛	1½	1½	1¾	2⅛	2⅞	1 3/16	1 7/16	1 9/16	1 9/16	1 13/16	2 3/16	3	7½	8¼	8¼	8¼	9½	10½	12	3
15/16	1¼	1⅝	1¾	2	2⅜	3¼	1 3/16	1 9/16	1 11/16	1 13/16	2 1/16	2 9/16	3 7/16	9	10	10	10¾	11½	12¼	14	4
1	1 7/16	1⅞	2⅛	2 7/16	3½	4½	1¼	1¾	1 15/16	2 3/16	2½	3⅝	4¾	11	12½	12½	14	15	15½	19	6
1⅛	1⅝	2⅛	2 7/16	2¾	3⅞	5¼	1⅜	1 15/16	2 3/16	2½	2 13/16	4 1/16	5 9/16	13½	15	15	16½	18½	19	21¾	8
1 3/16	1⅞	2⅜	2¾	3	4½	6¾	1 7/16	2 3/16	2 7/16	2 13/16	3 1/16	4 11/16	7 3/16	16	17½	17½	20	21½	23	26½	10
1¼	2	2½	2⅞	3⅜	5⅛	7½	1½	2 5/16	2 9/16	2 15/16	3 7/16	5 7/16	7 15/16	19	20½	20½	22	24	26½	30	12
1⅜	2⅛	2⅝	3	3⅝	5½	...	1⅝	2 7/16	2 11/16	3 1/16	3 13/16	5⅞	...	21	23	23	23¾	25¼	29½	...	14
1 7/16	2¼	2¾	3¼	3¾	6	...	1 11/16	2 9/16	2 13/16	3 5/16	3 15/16	6 7/16	...	23½	25½	25½	27	27¾	32½	...	16
1 9/16	2⅜	2⅞	3½	4¼	6⅝	...	1 13/16	2 11/16	2 15/16	3 9/16	4½	7 1/16	...	25	28	28	29¼	31	36	...	18
1 11/16	2½	3	3¾	4½	7¼	...	1 15/16	2⅞	3⅛	3⅞	4¾	7 11/16	...	27½	30½	30½	32	33¾	38¾	...	20
1⅞	2¾	3¼	4¼	5¾	8¼	...	2⅛	3 3/16	3 7/16	4 7/16	6⅛	8 13/16	...	32	36	36	37	41	46	...	24
2	3⅛	3¾	4½	5¾	...	...	...	3⅝	4	4¾	6 3/16	...	...	34¼	38¼	38¼	40	42¾	...	...	26
2 1/16	3⅜	4	4⅝	5⅞	...	...	...	3⅞	4¼	4⅞	6 5/16	...	...	36½	40¾	40¾	42¼	46	...	...	28
2⅛	3⅝	4¼	4¾	6⅛	...	...	...	4⅛	4½	5	6 9/16	...	...	38¾	43	43	44½	48½	...	...	30
2¼	3⅞	4½	4⅞	6½	...	...	...	4 7/16	4 13/16	5 3/16	6 15/16	...	...	41¾	45¼	45¼	47	51¾	...	...	32
2 5/16	4	4⅝	5	6¾	...	...	...	4 9/16	4 15/16	5 5/16	7 5/16	...	...	43¾	47½	47½	49	55	...	...	34
2⅜	4⅛	4¾	5⅛	7	...	...	...	4 11/16	5 1/16	5 7/16	7 9/16	...	...	46	50	50	51¾	57½	...	...	36
2⅝	...	...	...	...	...	...	...	...	...	...	...	...	...	53	...	...	...	...	...	...	42

(4) Correspondence of flange diameters and drilling templates: 125 lb cast iron, 150 lb steel and 150 lb brass or bronze are same.
250 lb cast iron, 300 lb steel and 300 lb brass or bronze are same.

(5) 150 lb in sizes larger than 24″ is not an MSS–SP–44 standard, but is common manufacturer's practice. 150 lb in sizes 26″ and large match 125 lb cast iron flanges (ASA B16.1).

TABLE 3.29B

DIMENSIONS OF STEEL FLANGES—DRILLING TEMPLATE AND BOLTING, WEIGHTS

Based on ASA B16.5 up to 24″, MSS SP-44 for sizes larger than 24″.
(Tables reproduced by permission of Taylor Forge and Pipe Works.)

150 lb

Drilling Template and Bolting

Nominal Pipe Size	DRILLING			BOLTING			
	Bolt Circle Diam.	Number of Holes	Diam. of Holes	Diam. of Bolts	Machine Bolt Length — Raised Face	Stud Bolt Length + Raised Face	Stud Bolt Length + Ring Joint
½	2⅜	4	⅝	½	1¾	2¼	- -
¾	2¾	4	⅝	½	2	2¼	- -
1	3⅛	4	⅝	½	2	2½	3
1¼	3½	4	⅝	½	2¼	2½	3
1½	3⅞	4	⅝	½	2¼	2¾	3¼
2	4¾	4	¾	⅝	2¾	3	3½
2½	5½	4	¾	⅝	3	3¼	3¾
3	6	4	¾	⅝	3	3½	4
3½	7	8	¾	⅝	3	3½	4
4	7½	8	¾	⅝	3	3½	4
5	8½	8	⅞	¾	3¼	3¾	4¼
6	9½	8	⅞	¾	3¼	3¾	4¼
8	11¾	8	⅞	¾	3½	4	4½
10	14¼	12	1	⅞	3¾	4½	5
12	17	12	1	⅞	4	4½	5
14	18¾	12	1⅛	1	4¼	5	5½
16	21¼	16	1⅛	1	4½	5¼	5¾
18	22¾	16	1¼	1⅛	4¾	5¾	6¼
20	25	20	1¼	1⅛	5¼	6	6½
22†	27¼	20	1⅜	1¼	5½	6½	7
24	29½	20	1⅜	1¼	5¾	6¾	7¼
26	31¾	24	1⅜	1¼	6	7	- -
28	34	28	1⅜	1¼	6	7	- -
30	36	28	1⅜	1¼	6¼	7¼	- -
32	38½	28	1⅝	1½	6¾	8	- -
34	40½	32	1⅝	1½	7	8	- -
36	42¾	32	1⅝	1½	7	8¼	- -
42	49½	36	1⅝	1½	7½	8¾	- -

Weights

Nominal Pipe Size	APPROXIMATE WEIGHT EACH—POUNDS			
	Welding Neck	Slip-on and Thr'd	Lap Joint	Blind
½	2	1	1	1
¾	2	2	2	2
1	3	2	2	2
1¼	3	3	3	3
1½	4	3	3	4
2	6	5	5	5
2½	8	7	7	7
3	10	8	8	9
3½	12	11	11	13
4	15	13	13	17
5	19	15	15	20
6	24	19	19	26
8	39	30	30	45
10	52	43	43	70
12	80	64	64	110
14	110	90	105	140
16	140	98	140	180
18	150	130	160	220
20	180	165	195	285
22	225	185	245	355
24	260	220	275	430
26	300	250	- -	525
28	315	285	- -	620
30	360	315	- -	720
32	435	395	- -	870
34	465	420	- -	990
36	520	480	- -	1125
42	750	680	- -	1625

† 22 in. is a manufacturer's standard.

+ Stud bolt length does not include height of points.

TABLE 3.29B (*Continued*)
DIMENSIONS OF STEEL FLANGES

300 lb

Drilling Template and Bolting

Nominal Pipe Size	DRILLING			BOLTING			
	Bolt Circle Diam.	Number of Holes	Diam. of Holes	Diam. of Bolts	Machine Bolt Length	Stud Bolt Length +	
					Raised Face	Raised Face	Ring Joint
½	2⅝	4	⅝	½	2	2½	3
¾	3¼	4	¾	⅝	2½	2¾	3¼
1	3½	4	¾	⅝	2½	3	3½
1¼	3⅞	4	¾	⅝	2¾	3	3½
1½	4½	4	⅞	¾	3	3½	4
2	5	8	¾	⅝	3	3¼	4
2½	5⅞	8	⅞	¾	3¼	3¾	4½
3	6⅝	8	⅞	¾	3½	4	4¾
3½	7¼	8	⅞	¾	3¾	4¼	5
4	7⅞	8	⅞	¾	3¾	4¼	5
5	9¼	8	⅞	¾	4	4½	5¼
6	10⅝	12	⅞	¾	4¼	4¾	5½
8	13	12	1	⅞	4¾	5¼	6
10	15¼	16	1⅛	1	5¼	6	6¾
12	17¾	16	1¼	1⅛	5¾	6½	7¼
14	20¼	20	1¼	1⅛	6	6¾	7½
16	22½	20	1⅜	1¼	6½	7¼	8
18	24¾	24	1⅜	1¼	6¾	7½	8¼
20	27	24	1⅜	1¼	7	8	8¾
22†	29¼	24	1⅝	1½	7½	8¾	9¾
24	32	24	1⅝	1½	7¾	9	10
26	34½	28	1¾	1⅝	8¾	10	11
28	37	28	1¾	1⅝	9¼	10½	11½
30	39¼	28	1⅞	1¾	10	11¼	12¼
32	41½	28	2	1⅞	10½	12	13¼
34	43½	28	2	1⅞	10¾	12¼	13½
36	46	32	2⅛	2	11¼	12¾	14

Weights

Nominal Pipe Size	APPROXIMATE WEIGHT EACH—POUNDS			
	Welding Neck	Slip-on and Thr'd	Lap Joint	Blind
½	2	2	2	2
¾	3	3	3	3
1	4	3	3	3
1¼	5	4	4	4
1½	7	6	6	6
2	9	7	7	8
2½	12	10	10	12
3	15	13	13	16
3½	18	17	17	21
4	25	22	22	27
5	32	28	28	35
6	42	39	39	50
8	67	58	58	81
10	91	81	91	125
12	140	115	140	185
14	180	165	190	250
16	250	190	250	295
18	320	250	295	395
20	400	315	370	505
22	465	370	435	640
24	580	475	550	790
26	670	570	- -	1050
28	810	720	- -	1275
30	930	810	- -	1500
32	1025	890	- -	1775
34	1200	1075	- -	2025
36	1300	1200	- -	2275

† 22 in. is a manufacturer's standard.

+ Stud bolt length does not include height of points.

TABLE 3.29B (*Continued*)
DIMENSIONS OF STEEL FLANGES

400 lb

Drilling Template and Bolting

| Nominal Pipe Size | DRILLING | | | BOLTING | | | |
| | Bolt Circle Diam. | Number of Holes | Diam. of Holes | Diam. of Bolts | Stud Bolt Length + | | |
					¼" Raised Face	Male-Female Tongue-Groove	Ring Joint
½	2⅝	4	⅝	½	3	2¾	3
¾	3¼	4	¾	⅝	3¼	3	3¼
1	3½	4	¾	⅝	3½	3¼	3½
1¼	3⅞	4	¾	⅝	3¾	3½	3¾
1½	4½	4	⅞	¾	4	3¾	4
2	5	8	¾	⅝	4	3¾	4¼
2½	5⅞	8	⅞	¾	4½	4¼	4¾
3	6⅝	8	⅞	¾	4¾	4½	5
3½	7¼	8	1	⅞	5¼	5	5½
4	7⅞	8	1	⅞	5¼	5	5½
5	9¼	8	1	⅞	5½	5¼	5¾
6	10⅝	12	1	⅞	5¾	5½	6
8	13	12	1⅛	1	6½	6¼	6¾
10	15¼	16	1¼	1⅛	7¼	7	7½
12	17¾	16	1⅜	1¼	7¾	7½	8
14	20¼	20	1⅜	1¼	8	7¾	8¼
16	22½	20	1½	1⅜	8½	8¼	8¾
18	24¾	24	1½	1⅜	8¾	8½	9
20	27	24	1⅝	1½	9½	9¼	9¾
22†	29¼	24	1¾	1⅝	10	9¾	10½
24	32	24	1⅞	1¾	10½	10¼	11
26	34½	28	1⅞	1¾	11½	- -	12
28	37	28	2	1⅞	12¼	- -	12¾
30	39¼	28	2⅛	2	13	- -	13½
32	41½	28	2⅛	2	13½	- -	14¼
34	43½	28	2⅛	2	13¾	- -	14½
36	46	32	2⅛	2	14	- -	14¾

Weights

| Nominal Pipe Size | APPROXIMATE WEIGHT EACH—POUNDS | | | |
	Welding Neck	Slip-on and Thr'd	Lap Joint	Blind
½	2	2	2	2
¾	4	3	3	3
1	4	4	4	4
1¼	6	5	5	5
1½	8	7	7	8
2	12	9	9	10
2½	18	13	12	15
3	23	16	15	20
3½	26	21	20	29
4	35	26	25	33
5	43	31	29	44
6	57	44	42	61
8	89	67	64	100
10	125	91	110	155
12	175	130	150	225
14	230	180	205	290
16	295	235	260	370
18	350	285	315	455
20	425	345	385	585
22	505	405	455	720
24	620	510	570	890
26	750	650	- -	1125
28	880	780	- -	1425
30	1000	900	- -	1675
32	1150	1025	- -	1975
34	1300	1150	- -	2250
36	1475	1325	- -	2525

† 22 in. is a manufacturer's standard.

+ Stud bolt length does not include height of points.

TABLE 3.29B *(Continued)*
DIMENSIONS OF STEEL FLANGES

600 lb

Drilling Template and Bolting

| Nominal Pipe Size | DRILLING | | | BOLTING | | | |
| | Bolt Circle Diam. | Number of Holes | Diam. of Holes | Diam. of Bolts | Stud Bolt Length + | | |
					¼" Raised Face	Male-Female Tongue-Groove	Ring Joint
½	2⅝	4	⅝	½	3	2¾	3
¾	3¼	4	¾	⅝	3¼	3	3¼
1	3½	4	¾	⅝	3½	3¼	3½
1¼	3⅞	4	¾	⅝	3¾	3½	3¾
1½	4½	4	⅞	¾	4	3¾	4
2	5	8	¾	⅝	4	3¾	4¼
2½	5⅞	8	⅞	¾	4½	4¼	4¾
3	6⅝	8	⅞	¾	4¾	4½	5
3½	7¼	8	1	⅞	5¼	5	5½
4	8½	8	1	⅞	5½	5¼	5¾
5	10½	8	1⅛	1	6¼	6	6½
6	11½	12	1⅛	1	6½	6¼	6¾
8	13¾	12	1¼	1⅛	7½	7¼	7¾
10	17	16	1⅜	1¼	8¼	8	8½
12	19¼	20	1⅜	1¼	8½	8¼	8¾
14	20¾	20	1½	1⅜	9	8¾	9¼
16	23¾	20	1⅝	1½	9¾	9½	10
18	25¾	20	1¾	1⅝	10½	10¼	10¾
20	28½	24	1¾	1⅝	11¼	11	11½
22†	30⅝	24	1⅞	1¾	12	11¾	12½
24	33	24	2	1⅞	12¾	12½	13¼
26	36	28	2	1⅞	13¼	- -	13¾
28	38	28	2⅛	2	13¾	- -	14¼
30	40¼	28	2⅛	2	14	- -	14½
32	42½	28	2⅜	2¼	14¾	- -	15½
34	44½	28	2⅜	2¼	15	- -	15¾
36	47	28	2⅝	2½	15¾	- -	16½

Weights

| Nominal Pipe Size | APPROXIMATE WEIGHT EACH—POUNDS | | | |
	Welding Neck	Slip-on and Thr'd	Lap Joint	Blind
½	2	2	2	2
¾	4	3	3	3
1	4	4	4	4
1¼	6	5	5	5
1½	8	7	7	8
2	12	9	9	10
2½	18	13	12	15
3	23	16	15	20
3½	26	21	20	29
4	42	37	36	41
5	68	63	61	68
6	81	80	78	86
8	120	115	110	140
10	190	170	170	230
12	225	200	200	295
14	280	230	250	355
16	390	330	365	495
18	475	400	435	630
20	590	510	570	810
22	720	590	670	1000
24	830	730	810	1250
26	1025	950	- -	1525
28	1175	1075	- -	1750
30	1300	1175	- -	2000
32	1500	1375	- -	2300
34	1650	1500	- -	2575
36	1750	1600	- -	2950

† 22 in. is a manufacturer's standard.
+ Stud bolt length does not include height of points.

TABLE 3.29B *(Continued)*
DIMENSIONS OF STEEL FLANGES

900 lb

Drilling Template and Bolting

| Nominal Pipe Size | DRILLING | | | BOLTING | | | |
| | Bolt Circle Diam. | Number of Holes | Diam. of Holes | Diam. of Bolts | Stud Bolt Length + | | |
					¼" Raised Face	Male-Female Tongue-Groove	Ring Joint
½	3¼	4	⅞	¾	4	3¾	4
¾	3½	4	⅞	¾	4¼	4	4¼
1	4	4	1	⅞	4¾	4½	4¾
1¼	4⅜	4	1	⅞	4¾	4½	4¾
1½	4⅞	4	1⅛	1	5¼	5	5¼
2	6½	8	1	⅞	5½	5¼	5¾
2½	7½	8	1⅛	1	6	5¾	6¼
3	7½	8	1	⅞	5½	5¼	5¾
4	9¼	8	1¼	1⅛	6½	6¼	6¾
5	11	8	1⅜	1¼	7¼	7	7½
6	12½	12	1¼	1⅛	7½	7¼	7½
8	15½	12	1½	1⅜	8½	8¼	8¾
10	18½	16	1½	1⅜	9	8¾	9¼
12	21	20	1½	1⅜	9¾	9½	10
14	22	20	1⅝	1½	10½	10¼	11
16	24¼	20	1¾	1⅝	11	10¾	11½
18	27	20	2	1⅞	12¾	12½	13¼
20	29½	20	2⅛	2	13½	13¼	14
24	35½	20	2⅝	2½	17	16¾	17¾
26	37½	20	2⅞	2¾	17½	- -	18¾
28	40¼	20	3⅛	3	18¼	- -	19½
30	42¾	20	3⅛	3	18¾	- -	20
32	45½	20	3⅜	3¼	20	- -	21¼
34	48¼	20	3⅝	3½	21	- -	22½
36	50¾	20	3⅝	3½	21½	- -	23

Weights

| Nominal Pipe Size | APPROXIMATE WEIGHT EACH—POUNDS | | | |
	Welding Neck	Slip-on and Thr'd	Lap Joint	Blind
½	5	4	4	4
¾	6	5	5	6
1	9	8	8	8
1¼	10	9	9	9
1½	13	12	12	13
2	25	25	25	25
2½	36	36	35	35
3	31	26	25	29
4	51	53	51	54
5	86	83	81	87
6	110	110	105	115
8	175	170	190	200
10	260	245	275	290
12	325	325	370	415
14	400	400	415	520
16	495	425	465	600
18	680	600	650	850
20	830	730	810	1075
24	1500	1400	1550	2025
26	1575	1525	- -	2200
28	1850	1800	- -	2575
30	2150	2075	- -	3025
32	2575	2500	- -	3650
34	3025	2950	- -	4275
36	3450	3350	- -	4900

+ Stud bolt length does not include height of points.

TABLE 3.29B (*Continued*)
DIMENSIONS OF STEEL FLANGES

1500 lb

Drilling Template and Bolting

Nominal Pipe Size	DRILLING			BOLTING			
					Stud Bolt Length+		
	Bolt Circle Diam.	Number of Holes	Diam. of Holes	Diam. of Bolts	¼" Raised Face	Male-Female Tongue-Groove	Ring Joint
½	3¼	4	⅞	¾	4	3¾	4
¾	3½	4	⅞	¾	4¼	4	4¼
1	4	4	1	⅞	4¾	4½	4¾
1¼	4⅜	4	1	⅞	4¾	4½	4¾
1½	4⅞	4	1⅛	1	5¼	5	5¼
2	6½	8	1	⅞	5½	5¼	5¾
2½	7½	8	1⅛	1	6	5¾	6¼
3	8	8	1¼	1⅛	6¾	6½	7
4	9½	8	1⅜	1¼	7½	7¼	7¾
5	11½	8	1⅝	1½	9½	9¼	9¾
6	12½	12	1½	1⅜	10	9¾	10¼
8	15½	12	1¾	1⅝	11¼	11	11¾
10	19	12	2	1⅞	13¼	13	13½
12	22½	16	2⅛	2	14¾	14½	15¼
14	25	16	2⅜	2¼	16	15¾	16¾
16	27¾	16	2⅝	2½	17½	17¼	18½
18	30½	16	2⅞	2¾	19¼	19	20¼
20	32¾	16	3⅛	3	21	20¾	22¼
24	39	16	3⅝	3½	24	23¾	25½

Weights

Nominal Pipe Size	APPROXIMATE WEIGHT EACH—POUNDS			
	Welding Neck	Slip-on and Thr'd	Lap Joint	Blind
½	5	4	4	4
¾	6	5	5	6
1	9	8	8	8
1¼	10	9	9	9
1½	13	12	12	13
2	25	25	25	25
2½	36	36	35	35
3	48	48	47	48
4	73	73	75	73
5	130	130	140	140
6	165	165	170	160
8	275	260	285	300
10	455	435	485	510
12	690	580	630	690
14	940	- -	890	975
16	1250	- -	1150	1300
18	1625	- -	1475	1750
20	2050	- -	1775	2225
24	3325	- -	2825	3625

+ Stud bolt length does not include height of points.

TABLE 3.29B *(Concluded)*
DIMENSIONS OF STEEL FLANGES

2500 lb

Drilling Template and Bolting

Nominal Pipe Size	DRILLING			BOLTING			
	Bolt Circle Diam.	Number of Holes	Diam. of Holes	Diam. of Bolts	Stud Bolt Length+		
					¼" Raised Face	Male-Female Tongue-Groove	Ring Joint
½	3½	4	⅞	¾	4¾	4½	4¾
¾	3¾	4	⅞	¾	4¾	4½	4¾
1	4¼	4	1	⅞	5¼	5	5¼
1¼	5⅛	4	1⅛	1	5¾	5½	6
1½	5¾	4	1¼	1⅛	6½	6¼	6¾
2	6¾	8	1⅛	1	6¾	6½	7
2½	7¾	8	1¼	1⅛	7½	7¼	7¾
3	9	8	1⅜	1¼	8½	8¼	8¾
4	10¾	8	1⅝	1½	9¾	9½	10¼
5	12¾	8	1⅞	1¾	11½	11¼	12¼
6	14½	8	2⅛	2	13½	13¼	14
8	17¼	12	2⅛	2	15	14¾	15½
10	21¼	12	2⅝	2½	19	18¾	20
12	24⅜	12	2⅞	2¾	21	20¾	22

Weights

Nominal Pipe Size	APPROXIMATE WEIGHT EACH—POUNDS			
	Welding Neck	Slip-on and Thr'd	Lap Joint	Blind
½	7	7	7	7
¾	8	8	8	8
1	12	11	11	11
1¼	17	16	16	17
1½	25	22	22	23
2	42	38	37	39
2½	52	55	53	56
3	94	83	80	86
4	145	125	120	135
5	245	210	205	225
6	380	325	315	345
8	580	485	470	530
10	1075	930	900	1025
12	1525	1100	1100	1300

+ Stud bolt length does not include height of points.

TABLE 3.30
DIMENSIONS OF STEEL-WELDING FITTINGS
(Based on ASA B16.9 except as noted)
All dimensions in inches

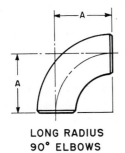

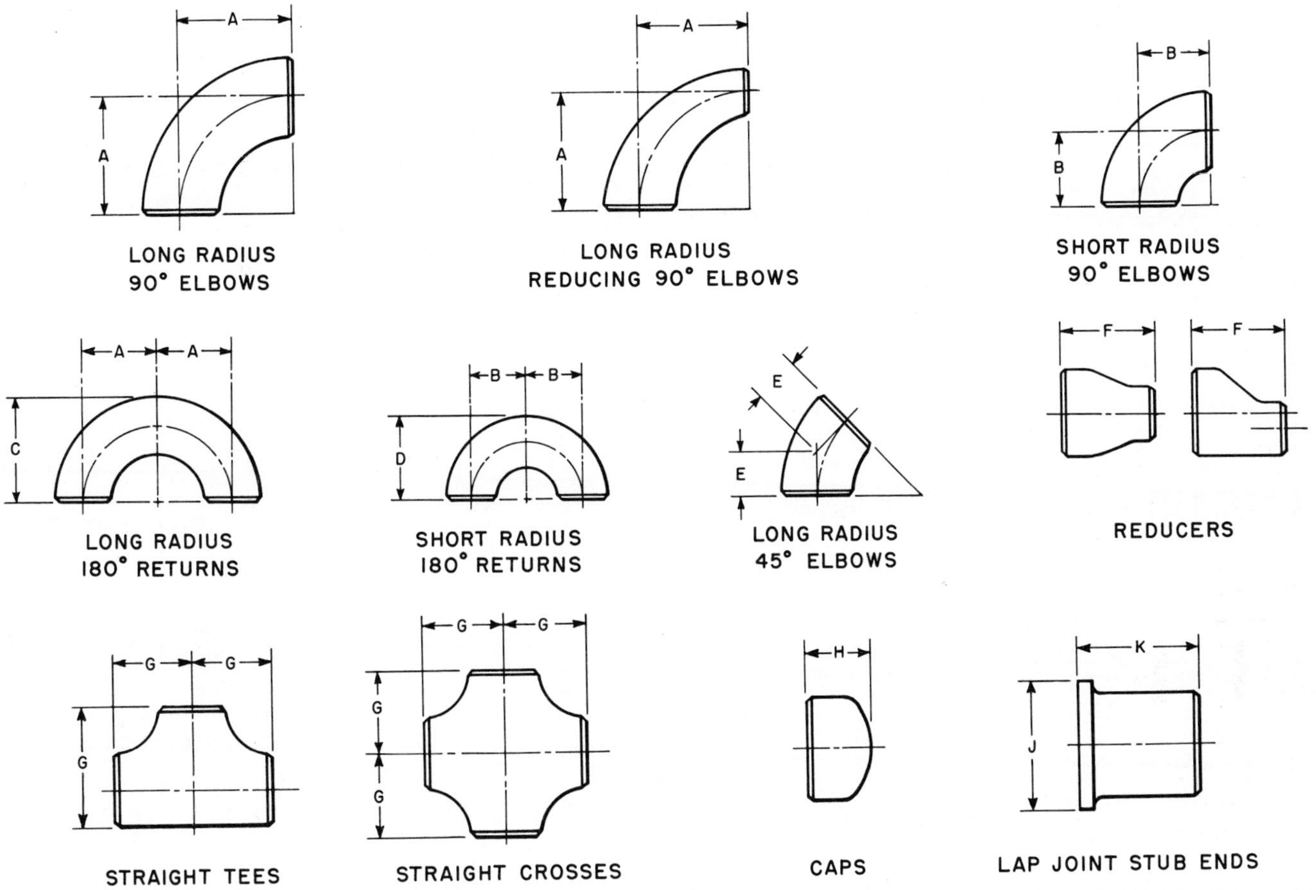

**LONG RADIUS
90° ELBOWS**

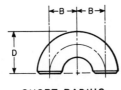

**LONG RADIUS
REDUCING 90° ELBOWS**

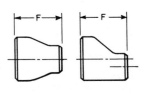

**SHORT RADIUS
90° ELBOWS**

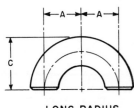

**LONG RADIUS
180° RETURNS**

**SHORT RADIUS
180° RETURNS**

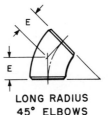

**LONG RADIUS
45° ELBOWS**

REDUCERS

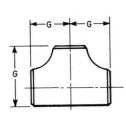

STRAIGHT TEES

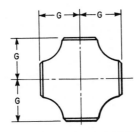

STRAIGHT CROSSES

CAPS

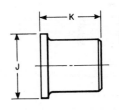

LAP JOINT STUB ENDS

Nominal Pipe Size	A	B	C	D	E	F	G	H[5]	J	K
1	1½		2³⁄₁₆	1⅝	⅞	2	1½	1½	2	4
1½	2¼		3¼	2⁷⁄₁₆	1⅛	2½	2¼	1½	2⅞	4
2	3		4³⁄₁₆	3³⁄₁₆	1⅜	3	2½	1½	3⅜	6
3	4½		6¼	4¾	2	3½	3⅜	2	5	6
4	6		8¼	6¼	2½	4	4⅛	2½	6³⁄₁₆	6
6	9		12⁵⁄₁₆	9⁹⁄₁₆	3¾	5½	5⅝	3½	8½	8
8	12		16⁵⁄₁₆	12⁵⁄₁₆	5	6	7	4	10⅝	8
10	15		20⅜	15⅜	6¼	7	8½	5	12¾	10
12	18		24⅜	18⅜	7½	8	10	6	15	10
14	21		28	21	8¾	13	11⁴	6½	16¼	12
16	24		32	23	10	14	12⁴	7	18½	12
18	27		36	25½	11¼	15	13½⁴	8	21	12
20	30		40	30½	12½	20	15⁴	9	23	12
24	36		48	34	15	20	17⁴	10½	27¼	12

For column A: (1.5)(Nominal Pipe Size). For column B: Same as Nominal Pipe Size.

(1) Dimensions for these fittings not an ASA standard but common commercial practice.

(2) For reducing tees, see ASA B16.9.

(3) For sizes larger than 24″, see MSS–SP–48.

(4) Center to end dimensions for outlet are not standardized in 14″ and larger. Dimensions given are in common use.

(5) For standard weight and extra strong. See ASA B16.9 for dimensions of other thicknesses.

TABLE 3.31
CENTER-TO-END AND END-TO-END DIMENSIONS OF SCREWED FITTINGS*

All dimensions in inches

Nominal Pipe Size	Elbows, Tees, and Crosses Center to End					Couplings End to End	Unions End to End		
	125 lb CI 150 lb MI	250 lb CI 300 lb MI	2000 lb, FS	3000 lb, FS	6000 lb, FS	3000 & 6000 lb, FS	150 lb MI	300 lb, MI 2000–3000 lb FS	6000 lb FS
1/8	. . .	. . .	1 3/16	1 3/16	3 1/32	1 1/4	. . .	1 1/2	. . .
1/4	1 1/16	1 5/16	1 3/16	3 1/32	1 1/8	1 3/8	1 3/8	1 3/8	2 1/16
1/2	1 1/8	1 1/4	1 1/8	1 5/16	1 1/2	1 7/8	1 7/8	1 15/16	2 7/8
3/4	1 5/16	1 7/16	1 5/16	1 1/2	1 3/4	2	2 1/8	2 1/4	3 3/8
1	1 1/2	1 5/8	1 1/2	1 3/4	2	2 3/8	2 3/8	2 7/16	3 3/8
1 1/2	1 15/16	2 1/8	2	2 3/8	2 1/2	3 1/8	2 15/16	3	4 3/16
2	2 1/4	2 1/2	2 3/8	2 1/2	3 1/4†	3 3/8	3 1/4	3 3/8	4 5/8
Standard	ASA B16.3, 4, and 19		MSS–SP–49			MSS–SP–49	No standard—common practice, but some variation with manufacturer		

* Extracted by permission: Manufacturers Standardization Society of the Valve and Fittings Industry, New York, and the American Society of Mechanical Engineers (publishers of the ASA Code), New York.

† Not part of standard but in general use.
CI = cast iron
FS = forged steel
MI = malleable iron

TABLE 3.32
CENTER-TO-END AND END-TO-END DIMENSIONS OF FORGED STEEL SOCKET-WELDED FITTINGS*
(Based on ASA B16.11)
All dimensions in inches

Nominal Pipe Size	Elbows, Tees, and Crosses Center to End		Couplings End to End
	2000 lb and 3000 lb	4000 lb and 6000 lb	All Classes
1/8	1 3/16	. . .	1
1/4	1 3/16	. . .	1
1/2	1 1/8	1 5/16	1 3/8
3/4	1 5/16	1 1/2	1 1/2
1	1 1/2	1 3/4	1 3/4
1 1/2	2	2 3/8	2
2	2 3/8	2 1/2	2 1/2

* 2000 lb is for use with schedule 40 pipe.
3000 lb is for use with schedule 80 pipe.
4000 lb is for use with schedule 160 pipe.
6000 lb is for use with double-extra strong pipe.

INDEX OF TABLES AND FIGURES OF
STANDARD PRESSURE-TEMPERATURE RATINGS
FOR FLANGES, VALVES, AND FITTINGS

TABLE 3.33
PRESSURE-TEMPERATURE RATINGS OF FORGED STEEL PIPE FLANGES AND FLANGED FITTINGS
(Including Flanged Values) (ASA B16.5)

Introductory Notes:

1. Product used within the jurisdiction of Section I, Power Boilers of the ASME Boiler and Pressure Vessel Code is subject to the same maximum temperature limitations placed upon the material in Table P7, 1956 edition thereof.

2. Product used within the jurisdiction of Section I, Power Piping of the ASA Code for Pressure Piping B31.1 is subject to the same maximum temperature limitations placed upon piping of the same general compositions in Table 2a, of ASA B31.1-1955.

3. The ratings at −20 to 100°F given for the materials covered in all classes shall also apply at lower temperatures. The ratings for low temperature service of the cast and forged materials listed in ASTM A352 and A350 shall be taken the same as the −20 to 100°F ratings for carbon steel.

Some of the materials listed in the rating tables undergo a decrease in impact resistance at temperatures lower than −20°F to such an extent as to be unable to safely resist shock loadings, sudden changes of stress or high stress concentrations. Therefore, products that are to operate at temperatures below −20°F shall conform to the rules of the applicable Codes under which they are to be used.

4. (*a*) Temperature limitations of materials:

Where welded construction is used consideration should be given to the possibility of graphite formation in the following:

    Carbon-steel above 775°F.
    Carbon-molybdenum Steel above 875°F.
    Chrome-molybdenum Steel (with chromium under 0.67) above 975°F.

(*b*) Consideration should be given to the possibility of excessive oxidation (scaling) on the following steels:

$$
\left.
\begin{array}{l}
1\ Cr\text{–}\tfrac{1}{2}\ Mo \\
1\tfrac{1}{4}\ Cr\text{–}\tfrac{1}{2}\ Mo \\
2\ Cr\text{–}\tfrac{1}{2}\ Mo \\
2\tfrac{1}{4}\ Cr\text{–}1\ Mo \\
3\ Cr\text{–}1\ Mo
\end{array}
\right\} \text{above } 1050°F
$$

5 Cr–½ Mo above 1100°F

5. (*a*) The pressure-temperature ratings in Tables 3.33.1 to 3.33.8 apply to flanges, flanged fittings and cast and forged steel flanged valves which, in other respects, merit these ratings.

(*b*) All ratings are the maximum allowable non-shock pressures (psig) at the tabulated temperatures (°F), which represent temperatures on the inside of the pressure retaining structure.

(*c*) These ratings assume proper gasketing and bolting. See Tables 3.34, 35 and 36 and ASA B16.5. They apply to ring-joint, small tongue and groove with any type gasket except flat solid metal and other facing (including raised face) with gasket combination that results in a bolt load or flange moment no greater than that for ring-joint or tongue and groove.

TABLE 3.33 (*Continued*). MATERIALS

Table 3.33.1. List of Material Specifications[1]

General Classification[3]	Applicable ASTM Specification	
	Forgings	Castings
Carbon Steel	A105-55T Grade I A105-55T Grade II A181-55T Grade I* A181-55T Grade II†	A216-53 Grade WCB
Carbon Steel (low temp.)	A350-55T Grade LF1‡	A352-55T Grade LCB‡
Carbon Moly	A182-55T Grade F1	A217-55 Grade WC1
Carbon Moly (low temp.)		A352-55T Grade LC1‡
½ Cr — ½ Mo	A335-55T Grade P2†	A217-55 Grade WC4
1 Cr — ½ Mo	A182-55T Grade F12	
1¼ Cr — ½ Mo	A182-55T Grade F11	A217-55 Grade WC5 and WC6
2 Cr — ½ Mo	A335-55T Grade P3b†	
2¼ Cr — 1 Mo	A182-55T Grade F22	A217-55 Grade WC9
3 Cr — 1 Mo	A335-55T Grade P21†	
5 Cr — ½ Mo	A182-55T Grade F5a	A217-55 Grade C5
5 Cr — ½ Mo-Si	A335-55T Grade P5b†	
9 Cr — 1 Mo	A182-55T Grade F9	A217-55 Grade C12
Type 304	A182-55T Grade F304	A351-52T Grade CF8
Type 310	A182-55T Grade F310	
Type 347	A182-55T Grade F347	A351-52T Grade CF8C
Type 321	A182-55T Grade F321	
Type 316	A182-55T Grade F316	A351-52T Grade CF8M
2 Ni		A352-55T Grade LC2‡
3½ Ni	A350-55T Grade LF3‡	A352-55T Grade LC3‡

Bolting[2]	
Bolts, Stud-Bolts	Nuts
ASTM A193-55T ASTM A307-55T-Grade B ASTM A320-55T ASTM A354-55T	ASTM A194-55T

[1] See Introductory Note 1.

[2] See ASA B16.5 for details. (Also see Table 3.34.)

[3] These classifications are used to group materials for pressure-temperature ratings in Tables 3.33.2 to 3.33.8 inclusive.

[4] For method of rating alloy steels not given in Tables 3.33.1–3.33.8, see Fig. 3.6.

* 150 lb. and 300 lb. classes only.

† ASTM Specifications directly covering forged flanges, flanged fittings and valves for these general material specifications do not presently exist. Flanges, flanged fittings and valves of these materials shall be specified to conform to the nearest grade in A182, except chemistry to conform to the ASTM specification listed above.

‡ Intended primarily for use for subzero service. See Note 3.

TABLE 3.33 (*Continued*)

Table 3.33.2. 150-Pound Pressure-Temperature Ratings

Note: These ratings are all subject to stipulations in Introductory Note 5 which form a part of this table. All pressures are in pounds per square inch gage (psig).

Service Temperature Deg. F	Material																Service Temperature Deg. F
	Carbon Steel	Carbon Moly	Cr-Mo ½-½	Cr-Mo 1-½	Cr-Mo 1¼-½	Cr-Mo 2-½	Cr-Mo 2¼-1	Cr-Mo 3-1	Cr-Mo 5-½	Cr-Mo 5½-Si	Cr-Mo 9-1	304	Types 347 & 321	316	310	
−20 to 100[3]							275									−20 to 100[3]
150							255									150
200							240									200
250							225									250
300							210									300
350							195									350
400							180									400
450							165									450
500							150									500
550							140									550
600							130									600
650							120									650
700							110									700
750							100									750
800							92									800
850	82[1]						82									850
875	75[1]	75[1]					75									875
900	70[1]	70[1]					70									900
925	60[1,2]	60[1,2]					60									925
950	55[1,2]	55[1,2]					55									950
975	50[1,2]	50[1,2]					50									975
1000	40[1,2]	40[1,2]					40									1000

Hydrostatic Shell Test Pressure 425

Notes: [1] See Introductory Note 1. [2] See Introductory Note 2. [3] See Introductory Note 3.

94

TABLE 3.33 (Continued)

Table 3.33.3. 300-Pound Pressure-Temperature Ratings

Note: These ratings are all subject to stipulations in Introductory Note 5 which form a part of this table. All pressures are in pounds per square inch gage (psig).

Service Temperature Deg. F	Carbon Steel	Carbon Moly	Cr-Mo ½-½	Cr-Mo 1-½	Cr-Mo 1¼-½	Cr-Mo 2-½	Cr-Mo 2¼-1	Cr-Mo 3-1	Cr-Mo 5-½	Cr-Mo 5½-Si	Cr-Mo 9-1	304	347 & 321	316	310	Service Temperature Deg. F
−20 to 100[3]	470					720						615		720		−20 to 100[3]
150	425					710						585		710		150
200	365					700						550		700		200
250						690						520		690		250
300						680						495		680		300
350						675						470		675		350
400						665						450		665		400
450						650						430		650		450
500						625						410		625		500
550						590						395		590		550
600						555						380		555		600
650						515						370		515		650
700	300[1]	480	480	485	485	480	485	480	485	480	485	355	495		490	700
750	260[1]	445	445	450	450	445	450	445	450	445	450	340	470		465	750
800	225[1]	410	410	415	415	410	415	410	415	410	415	330	450		440	800
850	190[1,2]	370	370	385	385	370	385	370	385	370	385	320	425		415	850
875	155[1,2]	355[1]	355	365	365	355	365	355	365	355	365	315	415		400	875
900	120[1,2]	335[1]	335	350	350	335	350	335	350	335	350	310	400		390	900
925	85[1,2]	320[1,2]	320	335	335	320	335	320	335	320	335	305	390		375	925
950		300[1,2]	300	315	315	300	315	300	315	300	315	305	380		365	950
975		280[1,2]	280	300	300	280	300	275	300	250	300	300	370		350	975
1000		215[1,2]	215	255	265	215	265	240	250	190	290	300	355		340	1000
1025				215	230[1]	180	235	215	215	155	240	295	345		325	1025
1050				170	190[1]	145	200	190	180	120	190	290	335		315	1050
1075				135	165[1]	120	170[1]	165	145	105	150	275	325		300	1075
1100				95	135[1]	95	145[1]	135	115	85	115	255	310		290	1100
1125				75[1,2]	110[1,2]	75	125[1]	115	95	75	95	225	300		270	1125
1150				55[1,2]	85[1,2]	60	105[1]	95	75	60	75	195	260	290	250	1150
1175				45[1,2]	65[1,2]	50	85[1,2]	70[2]	65	50	65	175	215	260	225	1175
1200				35[1,2]	40[1,2]	40	70[1,2]	50[2]	50	40	50	155	170	235	205	1200
1225												135	140	205	185	1225
1250												110	115	180	165	1250
1275												100	95	160	140	1275
1300												85	75	135	120	1300
1325												75	65	115	100	1325
1350												60	50	95	80	1350
1375												55	45	80	70	1375
1400												50	40	70	55	1400
1425												40	35	60	45	1425
1450												35	30	50	40	1450
1475												30	30	45	30	1475
1500												25	25	35	25	1500
Hydrostatic Shell Test Pressure							1100					925		1100		

Notes:
[1] See Introductory Note 1.
[2] See Introductory Note 2.
[3] See Introductory Note 3.

95

TABLE 3.33 (Continued)

Table 3.33.4. 400-Pound Pressure-Temperature Ratings

Note: These ratings are all subject to stipulations in Introductory Note 5 which form a part of this table. All pressures are in pounds per square inch gage (psig).

Material

Service Temperature Deg. F	Carbon Steel	Carbon Moly	Cr-Mo ½-½	Cr-Mo 1-½	Cr-Mo 1¼-½	Cr-Mo 2-½	Cr-Mo 2¼-1	Cr-Mo 3-1	Cr-Mo 5-½	Cr-Mo 5½-Si	Cr-Mo 9-1	304	Types 347 & 321	316	310	Service Temperature Deg. F
−20 to 100[3]						960						825		960		−20 to 100[3]
150						945						775		945		150
200						930						730		930		200
250						920						695		920		250
300						910						660		910		300
350						900						630		900		350
400						890						600		890		400
450						870						575		870		450
500						835						550		835		500
550						790						530		790		550
600						740						510		740		600
650						690						490		690		650
700	635	640	640	645	645	640	645	640	645	640	645	475	660		655	700
750	575	590	590	600	600	590	600	590	600	590	600	455	625		620	750
800	490	545	545	555	555	545	555	545	555	545	555	440	595		585	800
850	400[1]	495	495	510	510	495	510	495	510	495	510	425	565		550	850
875	350[1]	470[1]	470	490	490	470	490	470	490	470	490	420	550		535	875
900	295[1]	450[1]	450	465	465	450	465	450	465	450	465	415	535		520	900
925	250[1,2]	425[1,2]	425	445	445	425	445	425	445	425	445	410	520		500	925
950	205[1,2]	400[1,2]	400	420	420	400	420	400	420	400	420	405	505		485	950
975	160[1,2]	370[1,2]	370	400	400	370	400	365	400	330	400	405	490		470	975
1000	115[1,2]	285[1,2]	285	345	355	285	355	320	335	250	390	400	475		450	1000
1025				285	305[1]	240	310	285	285	205	320	395	460		435	1025
1050				230	250[1]	190	265	250	240	160	250	390	445		415	1050
1075				180	215[1]	160	230[1]	215	195	135	200	365	430	400	400	1075
1100				130	185[1]	125	190[1]	185	150	115	150	345	415		390	1100
1125				100[1,2]	150[1,2]	105	165[1]	155	125	100	125	305	400		360	1125
1150				70[1,2]	115[1,2]	80	135[1]	125	100	80	100	265	345	390	330	1150
1175				60[1,2]	85[1,2]	65	115[1,2]	95[2]	85	70	85	235	290	350	305	1175
1200				45[1,2]	55[1,2]	55	90[1,2]	70[2]	70	55	70	205	230	310	275	1200
1225												175	190	275	245	1225
1250												150	150	240	215	1250
1275												130	125	215	190	1275
1300												110	100	185	160	1300
1325												95	85	155	135	1325
1350												80	70	125	105	1350
1375												75	60	105	90	1375
1400												65	55	90	75	1400
1425												55	50	80	60	1425
1450												45	40	70	50	1450
1475												40	40	55	40	1475
1500												35	35	45	35	1500
Hydrostatic Shell Test Pressure						1450	1450					1250		1450		

Notes:
[1] See Introductory Note 1.
[2] See Introductory Note 2.
[3] See Introductory Note 3.

TABLE 3.33 (Continued)

Table 3.33.5. 600-Pound Pressure-Temperature Ratings

Note: These ratings are all subject to stipulations in Introductory Note 5 which form a part of this table. All pressures are in pounds per square inch gage (psig).

Service Temperature Deg. F	Carbon Steel	Carbon Moly	Cr-Mo ½-½	Cr-Mo 1-½	Cr-Mo 1¼-½	Cr-Mo 2-½	Cr-Mo 2¼-1	Cr-Mo 3-1	Cr-Mo 5-½	Cr-Mo 5½-Si	Cr-Mo 9-1	Type 304	Types 347 & 321	Type 316	Type 310
-20 to 100[3]						1440						1235		1440	
150						1420						1165		1420	
200						1400						1095		1400	
250						1380						1040		1380	
300						1365						985		1365	
350						1350						945		1350	
400						1330						900		1330	
450						1305						860		1305	
500						1250						825		1250	
550						1180						795		1180	
600						1110						765		1110	
650						1030						735		1030	
700	940	960	960	965	965	960	965	960	965	960	965	710	985	985	980
750	850	890	890	900	900	890	900	890	900	890	900	685	940	940	930
800	730	815	815	835	835	815	835	815	835	815	835	660	895	895	880
850	**600**[1]	745	745	765	765	745	765	745	765	745	765	640	850	850	830
875	525[1]	710[1]	710	735	735	710	735	710	735	710	735	630	825	825	805
900	445[1]	670[1]	670	700	700	670	700	670	700	670	700	620	805	805	780
925	375[1,2]	635[1,2]	635	665	665	635	665	635	665	635	665	615	780	780	755
950	310[1,2]	**600**[1,2]	**600**	635	635	**600**	635	**600**	635	**600**	635	610	760	760	725
975	240[1,2]	555[1,2]	555	**600**	**600**	555	**600**	550	**600**	495	**600**	605	735	735	700
1000	170[1,2]	430[1,2]	430	515	535	425	535	480	500	375	585	**600**	715	715	675
1025				430	455[1]	355	465	430	430	310	480	595	690	690	650
1050				345	375[1]	290	400	375	355	240	375	585	670	670	625
1075				265	325[1]	240	345[1]	325	290	205	300	550	645	645	**600**
1100				190	275[1]	190	290[1]	275	225	170	225	515	625	625	585
1125				150[1,2]	225[1,2]	155	245[1]	230	190	145	190	455	**600**	**600**	540
1150				105[1,2]	170[1,2]	120	205[1]	185	150	125	150	395	520	585	495
1175				85[1,2]	125[1,2]	100	170[1,2]	145[2]	125	105	125	350	430	525	455
1200				70[1,2]	80[1,2]	80	135[1,2]	105[2]	105	80	105	310	345	465	410
1225												265	285	415	370
1250												225	225	365	325
1275												195	190	320	285
1300												170	150	275	240
1325												145	125	230	200
1350												125	105	185	160
1375												110	95	160	135
1400												95	80	135	110
1425												80	70	120	95
1450												70	60	105	75
1475												60	55	85	65
1500												50	50	70	50
Hydrostatic Shell Test Pressure							2175					1875		2175	

Notes:
[1] See Introductory Note 1.
[2] See Introductory Note 2.
[3] See Introductory Note 3.

97

TABLE 3.33 (*Continued*)

Table 3.33.6. 900-Pound Pressure-Temperature Ratings

Note: These ratings are all subject to stipulations in Introductory Note 5 which form a part of this table. All pressures are in pounds per square inch gage (psig).

Service Temperature Deg. F	Carbon Steel	Carbon Moly	Cr-Mo ½-½	Cr-Mo 1-½	Cr-Mo 1¼-½	Cr-Mo 2-½	Cr-Mo 2¼-1	Cr-Mo 3-1	Cr-Mo 5-½	Cr-Mo 5½-Si	Cr-Mo 9-1	304	347 & 321	316	310
−20 to 100[3]						2160						1850		2160	
150						2130						1750		2130	
200						2100						1645		2100	
250						2070						1565		2070	
300						2050						1480		2050	
350						2025						1415		2025	
400						2000						1350		2000	
450						1955						1290		1955	
500						1875						1235		1875	
550						1775						1190		1775	
600						1660						1145		1660	
650						1550						1105		1550	
700	1410	1440	1440	1450	1450	1440	1450	1440	1450	1440	1450	1065	1480		1470
750	1275	1330	1330	1350	1350	1330	1350	1330	1350	1330	1350	1025	1410		1395
800	1100	1225	1225	1250	1250	1225	1250	1225	1250	1225	1250	985	1345		1320
850	900[1]	1115	1115	1150	1150	1115	1150	1115	1150	1115	1150	960	1275		1245
875	785[1]	1060[1]	1060	1100	1100	1060	1100	1060	1100	1060	1100	945	1240		1205
900	670[1]	1010[1]	1010	1050	1050	1010	1050	1010	1050	1010	1050	930	1205		1165
925	565[1,2]	955[1,2]	955	1000	1000	955	1000	955	1000	955	1000	920	1175		1130
950	465[1,2]	900[1,2]	900	950	950	900	950	900	950	900	950	915	1140		1090
975	360[1,2]	835[1,2]	835	900	900	835	900	825	900	745	900	905	1105		1050
1000	255[1,2]	645[1,2]	645	770	800	635	800	720	750	565	875	900	1070		1015
1025				645	685[1]	535	700	645	645	465	720	890	1035		975
1050				515	565[1]	430	595	565	535	360	565	875	1000		940
1075				400	490[1]	355	515[1]	490	435	310	455	825	970		900
1100				290	410[1]	285	430[1]	410	340	255	340	770	935		875
1125				225[1,2]	335[1,2]	230	370[1]	345	285	220	285	680	900		810
1150				160[1,2]	255[1,2]	180	310[1]	280	225	185	225	590	780	875	745
1175				130[1,2]	190[1,2]	150	255[1,2]	215[2]	190	155	190	525	650	785	680
1200				105[1,2]	125[1,2]	125	205[1,2]	155[2]	155	125	155	465	515	700	615
1225												400	425	620	555
1250												335	340	545	490
1275												295	285	480	425
1300												250	225	410	360
1325												220	190	345	300
1350												185	155	280	240
1375												165	140	240	205
1400												145	125	205	165
1425												125	110	180	140
1450												105	95	155	115
1475												90	85	130	95
1500												75	75	105	75
Hydrostatic Shell Test Pressure							3250					2775		3250	

Notes:
[1] See Introductory Note 1.
[2] See Introductory Note 2.
[3] See Introductory Note 3.

TABLE 3.33 (Continued)

Table 3.33.7. 1500-Pound Pressure-Temperature Ratings

Note: These ratings are all subject to stipulations in Introductory Note 5 which form a part of this table. All pressures are in pounds per square inch gage (psig).

Service Temp. Deg. F	Carbon Steel	Carbon Moly	Cr-Mo ½–½	Cr-Mo 1–½	Cr-Mo 1¼–½	Cr-Mo 2–½	Cr-Mo 2¼–1	Cr-Mo 3–1	Cr-Mo 5–½	Cr-Mo 5½–Si	Cr-Mo 9–1	Type 304	Types 347 & 321	Type 316	Type 310
−20 to 100[3]						3600						3085		3600	
150						3550						2915		3550	
200						3500						2740		3500	
250						3450						2605		3450	
300						3415						2470		3415	
350						3375						2360		3375	
400						3330						2245		3330	
450						3255						2150		3255	
500						3125						2055		3125	
550						2955						1985		2955	
600						2770						1910		2770	
650						2580						1845		2580	
700	2350	2400	2400	2415	2415	2400	2415	2400	2415	2400	2415	1775	2465		2455
750	2125	2220	2220	2250	2250	2220	2250	2220	2250	2220	2250	1710	2355		2325
800	1830	2040	2040	2080	2080	2040	2080	2040	2080	2040	2080	1645	2240		2200
850	1500[1]	1860	1860	1915	1915	1860	1915	1860	1915	1860	1915	1595	2125		2070
875	1305[1]	1770[1]	1770	1830	1830	1770	1830	1770	1830	1770	1830	1570	2070		2010
900	1115[1]	1680[1]	1680	1750	1750	1680	1750	1680	1750	1680	1750	1545	2010		1945
925	945[1,2]	1590[1,2]	1590	1665	1665	1590	1665	1590	1665	1590	1665	1535	1955		1880
950	770[1,2]	1500[1,2]	1500	1585	1585	1500	1585	1500	1585	1500	1585	1525	1900		1820
975	600[1,2]	1395[1,2]	1395	1500	1500	1390	1500	1370	1500	1245	1500	1510	1840		1755
1000	430[1,2]	1070[1,2]	1070	1285	1335	1065	1335	1200	1250	945	1455	1500	1785		1690
1025				1070	1140[1]	890	1165	1070	1070	770	1200	1485	1725		1625
1050				855	945[1]	720	995	945	890	600	945	1455	1670		1565
1075				670	815[1]	595	855[1]	815	730	515	755	1370	1615		1500
1100				480	685[1]	470	720[1]	685	565	430	565	1285	1555		1455
1125				375[1]	555[1,2]	385	615[1]	575	470	370	470	1135	1500		1350
1150				265[1]	430[1,2]	300	515[1]	465	375	310	375	985	1305	1455	1245
1175				220[1]	315[1,2]	255	430[1,2]	360[2]	315	255	315	880	1080	1310	1135
1200				170[1]	205[1,2]	205	345[1,2]	255[2]	255	205	255	770	855	1165	1030
1225												665	710	1035	920
1250												555	565	910	815
1275												490	470	795	705
1300												420	375	685	600
1325												365	315	575	500
1350												310	255	465	405
1375												275	230	405	340
1400												240	205	345	275
1425												205	180	300	230
1450												170	155	255	190
1475												150	140	215	160
1500												130	130	170	130
Hydrostatic Shell Test Pressure							5400					4650		5400	

Notes:
[1] See Introductory Note 1.
[2] See Introductory Note 2.
[3] See Introductory Note 3.

99

TABLE 3.33 (*Concluded*)

Table 3.33.8. 2500-Pound Pressure-Temperature Ratings

Note: These ratings are all subject to stipulations in Introductory Note 5 which form a part of this table.   All pressures are in pounds per square inch gage (psig).

Service Temperature Deg. F	Carbon Steel	Carbon Moly	Cr-Mo ½–½	Cr-Mo 1–½	Cr-Mo 1¼–½	Cr-Mo 2–½	Cr-Mo 2¼–1	Cr-Mo 3–1	Cr-Mo 5–½	Cr-Mo 5½–Si	Cr-Mo 9–1	304	347 & 321	316	310	Service Temperature Deg. F
–20 to 100[3]						6000						5145		6000		–20 to 100[3]
150						5915						4855		5915		150
200						5830						4565		5830		200
250						5750						4340		5750		250
300						5690						4115		5690		300
350						5625						3930		5625		350
400						5550						3745		5550		400
450						5430						3585		5430		450
500						5210						3430		5210		500
550						4925						3305		4925		550
600						4620						3180		4620		600
650						4300						3070		4300		650
700	3920	4000	4000	4025	4025	4000	4025	4000	4025	4000	4025	2960	4110		4090	700
750	3550	3700	3700	3745	3745	3700	3745	3700	3745	3700	3745	2850	3920		3875	750
800	3050	3400	3400	3470	3470	3400	3470	3400	3740	3400	3470	2745	3730		3665	800
850	2500[1]	3100	3100	3190	3190	3100	3190	3100	3190	3100	3190	2660	3540		3455	850
875	2180[1]	2950[1]	2950	3055	3055	2950	3055	2950	3055	2950	3055	2620	3445		3345	875
900	1855[1]	2800[1]	2800	2915	2915	2800	2915	2800	2915	2800	2915	2580	3330		3240	900
925	1570[1,2]	2650[1,2]	2650	2775	2775	2650	2775	2650	2775	2650	2775	2560	3260		3135	925
950	1285[1,2]	2500[1,2]	2500	2640	2640	2500	2640	2500	2640	2500	2640	2540	3165		3030	950
975	1000[1,2]	2320[1,2]	2320	2500	2500	2315	2500	2285	2500	2070	2500	2520	3070		2925	975
1000	715[1,2]	1785[1,2]	1785	2145	2230	1770	2230	2000	2085	1570	2430	2500	2975		2820	1000
1025				1785	1900[1]	1485	1945	1785	1785	1285	2000	2470	2880		2710	1025
1050				1430	1570[1]	1200	1655	1570	1485	1000	1570	2430	2785		2605	1050
1075				1115	1355[1]	995	1430[1]	1355	1215	855	1255	2285	2690		2500	1075
1100				800	1145[1]	785	1200[1]	1145	945	715	945	2145	2595		2430	1100
1125				620[1,2]	930[1]	645	1030[1]	955	785	615	785	1895	2500		2250	1125
1150				445[1,2]	715[1,2]	500	855[1]	770	630	515	630	1645	2170	2430	2070	1150
1175				365[1,2]	530[1,2]	420	715[1,2]	600[2]	530	430	530	1465	1800	2185	1895	1175
1200				285[1,2]	345[1,2]	345	570[1,2]	430[2]	430	345	430	1285	1430	1945	1715	1200
1225												1105	1185	1730	1535	1225
1250												930	945	1515	1355	1250
1275												815	785	1330	1180	1275
1300												700	630	1145	1000	1300
1325												605	530	955	835	1325
1350												515	430	770	670	1350
1375												455	385	670	565	1375
1400												400	345	570	455	1400
1425												345	300	500	385	1425
1450												285	255	430	315	1450
1475												250	235	355	265	1475
1500												215	215	285	215	1500
Hydrostatic Shell Test Pressure							9000					7725		9000		

Notes:
[1] See Introductory Note 1.
[2] See Introductory Note 2.
[3] See Introductory Note 3.

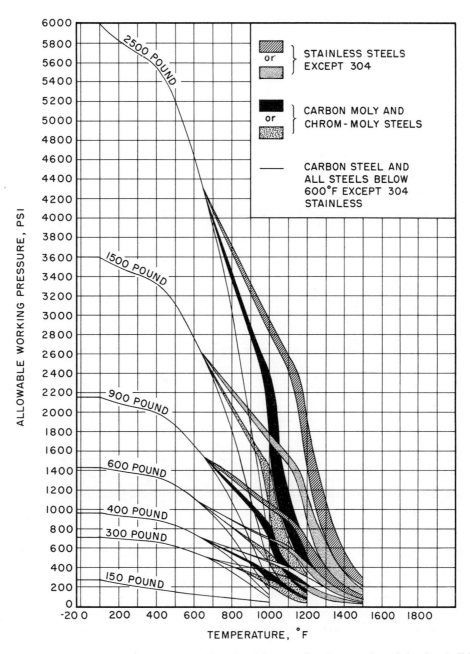

FIG. 3.5. Summary of pressure-temperature ratings of forged-steel flanges. (Based on complete tabular data in Table 3.33.)

FIG. 3.6. Method of rating alloy steels not given in Tables 3.33 (ASA B16.5). Notes:

Ratings of 150-lb pressure class flanges and flanged fittings for all alloy steels shall be taken to be the same as 150-lb pressure class ratings for carbon steel.

Ratings of 300-lb and higher pressure class flanges or flanged fittings made of alloy steels with allowable stresses given in Table P-7 of the ASME Boiler Construction Code other than those in Table 3.33 inclusive may be obtained by the following procedure:

1. *Definition of Symbols*

P = pressure ratings (psi) at desired temperature
S_c = Table P-7 allowable stress (psi) at that temperature
S_y = yield strength (psi) at that temperature
P_p = primary rating pressure (psi)

2. *Ratings*

(a) *Primary Rating Temperature.* Plot allowable stresses given in ASME Boiler Construction Code, Table P-7, for the alloy steel[1] against temperature. The temperature at which the curve intersects the 8750 psi stress line is the primary rating temperature. This temperature should be rounded to an even 25°F by adding not more than 5°, otherwise dropping to the next lower 25° interval.

(b) *Pressure-Temperature Ratings:*

Temperature	Pressure Ratings[2]
Up to 650°F	Ratings same as carbon steel Class A ratings
650°F to primary rating temperature	Ratings obtained by linear interpolation between pressure rating at 650°F and primary rating pressure at primary rating temperature
Above primary rating temperature	Ratings determined from formula $$P = \frac{S_c}{8750} P_p$$
All	Except:[3] Ratings shall not exceed those determined from formula: $$P = \frac{0.6 S_y}{8750} P_p$$

[1] The same ratings shall be used for cast and forged material. Where ASME Code gives different allowable stresses for the two forms of the same material, use the lower.

[2] Calculated ratings may be rounded to nearest 5 psi.

[3] This exception applies to materials with comparatively low yield strength at atmospheric and moderate temperatures, such as Type 304 austenitic stainless steel.

Example:

To obtain ratings of 300 LB ASA B16.5 flanges and flanged fittings made of SA-335, GR P7 (7Cr-1/2 Mol):

Pressure-temperature ratings:

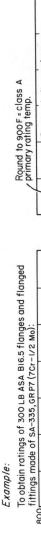
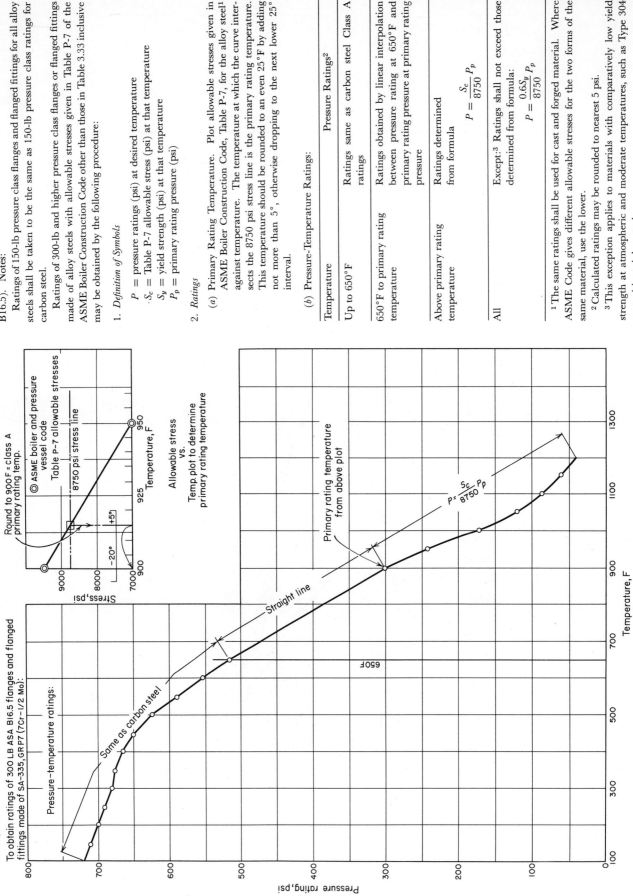

TABLE 3.34

RECOMMENDED BOLTING FOR FLANGES*

Bolting materials				Flange materials and ASA flange ratings							
ASTM		Temperature limit (2) °F	Notes	Cast iron		Carbon steel		Carbon ½ pct. mo. all	Chromium-molybdenum all	Austenitic nickel-chromium all	Nickel steel all
Specification	Grade			125 lb	250 lb	150 lb	300-2500 lb				
A307	B	450	—	A	A	B (4)	—	—	—	—	—
A261	BO	850	—	B (3)	B	A	B	B	B	—	B
A354	BB, BC, BD	750	—	B (3)	B	A	B	B	B	—	A
A193	B7	1000	—	—	—	A	A	A	A	A(5)	A
A193	B7a, B14, B16	1100	(6)	—	—	A	A	A	A (7)	—	—
A193	B5	1100	(8)	—	—	—	—	—	B	—	—
A193	B8,B8c, B8t	1500	—	—	—	—	—	—	—	A (7)	—
A193	B8F	—	—	—	—	—	—	—	—	(9)	—
Services below —20° F											
A320	L7 & L10	—150	—	—	—	A	A	—	—	A (10)	A
A320	L	—225	—	—	—	A	—	—	—	A (10)	—
A320	L8, L8c, L8t, L8F	—300	—	—	—	—	—	—	—	A (11)	—

Notes

(1) Recommended bolting is designated by A and B; A indicates preferred material, with B as acceptable alternate.
(2) The temperature limit shown includes the range of temperature from —20° F.
(3) Bolting conforming to this grade may be used with 125 lb ASA cast-iron flanges, provided a full-faced gasket is used or when a ring gasket is limited to soft gum rubber or material similar in behavior.
(4) Bolting conforming to this grade is not recommended when flanges are to be used with metal or metal-jacketed gaskets.
(5) This bolting is recommended for austenitic flanges at temperatures of 450° F and below to avoid the necessity of a special pre-tightening bolt.
(6) Grades B14 and B16 provide higher strength than Grade B7a at temperatures between 850° F and 1050° F. They should therefore be used for unusually severe service conditions.

(7) Bolting in elevated-temperature service should be selected so that the allowable stress at working temperature is not less than that permitted for the flange material. For operation at temperatures above 1100° F, bolting material should be special and of material similar to the flange.
(8) This material is not commonly used. Its selection depends primarily on the necessity to provide increased corrosion or oxidation resistance.
(9) The free-machining grades of austenitic bolting materials should be limited to low-temperature services until the Codes have sufficient data on mechanical properties to publish allowable stresses.
(10) These materials are preferred for use with austenitic flanges in low-temperature service within the temperature limits indicated.
(11) Austenitic bolting should be used only in the temperature range for which the ferritic grades cannot meet impact requirements.

* Reprinted by permission: Vogrin, C. M., F. S. G. Williams, and J. S. Worth. Paper No. 52-PET-7, 7th Pet. Div. Conf., Amer. Soc. Mech. Engrs. (1952). These recommendations are based on ASA B16.5 which restricts the use of carbon steel bolting to temperatures below 500°F and to pressures below 300 psi. For nonferrous flanges, see ASA B16.5.

TABLE 3.35

RING NUMBERS FOR RING JOINT FLANGES

(Based on ASA B16.5)

Nominal Pipe Size	½	¾	1	1¼	1½	2	2½	3	3½	4	5	6	8	10	12	14	16	18	20	24
150 LB.	...	...	R15	R17	R19	R22	R25	R29	R33	R36	R40	R43	R48	R52	R56	R59	R64	R68	R72	R76
*300 LB.-400 LB.-600 LB.	R11	R13	R16	R18	R20	R23	R26	R31	R34	R37	R41	R45	R49	R53	R57	R61	R65	R69	R73	R77
900 LB.	...	...	...	...	...	...	...	R31	...	R37	R41	R45	R49	R53	R57	R62	R66	R70	R74	R78
1500 LB.	R12	R14	R16	R18	R20	R24	R27	R35	...	R39	R44	R46	R50	R54	R58	R63	R67	R71	R75	R79
2500 LB.	R13	R16	R18	R21	R23	R26	R28	R32	...	R38	R42	R47	R51	R55	R60	...	...	...	...	...

* Reproduced by permission: Ladish Co.

TABLE 3.36
GASKET MATERIALS AND CONSTRUCTION (ASA B16.5)
(Based upon Table UA-47.1 of the ASME Unfired Pressure Vessels Code, 1956.
The design values and other details given in this table are suggested only and are not mandatory.)

Gasket Group Number	Gasket Material		Gasket† Factor m	Design Seating Stress Min.† y	Sketches
I	**Rubber** without fabric or a high percentage of asbestos fiber:				
	Below 75 Shore Durometer		0.50	0	
	75 or higher Shore Durometer		1.00	2.00	
	Asbestos with a suitable binder for the operating conditions	⅛″ Thick	2.00	1600	
		¹⁄₁₆″ Thick	2.75	3700	
	Rubber with cotton fabric insertion		1.25	400	
	Rubber with asbestos fabric insertion, with or without wire reinforcement	3-Ply	2.25	2200	
		2-Ply	2.50	2900	
		1-Ply	2.75	3700	
	Vegetable Fiber		1.75	1100	
	Spiral-Wound Metal, asbestos-filled	Carbon Steel	2.50	2900	
		Stainless or Monel	3.00	4500	
	Corrugated Metal, asbestos-inserted or **Corrugated Metal Jacketed**, asbestos-filled	Soft Aluminum	2.50	2900	
		Soft Copper or Brass	2.75	3700	
		Iron or Soft Steel	3.00	4500	
	Corrugated Metal	Soft Aluminum	2.75	3700	
		Soft Copper or Brass	3.00	4500	
II a & II b	**Asbestos** with a suitable binder for the operating conditions	¹⁄₃₂″ Thick	3.50	6500	
	Corrugated Metal, asbestos-inserted or **Corrugated Metal Jacketed**, asbestos-filled	Monel or 4–6% Chrome	3.25	5500	
		Stainless Steels	3.50	6500	
	Corrugated Metal	Iron or Soft Steel	3.25	5500	
		Monel or 4–6% Chrome	3.50	6500	
		Stainless Steels	3.75	7600	
	Flat Metal Jacketed, asbestos-filled	Soft Aluminum	3.25	5500	
		Soft Copper or Brass	3.50	6500	
		Iron or Soft Steel	3.75	7600	
		Monel	3.50	8000	
		4–6% Chrome	3.75	9000	
		Stainless Steels	3.75	9000	
	Grooved Metal	Soft Aluminum	3.25	5500	
		Soft Copper or Brass	3.50	6500	
		Iron or Soft Steel	3.75	7600	
		Monel or 4–6% Chrome	3.75	9000	
		Stainless Steels	4.25	10100	
	Solid Flat Metal	Soft Aluminum	4.00	8800	
III a & III b	**Solid Flat Metal**	Soft Copper or Brass	4.75	13000	
		* Iron or Soft Steel	5.50	18000	
		* Monel or 4–6% Chrome	6.00	21800	
		* Stainless Steels	6.50	26000	

* Same values apply for ring-joint gaskets.

† This table gives a list of many commonly used gasket materials and contact facings with suggested design values of m and y that have generally proved satisfactory in actual service when using effective gasket seating width b given in Table UA-47.2 of the ASME Unfired Pressure Vessel Code. Design value m is the ratio of gasket seating pressure to operating pressure. Design value y is gasket stress required to seat gasket under bolt-up conditions without internal pressure. Details of dimensions given in ASA B16.5.

Gasket contact widths: Group I ASA B16.5 slip-on raised face width
Group II ASA B16.5 large tongue width
Group III ASA B16.5 small tongue width
minus ½″ but not less than ³⁄₁₆″.

Gaskets with inside diameter equal to outside diameter of pipe: Group I, IIa, IIIa.

Gaskets with outside contact diameter equal to outside diameter of raised face: Group IIb and IIIb.

TABLE 3.37‡
PRESSURE-TEMPERATURE RATINGS FOR NON-FERROUS WROUGHT FLANGES

Maximum Non-Shock Service Pressure, psi

Service Temp. °F	Aluminum Alloy ASTM B247, 3003-0		Aluminum Alloy ASTM B247, 6061-T6†		Nickel, Annealed ASTM B160		Hastelloy B ASME Code Case 1173	
	150 lb Class	300 lb Class	150 lb Class	300 lb Class	150 lb Class	300 lb Class	150 lb Class	300 lb Class
100	40	105	275	720	120	310	350*	910*
150	40	100	270*	710	120	310	345*	900*
200	35	95	265*	700	120	310	335*	880*
250	35	95	260*	675	120	310	330*	865*
300	35	85	215*	565	120	310	320*	845*
350	30	80	155	410	120	310	315*	825*
400	25	60	100	265	120	310	305*	800*
450	...	...	...	...	120	310	295*	770*
500	...	...	...	...	120	310	285*	745*
550	...	...	...	...	120	310	280*	725*
600	...	...	...	...	120	310	275*	715*
650	...	...	...	...	...	...	265*	700*

† Slip-on flanges of 6061-T6 material are rated at two-thirds of the values shown because of the annealing effect of welding on 6061-T6 aluminum.

Maximum Non-Shock Service Pressure, psi

Service Temp. °F	Monel ASTM B164		Inconel ASTM B166		Low Carbon Nickel ASTM B160		Hastelloy C ASME Code Case 1194	
	150 lb Class	300 lb Class	150 lb Class	300 lb Class	150 lb Class	300 lb Class	150 lb Class	300 lb Class
100	195	515	235	615	80	205	345*	900*
150	180	475	225	595	75	200	335*	875*
200	175	455	220	580	75	200	320*	845*
250	165	435	220	575	75	195	310*	815*
300	160	420	215*	560	75	195	300*	790*
350	155	410	210*	555	75	190	290*	765*
400	155	410	210*	555	75	190	285*	740*
450	155	405	210*	555	75	190	280*	725*
500	155*	405	210*	555	75	190	275*	715*
550	155*	405	210*	555	75	190	265*	695*
600	155*	405	210*	555	75	190	260*	680*
650	155*	405	210*	555*	75	190	260*	680*
700	155*	405	210*	550*	70	190	260*	680*
750	155*	405	205*	540*	70	185	260*	680*
775	155*	405	205*	535*	70	185	260*	680*
800	155*	405	205*	530*	70	180	260*	680*
825	155*	405	200*	525*	70	180	260*	680*
850	145*	375	200*	520*	70	175	260*	680*
875	120*	310	195*	510*	65	175	260*	680*
900	105*	275	190*	505*	60	155	260*	680*
925	...	...	170*	445*	55	135	260*	680*
950	...	...	140*	360	50	125	260*	680*
975	...	...	110*	290	45	115	260*	680*
1000	...	...	90*	240	40	105	260*	680*
1025	...	...	80*	205	35	95	...	...
1050	...	...	65*	170	30	85	...	...
1075	...	...	55*	150	30	75	...	...
1100	...	...	40*	105	25	70	...	...
1125	...	...	40*	100	25	60	...	...
1150	...	...	35*	85	20	50	...	...
1175	...	...	30*	75	15	45	...	...
1200	...	...	25*	70	15	40	...	...

* Pressure-Temperature ratings for wrought non-ferrous flanges were tentatively established in the 1960 Addendum to ASA B16.5 and are reproduced as Addenda in the 1961 edition.

The use of these ratings requires care in selecting bolting material of adequate strength and gaskets conforming to the requirements of Introductory Note 6.10 of ASA B16.5.

For Pressure-Temperature ratings of higher pressure classes, the following factors may be used. Because of round-off of the basic pressure ratings, the use of these factors will give pressure ratings that are approximate.

For exact ratings, see ASA B16.5.

Factors to be Multiplied by 300-lb Rating to Obtain Class Rating

Class	300 lb	400 lb	600 lb	900 lb	1500 lb	2500 lb
Factor	1.000	1.333	2.000	3.000	5.000	8.333
Max Error	0	6 psi	7½ psi	10 psi	15 psi	24 psi

‡ Reproduced by permission: Cat. No. 311, Copyright: Chemetron Corporation, Tube-Turns Division.

TABLE 3.38
PRESSURE-TEMPERATURE RATINGS FOR FORGED STEEL SCREWED AND SOCKET WELD END FITTINGS[3]

(Reproduced from Vogt Catalog F-10 by permission of Henry Vogt Machine Co.)

Pressure Class / Service Temperature Degree F.	2000						3000					
	Carbon Steel A-105 Gr. 2	1 Cr. ½ Mo. A-182 Gr. F12	1¼ Cr. ½ Mo. A-182 Gr. F11	5 Cr. ½ Mo. Grade F5a	A-182 Grade F-304	A-182 Grade F-316	Carbon Steel A-105 Gr. 2	1 Cr. ½ Mo. A-182 Gr. F12	1¼ Cr. ½ Mo. A-182 Gr. F11	5 Cr. ½ Mo. A-182 Grade F5a	A-182 Grade F-304	A-182 Grade F-316
−20 to 100	2000	2000	2000	2000	1715	2000	3000	3000	3000	3000	2570	3000
150	1970	1970	1970	1970	1615	1970	2955	2955	2955	2955	2425	2955
200	1940	1940	1940	1940	1520	1940	2915	2915	2915	2915	2280	2915
250	1915	1915	1915	1915	1445	1915	2875	2875	2875	2875	2170	2875
300	1895	1895	1895	1895	1370	1895	2845	2845	2845	2845	2055	2845
350	1875	1875	1875	1875	1310	1875	2810	2810	2810	2810	1965	2810
400	1850	1850	1850	1850	1245	1850	2775	2775	2775	2775	1870	2775
450	1810	1810	1810	1810	1195	1810	2715	2715	2715	2715	1790	2715
500	1735	1735	1735	1735	1140	1735	2605	2605	2605	2605	1715	2605
550	1640	1640	1640	1640	1100	1640	2460	2460	2460	2460	1650	2460
600	1540	1540	1540	1540	1060	1540	2310	2310	2310	2310	1590	2310
650	1430	1430	1430	1430	1020	1430	2150	2150	2150	2150	1535	2150
700	1305	1340	1340	1340	985	1370	1960	2010	2010	2010	1480	2055
750	1180	1245	1245	1245	950	1305	1775	1870	1870	1870	1425	1960
800	1015	1155	1155	1155	915	1240	1525*	1735	1735	1735	1370	1865
850	830[1]	1060	1060	1060	885	1180	1250[1]	1595	1595	1595	1330	1770
875	725[1]	1015	1015	1015	870	1145	1090[1]	1525	1525	1525	1310	1720
900	615[1]	970	970	970	860	1115	925[1]	1455	1455	1455	1290	1675
925	520[1][2]	925	925	925	850	1085	785[1][2]	1385	1385	1385	1280	1630
950	425[1][2]	880	880	880	845	1055	640[1][2]	1320	1320	1320	1270	1580
975	330[1][2]	830	830	830	840	1020	500[1][2]	1250	1250	1250	1260	1535
1000	235[1][2]	715	740	695	830	990	355[1][2]	1070	1115	1040	1250	1485
1025	....	595	630[1]	595	820	960	....	890	950[1]	890	1235	1440
1050	....	475	520[1]	495	810	925	....	715	785[1]	740	1215	1390
1075	....	370	450[1]	405	760	895	....	555	675[1]	605	1140	1345
1100	....	265	380[1]	315	715	865	....	400	570[1]	470	1070	1295
1125	....	205[1][2]	310[1][2]	260	630	830	....	310[1][2]	465[1][2]	390	945	1250
1150	....	145[1][2]	235[1][2]	210	545	810	....	220[1][2]	355[1][2]	315	820	1215
1175	....	120[1][2]	175[1][2]	175	485	725	....	180[1][2]	265[1][2]	265	730	1090
1200	....	95[1][2]	115[1][2]	140	425	645	....	140[1][2]	170[1][2]	215	640	970

[1] Product used within jurisdiction of Section 1 (Power Boilers) of ASME Code is subject to maximum temperature limitations of that code.

[2] Product used within jurisdiction of Section 1 (Power Piping) of ASA B31.1 is subject to maximum temperature limitation of that code.

[3] The carbon steel ratings at 2000, 3000, and 6000 are in accordance with MSS-SP-49. Other ratings are typical manufacturer's (slight variations with manufacturers exist). Socket welding fittings are customarily rated the same as screwed fittings although ASA B16.11 permits pressure ratings equal to those of pipe of the same schedule number and materials.

TABLE 3.38 (*Concluded*)
PRESSURE-TEMPERATURE RATINGS*

Pressure Class	4000						6000					
Service Temperature Degree F.	Carbon Steel A-105 Gr. 2	1 Cr. ½ Mo. A-182 Gr. F12	1¼ Cr. ½ Mo. A-182 Gr. F11	5 Cr. ½ Mo. Grade F5a	A-182 Grade F-304	A-182 Grade F-316	Carbon Steel A-105 Gr. 2	1 Cr. ½ Mo. A-182 Gr. F12	1¼ Cr. ½ Mo. A-182 Gr. F11	5 Cr. ½ Mo. Grade F5a	A-182 Grade F-304	A-182 Grade F-316
−20 to 100	4000	4000	4000	4000	3930	4000	6000	6000	6000	6000	5145	6000
150	3940	3940	3940	3940	3235	3940	5915	5915	5915	5915	4855	5915
200	3885	3885	3885	3885	3040	3885	5830	5830	5830	5830	4565	5830
250	3830	3830	3830	3830	2890	3830	5750	5750	5750	5750	4340	5750
300	3790	3790	3790	3790	2740	3790	5690	5690	5690	5690	4115	5690
350	3750	3750	3750	3750	2620	3750	5625	5625	5625	5625	3930	5625
400	3700	3700	3700	3700	2495	3700	5550	5550	5550	5550	3745	5550
450	3620	3620	3620	3620	2390	3620	5430	5430	5430	5430	3585	5430
500	3470	3470	3470	3470	2285	3470	5210	5210	5210	5210	3430	5210
550	3280	3280	3280	3280	2200	3280	4925	4925	4925	4925	3305	4925
600	3080	3080	3080	3080	2120	3080	4620	4620	4620	4620	3180	4620
650	2865	2865	2865	2865	2045	2865	4300	4300	4300	4300	3070	4300
700	2610	2680	2680	2680	1970	2740	3920	4025	4025	4025	2960	4110
750	2365	2495	2495	2495	1900	2610	3550	3745	3745	3745	2850	3920
800	2030	2310	2310	2310	1830	2485	3050	3470	3470	3470	2745	3730
850	1665[1]	2125	2125	2125	1770	2360	2500[1]	3190	3190	3190	2660	3540
875	1450[1]	2035	2035	2035	1745	2295	2180[1]	3055	3055	3055	2620	3445
900	1235[1]	1940	1940	1940	1720	2230	1855[1]	2915	2915	2915	2580	3350
925	1045[1][2]	1850	1850	1850	1705	2170	1570[1][2]	2775	2775	2775	2560	3260
950	855[1][2]	1760	1760	1760	1690	2110	1285[1][2]	2640	2640	2640	2540	3165
975	665[1][2]	1665	1665	1665	1680	2045	1000[1][2]	2500	2500	2500	2520	3070
1000	475[1][2]	1430	1485	1390	1665	1980	715[1][2]	2145	2230	2085	2500	2975
1025	....	1190	1265[1]	1190	1645	1920	....	1785	1900	1785	2470	2880
1050	....	950	1045[1]	990	1620	1855	....	1430	1570	1485	2430	2785
1075	....	740	900[1]	810	1520	1790	....	1115	1355	1215	2285	2690
1100	....	530	760[1]	630	1430	1730	....	800	1145	905	2145	2595
1125	....	410[1][2]	620[1][2]	520	1260	1665	....	620[1][2]	930	785	1895	2500
1150	....	295[1][2]	475[1][2]	420	1095	1620	....	445[1][2]	715	630	1645	2430
1175	....	240[1][2]	350[1][2]	350	975	1455	....	365[1][2]	530	530	1465	2185
1200	....	190[1][2]	230[1][2]	285	855	1295	....	285[1][2]	345	430	1285	1945

* See notes on previous page.

Equivalent Schedule Numbers for Socket Weld Fittings

Class	Schedule Number
2000	40
3000	80
4000	160
6000	Double extra strong

TABLE 3.39
RATINGS OF OTHER SCREWED FITTINGS*

Bronze and Brass

			Temperature, °F	
Class	Service	Pressure, psig	Red Brass ASTM B62	ASTM B16 B140
125	steam	125	400	400
	gas or liquid	175	150	150
				ASTM B61
250	steam	250	406	550
	gas or liquid	400	150	150

Cast Iron

Class	Service	Pressure, psig	Temperature, °F
125	steam	125	353
	liquid or gas	175	150
250	steam	250	406
	liquid or gas	400	150

Malleable Iron

Class	Service	Pressure, psig	Temperature, °F
150	steam	150	366
	liquid or gas	300	150
300	steam and oil	300	550 (¼″ to 3″ incl.)
	liquid or gas†	2000 ¼ to 1″	
		1500 1¼ to 2″	150
		1000 2½ to 3″	

* ASTM ratings are quoted by permission: American Society for Testing Materials, Philadelphia, Pa.
† Street elbow not recommended for pressures over 600 psig.

LIST OF STANDARD MATERIAL SPECIFICATIONS
(Physical and Chemical Properties of Materials and Finished Products)

FERROUS | Designations*

			Designations*

PIPE

Carbon Steel

Welded and seamless for coiling, bending, and flanging	A53
Seamless for high-temperature service	A106
Welded and seamless for ordinary uses and galvanized	A120
Electric-fusion (arc) welded (16 in. and over)	A134
Electric-resistance welded, general purpose	A135
Electric-fusion (arc) welded (4 in. and over)	A139
Electric-fusion welded for high-temperature service	A155
Spiral-welded steel or iron	A211
Welded and seamless hearth-iron pipe	A253
Welded and seamless for low-temperature service (carbon steel and alloy)	A333
Metal-arc welded, for high-pressure transmission service	A381
Pipeline	API 5L
High-test line pipe	API 5LX

Alloy Steel

Electric-fusion welded for high-temperature service (carbon steel and carbon-moly, and chrome-moly)	A155
Seamless and welded austenitic stainless	A312
Seamless and welded for low-temperature service (carbon steel and nickel steel)	A333
Seamless ferritic† alloy for high-temperature service	A335
Electric-fusion welded austenitic chromium-nickel alloy for high-temperature service	A358
Iron-chromium and iron-chromium-nickel alloy tabular centrifugal castings for general service	A362
Ferritic alloy forged and bored pipe for high-temperature service	A369
Seamless ferritic alloy, specially heat-treated for high-temperature service	A405

Welded large-diameter light-wall austenitic chromium-nickel alloy for corrosive or high-temperature service	A409
Centrifugally cast ferritic alloy, for high-temperature service	A426
Austenitic, forged and bored, for high-temperature service	A430

Wrought Iron

Welded	A72
Electric-fusion welded (16 in. and over)	A419
Line pipe	API 5L

Cast Iron

Soil pipe and fittings	A74
Culvert pipe	A142
Pressure pipe (refers to ASA specifications A21, 22, 23, 26, 27, 28, and 29)	A377

TUBING

There are 22 specifications for steel boiler, exchanger, and pipe-still tubing having chemical and physical properties similar to those given in the various pipe specifications already listed. Of these, tubing specifications ASTM A83 is often used as a substitute for seamless ASTM A53 or A106 pipe in sizes 1½ in. and smaller when these cannot be obtained. In such cases the tubing is ordered to correspond to iron-pipe sizes.

FITTINGS AND VALVES

Steel Forged or Rolled Pipe Flanges, Forged Fittings, and Valves and Parts

Carbon steel for high-temperature service	A105
Carbon steel and austenitic alloy for general service	A181
Ferritic steel for high-temperature service	A182
Carbon and alloy (nickel and nickel-copper-chromium-aluminum) steel for low-temperature service	A350
Chrome-moly alloy specially heat-treated for high-temperature service	A404

* Unless otherwise noted, these designations refer to ASTM specifications.

† Ferritic alloys in this listing refer to carbon-moly or chrome-moly alloys.

	Designations*		Designations*
Steel Welding Fittings (factory made)		Seamless copper pipe (for plumbing, etc.)	B42
Wrought carbon steel and ferritic alloy	A234	Seamless copper tube, bright-annealed (dirt and scale free)	B68
Wrought austenitic steel	A403	Seamless copper tube (general use)	B75
Wrought carbon steel and alloy (nickel and nickel-copper-chromium-aluminum) steel for low-temperature service (seamless or welded)	A420	Seamless copper water tube (Types K, L, & M)	B88
		Copper and copper-alloy seamless condenser tubes	B111
Castings for Valves, Flanges, and Fittings		Threadless copper pipe for brazed-joint fittings	B302
Steel		Red brass pipe (for plumbing, etc.)	B43
Carbon steel suitable for fusion welding, for high-temperature service	A216	Brass tube, seamless	B135
		Copper-silicon alloy pipe and tube	B315
Low alloy steel (carbon-moly, nickel-chrome-moly and chrome-moly) for high-temperature service	A217	*Nickel and Nickel Alloy*	
		Nickel and low-carbon nickel seamless pipe and tube	B161
Ferritic (high chrome) and austenitic steel, for high-temperature service alloys	A351	Nickel-copper alloy seamless pipe and tube	B165
		Nickel-chromium-iron pipe and tube	B167
Ferritic steel for low-temperature service (carbon steel, carbon-moly, and nickel-steel)	A352	*Aluminum and Aluminum Alloy*	
		Aluminum alloy pipe	B241
Alloy steel, specially heat-treated for high-temperature service (chrome-moly)	A389	Aluminum alloy drawn seamless tube	B210
Malleable Iron		Aluminum alloy extruded tubes	B235
Including flanges	A47 and A338		
Cast Iron		CASTINGS FOR VALVES, FITTINGS, AND FLANGES	
Gray iron	A126		
Bolting materials		*Copper and Copper Alloy*	
Alloy steel bolting (ferritic and austenitic) for high-temperature service	A193	Copper-base alloys for casting	B30 and B45
		Bronze castings	B61
Carbon and alloy steel (ferritic and austenitic) nuts for high-pressure and high-temperature service	A194	Composition brass (copper, tin, lead, and zinc)	B62
		Leaded high-strength yellow brass and castings	B132
Heat-treated carbon-steel bolting	A261	Nickel-tin bronze	B292
Alloy steel (ferritic and austenitic) bolting for low-temperature service	A320	Specifications listed include those commonly used for valves, fittings, and flanges. Specifications for other casting materials which can be applied include B143 through 149, 179, 198, and 271.	
Low-carbon steel externally and internally threaded standards fasteners (Grade B used for flange bolts when one or both flanges are cast iron)	A307	*Nickel Alloy*	
		Nickel-molybdenum and nickel-molybdenum-chromium alloy castings	B332
Quenched and tempered alloy-steel bolts and studs with suitable nuts (for normal atmospheric temperatures requiring high strength)	A354	*Aluminum Alloy*	
		Aluminum-base alloys in ingot form for casting (See also B26, B85, and B108)	B179

NON-FERROUS METALS

PIPE AND TUBE

Copper and Copper Alloy

	Designations*
General requirements (dimensional standards included)	B251

FORGINGS

Copper and Copper Alloy

	Designations*
Forging rod, bar and shapes	B124

Aluminum Alloys

	Designations*
Die forgings	B247

	Designations*
BOLTING	

Specific material specifications for bolting composed of these materials are not available. General material specifications from which bolts can be designed include:

Copper and copper alloy	B12 and B98
Nickel and nickel alloy	B160, B164 & B166
Aluminum and aluminum alloy	B211
(Carbon steel bolts galvanized or aluminum painted may be used)	

MISCELLANEOUS NON-METALLIC PIPE

ASBESTOS-CEMENT

Pressure pipe	C296

CLAY PIPE

Drain tile	C4
Standard strength sewer pipe (salt-glazed)	C13
Standard strength sewer pipe (ceramic, glazed or unglazed)	C261

	Designations*
Extra strength (salt glazed)	C200
Extra strength (ceramic glazed or unglazed)	C278
Perforated, standard strength	C211
Vitrified clay pipe joints	C425

CONCRETE

Drain tile	C412
Sewer pipe	C14
Reinforced culvert, storm drain and sewer	C76
Low-head pressure pipe	C361
Low-head internal pressure sewer pipe	C362

PLASTIC

Solvent-welded cellulose acetate butyrate pipe (SWP size)	D1503
Non-rigid polyvinyl (chloride) tubing (up to 2 in.—primarily for electrical insulation)	D922

TABLE 3.40
PIPE, FITTING, AND FLANGE MATERIALS
(Reproduced by permission: Cat. No. 311, Copyright 1962, Chemetron Corporation, Tube-Turns Division)

Low and Intermediate Alloy Steels

| Material | ASTM Specifications[1] | | | Chemistry[2] | | | | | | | ASTM Specified | |
	Number	Grade	Form	Carbon per cent	Manganese per cent	Silicon per cent	Chromium per cent	Nickel per cent	Moly per cent	Other	Minimum U.T.S. psi	Minimum Yield psi
Carbon Moly	A 182	F 1	Flanges	.20/.30	.60/.90	.20/.35			.40/.60	a	70,000	40,000
	A 335	P 1	Pipe	.10/.20	.30/.80	.10/.50			.44/.65	b	55,000	30,000
	A 234	WP 1	Fittings	Covers Manufactured Fittings—For chemistries other specifications apply								
½ Cr ½ Moly	A 335	P 2	Pipe	.10/.20	.30/.61	.10/.30	.50/.81		.44/.65	b	55,000	30,000
1 Cr ½ Moly	A 234 A 335	WP 12 P 12	Fittings Pipe	.15 Max	.30/.61	.50 Max	.80/1.25		.44/.65	b	60,000	30,000
1 Cr ½ Moly	A 182	F 12	Flanges	.10/.20	.30/.80	.10/.60	.85/1.20		.45/.65	a	70,000	40,000
1¼ Cr ½ Moly	A 234 A 335	WP 11 P 11	Fittings Pipe	.15 Max	.30/.60	.50/1.00	1.0/1.5		.44/.65	c	60,000	30,000
1¼ Cr ½ Moly	A 182	F 11	Flanges	.10/.20	.30/.80	.50/1.00	1.0/1.5		.45/.65	a	70,000	40,000
2¼ Cr 1 Moly	A 234 A 335	WP 22 P 22	Fittings Pipe	.15 Max	.30/.60	.50 Max	1.90/2.60		.87/1.13	c	60,000	30,000
2¼ Cr 1 Moly	A 182	F 22	Flanges	.15 Max	.30/.60	.50 Max	2.00/2.50		.90/1.10	a	70,000	40,000
3 Cr 1 Moly	A 335	P 21	Pipe	.15 Max	.30/.60	.50 Max	2.65/3.35		.80/1.06	c	60,000	30,000
5 Cr ½ Moly	A 234 A 335	WP 5 P 5	Fittings Pipe	.15 Max	.30/.60	.50 Max	4.00/6.00		.45/.65	c	60,000	30,000
5 Cr ½ Moly	A 182	F 5	Flanges	.15 Max	.30/.60	.50 Max	4.00/6.00	.50 Max	.45/.65	c	60,000	30,000
7 Cr ½ Moly	A 335	P 7	Pipe	.15 Max	.30/.60	.50/1.00	6.00/8.00		.44/.65	c	60,000	30,000
7 Cr ½ Moly	A 182	F 7	Flanges	.15 Max	.30/.60	.50/1.00	6.00/8.00		.45/.65	c	60,000	36,000
9 Cr 1 Moly	A 335	P 9	Pipe	.15 Max	.30/.60	.25/1.00	8.00/10.00		.90/1.10	c	60,000	30,000
9 Cr 1 Moly	A 182	F 9	Flanges	.15 Max	.30/.60	.50/1.00	8.00/10.00		.90/1.10	c	100,000	70,000
3½ Nickel	A 420 A 333	WPL 3 3	Fittings Pipe	.19 Max	.31/.64	.18/.37		3.18/3.82		d	65,000	35,000
3½ Nickel	A 350	LF 3	Flanges	.20 Max	.30/.60	.15/.35		3.25/3.75		a	70,000	40,000
5 Nickel	A 420 A 333	WPL 5 5	Fittings Pipe	.19 Max	.20/.64	.18/.37		4.68/5.32		d	65,000	35,000
Cr Cu Ni	A 420 A 333	WPL 4 4	Fittings Pipe	.12 Max	.50/1.05	.08/.37	.44/1.01	.47/.98		a, e	60,000	30,000
Cr Cu Ni	A 350	LF 4	Flanges	.12 Max	.55/1.00	.10/.35	.50/.95	.50/.95		a, e	60,000	30,000

[1] For specifications on bolting materials for use with flanges of above materials, refer to Table 3.34.

[2] Where maximum values are used, the minimum content is determined by strength requirements. Determinations of chemistry are by Check Analysis.

(a) Phosphorus .040 per cent max, Sulfur .040 per cent max.
(b) Phosphorus .045 per cent max, Sulfur .045 per cent max.
(c) Phosphorus .030 per cent max, Sulfur .030 per cent max.
(d) Phosphorus .050 per cent max, Sulfur .050 per cent max.
(e) Aluminum .04/.30, Copper .40/.75.

TABLE 3.40 (*Continued*)

Stainless Steels

Material	ASTM Specifications[1]			Chemistry[2]							ASTM Specified	
	Number	Grade	Form	Max Carbon per cent	Manganese per cent	Silicon per cent	Chromium per cent	Nickel per cent	Moly per cent	Other	Minimum U.T.S. psi	Minimum Yield psi
18-8 Cr Ni Type 304*	A 182	F 304	Flanges	.08	2.0 Max	1.00 Max	18.0/20.0	8.0/11.0		f	75,000	30,000
	A 312	TP 304	Pipe	.08	2.0 Max	.75 Max	18.0/20.0	8.0/11.0		f	75,000**	30,000
	A 403	WP 304	Fittings	Covers Manufactured Fittings—For chemistries other specifications apply								
18-8 Cr Ni Type 304L	A 403 / A 312	WP 304L / TP 304L	Fittings / Pipe	.035	2.0 Max	.75 Max	18.0/20.0	8.0/13.0		f	70,000	25,000
	A 182	F 304L	Flanges	.035	2.0 Max	1.00 Max	18.0/20.0	8.0/13.0		f	65,000	25,000
25-R Cr Ni Type 309	A 403 / A 312	WP 309 / TP 309	Fittings / Pipe	.15	2.0 Max	.75 Max	22.0/24.0	12.0/15.0		f	75,000	30,000
25-20 Cr Ni Type 310	A 403 / A 312	WP 310 / TP 310	Fittings / Pipe	.15	2.0 Max	.75 Max	24.0/26.0	19.0/22.0		f	75,000	30,000
	A 182	F 310	Flanges	.15	2.0 Max	1.00 Max	24.0/26.0	19.0/22.0		f	75,000	30,000
18-8 Moly Type 316*	A 403 / A 312	WP 316 / TP 316	Fittings / Pipe	.08	2.0 Max	.75 Max	16.0/18.0	11.0/14.0	2.0/3.0	f	75,000	30,000
	A 182	F 316	Flanges	.08	2.0 Max	1.00 Max	16.0/18.0	10.0/14.0	2.0/3.0	f	75,000	30,000
18-8 Moly Type 316L	A 403 / A 312	WP 316L / TP 316L	Fittings / Pipe	.035	2.0 Max	.75 Max	16.0/18.0	10.0/15.0	2.0/3.0	f	70,000	25,000
	A 182	F 316L	Flanges	.035	2.0 Max	1.00 Max	16.0/18.0	10.0/15.0	2.0/3.0	f	65,000	25,000
19-9 Moly Type 317	A 403 / A 312	WP 317 / TP 317	Fittings / Pipe	.08	2.0 Max	.75 Max	18.0/20.0	11.0/14.0	3.0/4.0	f	75,000	30,000
18-8 Ti Type 321*	A 403 / A 312	WP 321 / TP 321	Fittings / Pipe	.08	2.0 Max	.75 Max	17.0/20.0	9.0/13.0		f,g	75,000	30,000
	A 182	F 321	Flanges	.08	2.5 Max	.85 Max	17.0 Min	9.0 Min		g,h	75,000	30,000
18-8 Cb Type 347*	A 403 / A 312	WP 347 / TP 347	Fittings / Pipe	.08	2.0 Max	.75 Max	17.0/20.0	9.0/13.0		f,i	75,000	30,000
	A 182	F 347	Flanges	.08	2.0 Max	1.00 Max	17.0/20.0	9.0/13.0		c,i	75,000	30,000
18-8 Cb Type 348*	A 403 / A 312	WP 348 / TP 348	Fittings / Pipe	.08	2.0 Max	.75 Max	17.0/20.0	9.0/13.0		f,i,j	75,000	30,000
	A 182	F 348	Flanges	.08	2.0 Max	1.00 Max	17.0/20.0	9.0/13.0		c,i,j	75,000	30,000
12 Cr Type 410	A 268	TP 410	Tubing	.15	1.0 Max	.75 Max	11.5/13.5	.50 Max		f	60,000	30,000
	A 182	F 6	Flanges	.12	1.0 Max	1.00 Max	11.5/13.5	.50 Max		f	85,000	55,000
17 Cr Type 430	A 268	TP 430	Tubing	.12	1.0 Max	.75 Max	14.0/18.0	.50 Max		f	60,000	35,000

* 304, 316H, 321H, 347H and 348H Grades are available differing from the regular grades in that the carbon content is 0.04 to 0.10 and that the Columbium, Tantalum and/or Titanium limits for 321H, 347H and 348H are already modified.

** For TP304 Schedule 140 in sizes 8″ and larger, the U.T.S. is 70,000 psi minimum.

[1] For specifications on bolting materials for use with flanges of above materials, refer to Table 3.34.

[2] Where maximum values are used, the minimum content is determined by strength requirements. Determinations of chemistry are by Check Analysis.

(f) Phosphorus .040 per cent max, Sulfur .030 per cent max.

(g) Titanium content shall be not less than five times the carbon content and not more than 0.60 per cent.

(h) Phosphorus .035 per cent max, Sulfur .030 per cent max.

(i) Columbium plus Tantalum content shall not be less than ten times the carbon content and not more than 1.00 per cent.

(j) Tantalum 0.10 per cent max.

Aluminum and Aluminum Alloys

Types of Aluminum	Alloy	Forms Available†	Chemical Composition, per cent‡									Others		ASME* Minimum Strengths, psi			
			Aluminum	Copper	Iron	Silicon	Manganese	Magnesium	Zinc	Chromium	Titanium	Each	Total	Unwelded U.T.S	Unwelded Yield	Welded U.T.S.	Welded Yield
Non-Heat-Treatable Aluminum Alloys																	
Pure(a)	1060	2,3,4	99.6 Min	.05	.35	.25	.03	.03	.05	...	.03	.03	...	10,000	4,000	9,500	2,500
	1100	3,4,5	99.0 Min	.20	(1.0 Total)		.05	...	.10	...	...	.05	.15	12,000	5,000	11,000	3,500
Manganese Alloy	3003	1,2,3,4,5	Remainder	.20	.7	.6	1.0/1.5	...	.10	...	...	.05	.15	14,500	6,000	14,000	5,000
Magnesium Alloys(b)	5052	2,3,4	Remainder	.10	(.45 Total)		.10	2.2/2.8	.20	.15/.35	...	.05	.15	25,000	9,500	25,000	9,500
	5154	2,3,4	Remainder	.10	(.45 Total)		.10	3.1/3.9	.20	.15/.35	.20	.05	.15	30,000	11,000	30,000	11,000
Heat-Treatable Aluminum Alloys																	
Magnesium Silicon Alloys(c)	6061	1,2,3,4,5	Remainder	.15/.40	.7	.40/.8	.15	.8/1.2	.25	.15/.35	...	.05	.15	38,000	35,000	24,000	...
	6062	1,2,4	Remainder	.15/.40	.7	.40/.8	.15	.8/1.2	.25	.04/.14	.15	.05	.15	...	...	...	...
	6063	1,2,4	Remainder	.10	.35	.20/.6	.10	.45/0.9	.10	.10	.10	.05	.15	30,000	25,000	17,000	...

* Values are from Table UNF-23 of Section VIII of the ASME Boiler and Pressure Vessel Code, 1959 Edition, with tempers of H112 (unwelded) and O (welded) for the non-heat-treatable alloys and T6 and T6-welded for the heat-treatable alloys.

† 1 = Pipe, 2 = Tube, 3 = Plate, 4 = Bar and 5 = Forging.

‡ Single values are maximum amounts permitted.

Fittings are made to ASTM B361 which includes permissible raw materials to: Pipe-ASTM B241; Tube-ASTM B210, B234 and B235; Plate-ASTM B209; Bar-ASTM B211 and B221; Forgings-ASTM B247.

(a) Alloys 1160, 1260, and 1360 also are available.

(b) Alloys 5652 and 5254 also are available.

(c) Alloys 6053 and 6363 also are available.

TABLE 3.40 (*Continued*)

Carbon Steels and Wrought Iron

Material	ASTM Specifications[1]			Chemistry[2]					Service Temperature Limits Deg F[3]	Welding		
	Number	Grade	Form	Max Carbon per cent	Manganese per cent	Phosphorus per cent	Sulphur per cent	Silicon per cent		Filler Metal	Preheat Deg F[4]	Stress Relief Deg F[6]
Carbon Steel	A 53	Furnace Welded	Pipe	Max Phosphorus: .08 Open Hearth(OH); .13 Acid Bessemer(AB)					−20 to 750	E6010 GA60	No	No
	A 53	A or B	Pipe	Max Phosphorus: .048 Seamless(OH); .11 Seamless (AB); .050 Electric Resistance Welded (OH)					−20 to 1100	E6010 GA60	No	No
	A 105	I	Flanges	.35	.90 Max	.055 Max	.055 Max	.35 Max	−20 to 1000	E6010 GA60	No	No
	A 105	II	Flanges	.35	.90 Max	.055 Max	.055 Max	.35 Max	−20 to 1000	E6010 GA60	250F[5]	No
	A 106	A	Pipe	.25	.27/.93	.048 Max	.058 Max	.10 Min	−20 to 1100	E6010 GA60	No	No
	A 106	B	Pipe	.30	.29/1.06	.048 Max	.058 Max	.10 Min	−20 to 1100	E6010 GA60	No	No
	A 120		Pipe	No Chemistry Specified					−20 to 450	E6010 GA60	No	No
	A 135	A or B	Pipe			.050 Max	.060 Max		−20 to 1100	E6010 GA60	No	No
	A 181	I	Flanges	.35	.90 Max	.055 Max	.055 Max	.35 Max	−20 to 1000	E6010 GA60	No	No
	A 181	II	Flanges	.35	.90 Max	.055 Max	.055 Max	.35 Max	−20 to 1000	E6010 GA60	250F[5]	No
	A 234	WPA	Fittings	Covers Manufactured Fittings—For chemical analyses, other specifications apply					−20 to 1000	E6010 GA60	No	No
	A 234	WPB	Fittings	Covers Manufactured Fittings—For chemical analyses, other specifications apply					−20 to 1000	E6010 GA60	No	No
	A 333	C	Pipe	.25	.64/1.06	.050 Max	.060 Max		−50 to 1000	E6010 GA60	No	No
	A 350	LFI	Flanges	.30	1.06 Max	.040 Max	.050 Max		−50 to 750	E6010 GA60	No	1100/1200
Wrought Iron	A 72		Pipe		.06 Max	Essentially Iron—Approx. 1.5% Slag			−20 to 750	E6010 GA60	No	No

[1] For specifications on bolting materials for use with flanges of above materials, refer to ASTM Specifications A193, A194, A261 and A307.

[2] Where maximum values are used, the minimum content is determined by strength requirements. Determinations of chemistry are by Check Analysis.

[3] Permissible temperature will be governed by service conditions. For most materials shown allowable working stresses have been established by Codes governing piping design. Refer to American Standard Code for Pressure Piping (ASA B31.1) and ASME Boiler and Pressure Vessel Code, Section I and Section VIII. In some cases the maximum temperatures for which codes have established working stresses are lower than the limiting temperatures shown herein. *The codes are not to be interpreted as implying that the materials may be safely used under all service conditions within the temperature range they give,* nor that they cannot be safely employed beyond those temperature ranges. In selecting materials the design engineer must take into account all operating conditions affecting the adequacy of the piping material.

For example, to avoid *graphitization*, carbon steel is conservatively limited to 775F max.

[4] Preheating of these steels is not required but is recommended when (A) the ambient temperature is below 32F in which case local preheating to a hand-hot temperature condition is recommended, (B) the nominal thickness exceeds ½″ and the carbon content exceeds .20%, or when the nominal thickness exceeds ¾″ regardless of analysis, in which case preheating to approximately 400F is advisable.

[5] Preheating is not required but experience has indicated that preheating to 250F is advisable, regardless of ambient temperature, when welding these materials.

[6] ASA B31.1, Section 6, requires stress relieving at 1100F or over for carbon steels thicker than ¾″. ASA B31.8 requires stress relieving at 1100F or over for carbon steels having a carbon content in excess of .32% (ladle analysis) or a carbon equivalent (C + ¼ Mn) in excess of .65% (ladle analysis) and all carbon steels heavier than 1¼″.

Nickel and Nickel Alloys

TABLE 3.40 (*Concluded*)

Material	ASTM Specifications[1] Number	ASTM Specifications[1] Form	Max Carbon per cent	Max Manganese per cent	Max Silicon per cent	Nickel per cent	Chromium per cent	Copper per cent	Iron per cent	Moly per cent	Other per cent	ASME* Specified Minimum U.T.S. psi	ASME* Specified Minimum Yield psi
Nickel	B160 B161 B162	Bars, Rods Pipe Plate	.15	.35	.35	99.0 Min		.25 Max	.40 Max		S .01 Max	55,000	15,000
Nickel Low C	B160 B161 B162	Bars, Rods Pipe Plate	.02	.35	.35	99.0 Min		.25 Max	.40 Max		S .01 Max	50,000	12,000
Monel	B127 B164 B165	Plate Bars, Rods Pipe	.30	1.25(Plate) 2.00	.50	63.0/70.0		Remainder	2.50 Max		Al .50 Max S .024 Max	70,000	28,000
Inconel	B166 B167 B168	Bars, Rods Pipe Plate	.15	1.00	.50	72.0 Min	14.0/17.0	.50 Max	6.0/10.0		S .015 Max	80,000	30,000
Incoloy			.10	1.50	1.00	30.0/34.0	19.0/22.0	.50 Max	Remainder		S .03 Max	(75,000)	(30,000)
Inconel X			.08	.30/1.00	.50	70.0 Min	14.0/16.0	.20 Max	5.0/9.0		Ti 2.25/2.75 Cb+Ta .7/1.2 Al .4/1.0 S .01 Max	(160,000)	(100,000)
Hastelloy B Wrought	B335 B333	Rods Plate	.05	1.00	1.00	Remainder	1.00 Max		4.0/6.0	26.0/30.0	Co 2.5 Max V .2/.4 P .025 Max S .03 Max	100,000	45,000
Hastelloy C Wrought	B336 B334	Rods Plate	.08	1.00	1.00	Remainder	14.5/16.5		4.0/7.0	15.0/17.0	W 3.0/4.5 Co 2.5 Max V .35 Max P .04 Max S .03 Max	100,000	45,000
Hastelloy F Wrought			.05	1.00/2.00	1.00	Remainder	21.0/23.0		13.5/17.0	5.5/7.5	W 1.00 Max Co 2.5 Max Cb+Ta 1.75/2.50 P .04 Max S .03 Max		

* Values from Table UNF-23 of Section VIII of the ASME Boiler and Pressure Vessel Code, 1959 Edition, for the annealed condition. Values shown in parentheses are from the alloy manufacturer's published data.
[1] For specifications on bolting materials for use with flanges of above materials refer to ASTM Specifications B160, B164 and B166.

[2] Where maximum values are used, the minimum content is determined by strength requirements. Determinations of chemistry are by Check Analysis.

Copper and Copper Alloys

Material	ASTM Specifications[1] Number	ASTM Specifications[1] Grade	ASTM Specifications[1] Form	Copper per cent	Tin per cent	Zinc per cent	Nickel per cent	Silicon per cent	Iron per cent	Max Lead	Other per cent	ASME* Specified Minimum U.T.S. psi	ASME* Specified Minimum Yield psi
Admiralty	B111	A B C D	Tubing	70.0/73.0	.90/1.20	Remainder			.06 Max	.07	As .02/.10 (B) Sb .02/.10 (C) P .02/.10 (D)	45,000	15,000
Red Brass	B36 / B43	3	Plate / Pipe	84.0/86.0		Remainder			.05 Max	.05 / .06		40,000	12,000
Deoxidized Copper	B42		Pipe	99.9 Min							P .04 Max	30,000	9,000
Copper Nickel 70/30	B111 / B122	70/30 Cu Ni / Alloy 5	Tubing / Plate	65.0 Min		1.0 Max	29.0/33.0		.40/.70 / .70 Max	.05	Mn 1.0 Max	52,000	18,000
Copper Nickel 90/10	B111	90/10 Cu Ni	Tubing	86.5 Min		1.0 Max	9.0/11.0		.5/2.0	.05	Mn 1.0 Max	40,000	15,000
Silicon Bronze	B96 B124 / B97	A Alloy 7 / B	Plate Forgings / Plate	94.8 Min / 96.0 Min	1.5 Max		.06	2.8/3.5 / .8/2.0	1.6 Max / .8 Max	.05 / .05	Mn 1.5 Max / Mn .7 Max	50,000	18,000
Aluminum Bronze	B169	D	Plate	88.0/92.5					1.5/3.5		Al 6.0/8.0	70,000	30,000

* Values are from Table UNF-23 of Section VIII of the ASME Boiler and Pressure Vessel Code, 1959 Edition, for the annealed condition.
[1] For specifications on bolting materials for use with flanges of above materials, refer to ASTM Specifications B12 and B98.

4

Pressure drop in piping systems and line sizing

In this chapter various charts and tables are included which can be used to solve any practical problem in the sizing of lines or the calculation of pressure drops in lines. Examples of many types of problems as well as their solutions are also given. Intelligent users of such information, however, want and need to know its theoretical basis and limitations. Hence, the data section is preceded by a brief review of fluid-flow theory.

BASIC RELATIONSHIPS

Definition of Force and Its Units

The definition of force and its units must be clearly understood in order to avoid confusion which may affect decisions on even the most practical problems. Force is defined by Newton's law as:

$$\mathscr{F} = ma \qquad (1)$$

	English Units	Metric Units
where $\mathscr{F}$ is force	lb-ft/sec², called poundals	gr-cm/sec², called dynes
m is mass	pounds	grams
a is acceleration	ft/sec²	cm/sec²

It was found convenient to express a standard unit of force in terms of the force exerted by gravity on a unit mass at sea level and 45° latitude. The acceleration due to gravity, g, at these conditions is 32.174 ft/sec² or 980.665 cm/sec². This standard force is called a pound of force or a gram of force.

$1\ \text{lb}_f = (1\ \text{lb}_m)(32.174\ \text{ft/sec}^2) = 32.174\ \text{lb}_m\text{-ft/sec}^2$
$1\ \text{gram}_f = (1\ \text{gram}_m)(980.665\ \text{cm/sec}^2)$
$\qquad\qquad = 980.665\ \text{gram}_m\text{-cm/sec}^2$

where lb_f = pounds of force
$\quad \text{gram}_f$ = grams of force
$\quad\quad \text{lb}_m$ = pounds of mass
gram_m = grams of mass

Since it is customary to express pressure in terms of pounds or kilograms of force per unit area, working equations are converted from the basic absolute system of force (lb_m-ft/sec² or gram_m-cm/sec²) to the gravitational system of pounds of force or kilograms of force. The conversion factor, called g_c, is:

$$32.174\ \frac{(\text{lb}_m)(\text{ft/sec}^2)}{\text{lb}_f} \qquad \text{or}$$

$$980.665\ \frac{(\text{gram}_m)(\text{cm/sec}^2)}{\text{gram}_f} \qquad \text{or}$$

$$980665.0\ \frac{(\text{gram}_m)(\text{cm/sec}^2)}{\text{kg}_f}$$

It is essential for the designer to realize that pounds and grams of force are quite different from pounds and grams of mass and, therefore, cannot be equated.

Basic Energy Equation for Flow

From the first law of thermodynamics (conservation of energy) a useful equation can be derived which expresses the pressure change in a flowing system. In terms of differential quantities, this equation is

$$-\frac{dP}{\rho} = \frac{d(u^2)}{2g_c} + \frac{g}{g_c}dZ + dF \qquad (2)$$

	English Units	Metric Units
where ρ = density	lb_m/ft^3	$gram_m/cm^3$
P = pressure	lb_f/ft^2	kg_f/cm^2*
u = velocity	ft/sec	cm/sec
g_c = conversion factor	$32.17 \dfrac{(lb_m)(ft/sec^2)}{lb_f}$	$980665 \dfrac{(gram_m)(cm/sec^2)}{kg_f}$
g = acceleration due to gravity	ft/sec^2	cm/sec^2
Z = height above datum	ft	cm
F = energy loss due to friction	$\dfrac{(ft)(lb_f)}{lb_m}$	$\dfrac{(cm)(kg_f)}{gram_m}$

When density changes are small, Eq. (2) can be integrated between a downstream point D and an upstream point U.

$$\frac{P_U - P_D}{\rho} = \frac{u_D^2 - u_U^2}{2g_c} + (Z_D - Z_U)\frac{g}{g_c} + F \quad (3)$$

This equation is composed of so-called head terms common to piping design. Each term has the units ft-lb_f/lb_m and since some users of the equation overlook the difference between lb_f and lb_m, we often hear the expression "feet of head." It should be realized, however, that lb_f and lb_m do not cancel and the units are actually energy per pound of mass (ft-lb_f/lb_m). The terms are designated as follows:

$\dfrac{P_U - P_D}{\rho}$ is the change in pressure head

$\dfrac{u_D^2 - u_U^2}{2g_c}$ is the change in velocity head (kinetic energy)

$(Z_D - Z_U)\dfrac{g}{g_c}$ is the change in static head

F is the loss in head due to friction

Equation (3) states in mathematical form a fact that is well-known to most practical men; that is, the loss in pressure head is composed of the change in velocity and static heads and the energy loss due to friction. More concretely, if water at 60°F, flowing at 5 ft/sec at point U and at a pressure of 100 psig loses 20 ft-lb_f/lb_m due to friction as it flows to a point D, 80 ft above point U, where it flows in a smaller pipe at 10 ft/sec, the final pressure P_D is from Eq. (3):

$$\frac{P_U - P_D}{\rho} = \frac{(10)^2 - (5)^2}{2g_c} + (80 - 0)\frac{g}{g_c}$$

$$+ 20 = 1.2 + 80 + 20 = 101.2$$

$\dfrac{g}{g_c}$ is approximately equal to 1.

$$P_D = \frac{(100)(144)}{62.4} - 101.2 = 129.8 \text{ ft-}lb_f/lb_m$$

or $$P_D = (129.8)(62.4) = 8090 \text{ } lb_f/ft^2$$

or $$\frac{8090}{144} = 56.3 \text{ psi}$$

Many designers prefer to solve such a problem based on their own knowledge of the facts rather than setting it into a formal equation. The fluid possesses a certain energy as pressure head at some point where the pressure is known which can be designated as the starting point. The changes in head are then followed as in this tabulation:

Head	Energy Consumed (Pressure Head is Decreased)	Energy Gained (Pressure Head is Increased)
Static head: Change in position of piping	(a) Higher than at starting point	(b) Lower than at starting point
Velocity head: Change in velocity in pipe	(a) Higher than at starting point	(b) Lower than at starting point
Friction head: Losses due to friction	(a) In direction of flow	(b) Counter to direction of flow

Thus, instead of relying on remembering a specific equation, a solution is readily obtained by reasoning for each form of energy whether it is gained or lost. Before adding or subtracting energy terms, however, each term must be in the same units.

Example. Pressure at Point U is 100 psi or

$$\frac{(100)(144)}{62.4} = 231 \text{ ft-}lb_f/lb_m$$

Static Head: (The elevation of piping at Point D is 80 ft above that at U and energy is consumed in moving the fluid to this higher level.)

$$-80\frac{g}{g_c} = -80 \text{ ft-}lb_f/lb_m$$

---

* Pressure expressed in kg_f/cm^2 is approximately equal to atmospheres (1 $kg_f/cm^2 = 1.03$ atmospheres).

Velocity Head: (The velocity at D is higher than at point U and thus the kinetic energy is higher. This energy must come from the pressure energy.)

$$-\frac{10^2 - 5^2}{2g_c} = -1.2 \text{ ft-lb}_f/\text{lb}_m$$

Friction Head: (This is always energy lost.)

$$-20 \text{ ft-lb}_f/\text{lb}_m$$

Net energy at Point D: $231 - 101.2 = 129.8$ ft-lb$_f$/lb$_m$ or in terms of pressure:

$$\frac{(129.8)(62.4)}{144} = 56.3 \text{ psi}$$

In many process lines pipe sizes remain constant or the change is such that the velocity head difference is negligible as in the preceding example.

Basic Friction-Loss Relationships (Friction Head)

Experimental evidence based on studies of drag exerted by the pipe wall on flowing fluid has resulted in the Fanning equation:

$$F = \frac{fL}{D}\frac{u^2}{2g_c} \qquad (4)$$

		English Units	Metric Units
where	F = friction loss	$\dfrac{\text{ft-lb}_f}{\text{lb}_m}$	$\dfrac{\text{cm-gram}_f}{\text{gram}_m}$
	f = dimensionless constant called the friction factor		
	L = length of pipe in	ft	cm
	g_c = unit conversion factor	$32.17 \dfrac{\text{ft-lb}_m}{(\text{lb}_f)(\text{sec}^2)}$	$980.667 \dfrac{\text{cm-gram}_m}{(\text{sec}^2)(\text{gram}_f)}$
	D = inside diameter in	ft	cm

Equation (4) can be converted into a number of different forms to suit the user. For instance, if it is desired to express the flow rate in gpm (Q), diameter in inches (d), and F in terms of Δp in psi, the equation becomes

$$\Delta p = 0.000217 fLQ/d^5 \qquad (5)$$

Care should be exercised, however, in using equations found in handbooks written in various forms. The basic Fanning equation has been presented by numerous authors, and the factor f has not always been defined in the same manner. In some equations, in place of f as given in Eq. 4, a factor $4f_1$ is substituted. If such equations are to be used, they must be used with the partic-

ular friction factor chart upon which they are based. Equation (4) is that approved by the Hydraulic Institute and is most generally accepted.

THE FRICTION FACTOR

The friction factor f has been shown to depend upon a dimensionless Reynolds number $Re = D\rho u/\mu$ (μ is the viscosity of the fluid) and a dimensionless term expressing the roughness of the pipe (ϵ/D where ϵ is the depth of irregularities in the pipe). The relationship as given by Moody[16] is shown in Figs. 4.1 and 4.2. As will be demonstrated, these curves can be placed in more convenient form for routine use. The limitations of these data, however, can best be discussed by considering the original form of presentation.

By reference to Fig. 4.2 it will be noted that the relative roughness for a given type of pipe is based on a single value of roughness. This must necessarily be some average value in order to simplify the problem. Pigott[18] has shown that, before more reliable friction factor data can be obtained, a better understanding of roughness is needed as well as improved methods for measuring roughness.

In addition to the problem of knowing the roughness of new pipe, the effects of service life on the wall characteristics are not known well enough to express them quantitatively. The data as plotted on the Moody charts, although among the best (Moody estimates a probable variation in the correlation of ±10%), should be thought of as design data and used with an adequate safety factor, as will be demonstrated in the example calculations.

It should further be noted that the flow in Fig. 4.1 is divided into three distinct regions, laminar (below 2100), transition (2100 to 3000), and turbulent (above 3000). Because of the uncertainty of the relationship in the transition region it is preferable to use values of f for that region extrapolated from the turbulent region.

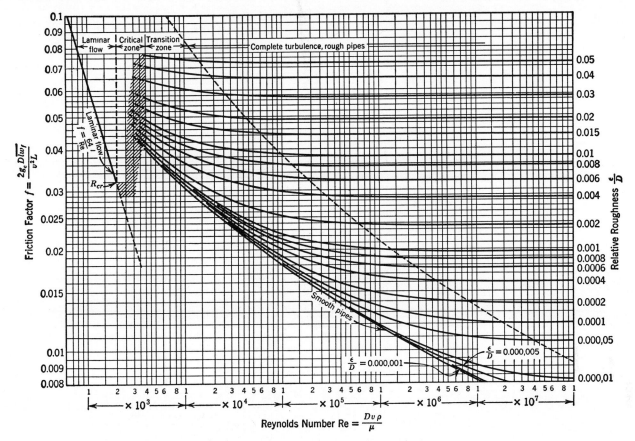

FIG. 4.1. Moody friction-factor chart. [Reproduced by permission from L. F. Moody, *Trans. ASME,* **66,** 671 (1944).]
(Note: See Fig. 4.2, page 120, for ϵ/D.)

FRICTION LOSS RELATIONSHIP FOR GASES AND VAPORS

Equation (2) is basic and applies to all fluids. The integrated form (Eq. 3) was obtained, however, by assuming an average density which is satisfactory for liquids but not for gases and vapors unless the change in pressure is small, as in short lines. For longer runs, the pressure drop is high enough to cause an appreciable variation in gas density.

Static head changes can be neglected for most process gas and vapor lines since gas or vapor density is so low that even a 200-ft change in height represents only a small change in pressure. Velocity head changes are also negligible, and the pressure drop in the line is represented almost entirely by the friction loss which is in differential form:

$$dF = \frac{fu^2 dL}{2g_c D} \qquad (6)$$

By representing density in terms of the actual gas law ($\rho = zRT/PM$) Weymouth's equation can be derived. This equation is applicable for gases and vapors undergoing large pressure drops and, thus, changing densities.

$$p_1{}^2 - p_2{}^2 = \frac{16fLW^2zRT}{g_c M\pi^2 D^5(144)^2} \qquad (7)$$

where W = lb/sec
D = inside diameter in ft
p = pressure in psia
R = 1544 (lb$_f$/sq ft)(cu ft)/(lb-mole)($^\circ R$)
M = molecular weight
T = temperature in $^\circ R$ ($^\circ$F + 460)
z = compressibility factor (see Fig. A-9 in Appendix)

When it is not possible to neglect the change in pressure caused by differences in elevation, as in deep gas wells, more complex equations result.

FRICTION-LOSS RELATIONSHIP FOR NON-NEWTONIAN FLUIDS

The relations for losses due to friction described thus far are applicable to all fluids whose flow properties can be completely characterized by the viscosity of the fluid. Such fluids are called Newtonian and include all gases and all low molecular-weight liquids and solutions. Solutions or melts of polymeric materials, such as melted plastics, and slurries of high concentration, especially those composed of solids which swell, such as clay, are non-Newtonian in character. Their flow properties must be described by factors related to particle size and

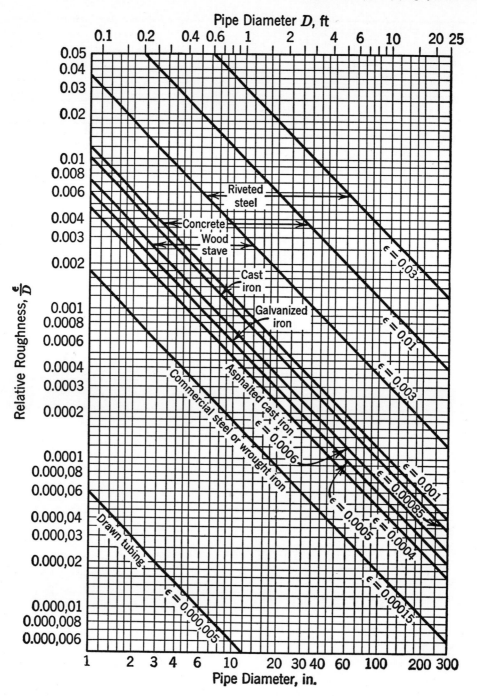

FIG. 4.2. Relative roughness for pipe. [Reproduced by permission from L. F. Moody, *Trans. ASME,* **66,** 671 (1944).]

shape, particle flexibility, presence and magnitude of electrical charges on the particles, and other colloidal properties.[14]

Metzner and Reed[15] have proposed a generalized Reynolds number such that in the laminar region (below $Re = 2100$) experimental data for both Newtonian and non-Newtonian fluids follow the same curve

$(f = 64/Re)$. This generalized Reynolds number is:

$$N_{Re}' = \frac{D^n u^{2-n} \rho}{\gamma} \qquad (8)$$

where D = inside pipe diameter, ft
  u = average bulk velocity, ft/sec
  ρ = density, lb_m/cu ft

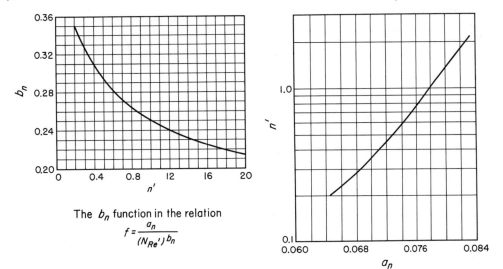

FIG. 4.3. Empirical constants for non-Newtonian friction-factor Equation 9. [Reproduced by permission from D. W. Dodge and A. B. Metzner, *AIChE Journal*, **5**, 189 (1959).]

γ = generalized viscosity coefficient $\text{lb}_m/(\text{ft})(\text{sec})$ defined as $= g_c K' 8^{n-1}$ (See Table 4.1)

K' = experimentally determined flow constant related to fluid consistency, $K' = \mu/g_c$ for a Newtonian fluid

n = experimentally determined constant related to the degree of non-Newtonian behavior (flow-behavior index). $n = 1.0$ for a Newtonian fluid

Since $n = 1.0$ and $K' = \mu/g_c$ for a Newtonian fluid, the generalized Reynolds number reduces to the familiar $Du\rho/\mu$ for Newtonian fluids.

For Reynolds numbers above 2100 and up to 100,000 the friction factor can be estimated from the empirical equation

$$f = \frac{a_n}{(N_{Re}')b_n} \qquad (9)$$

where N_{Re}' = generalized Reynolds number
a_n, b_n = empirical constants given in Fig. 4.3

Typical values of n and γ for some non-Newtonian systems are given in Table 4.1. Data for fluids not reported in the literature must be obtained from viscosimeter measurements.[15]

TABLE 4.1

RHEOLOGICAL CONSTANTS FOR SOME TYPICAL NON-NEWTONIAN FLUIDS*

(See Equation 8 for definition of terms)

Composition of fluid	Rheological constants		Composition of fluid	Rheological constants	
	n	γ		n	γ
23.3% Illinois yellow clay in water	0.229	0.863	18.6% Solids, Mississippi clay in water	0.022	0.105
0.67% Carboxy-methyl-cellulose (CMC) in water	0.716	0.121	14.3% Clay in water	0.350	0.0344
			21.2% Clay in water	0.335	0.0855
1.5% CMC in water	0.554	0.920	25.0% Clay in water	0.185	0.204
3.0% CMC in water	0.566	2.80	31.9% Clay in water	0.251	0.414
33% Lime water	0.171	0.983	36.8% Clay in water	0.176	1.07
10% Napalm in Kerosene	0.520	1.18	40.4% Clay in water	0.132	2.30
4% Paper pulp in water	0.575	6.13	23% Lime in water	0.178	1.04
54.3% Cement rock in water	0.153	0.331			

* Reproduced by permission: A. B. Metzner and J. C. Reed, *AICHE Journal*, **1**, 434 (1955).

WORKING CHARTS AND EQUATIONS
FOR FRICTION-LOSS CALCULATIONS

Valves and Fittings

Figures 4.1, 4.2, and related correlations are used for predicting friction loss in straight runs of pipe. The flow through various valves and fittings, however, is much too complex to correlate in the same generalized way. Perhaps the simplest and most accurate method for representing losses in valves and fittings is by some fraction of the kinetic energy (velocity head $u^2/2g_c$) of the fluid entering the valve or fitting.

$$F = K \frac{u^2}{2g_c} \qquad (10)$$

where K is constant called resistance coefficient for fitting or valve

u is velocity at entrance to fitting or valve.

This procedure is more logical than the widely used equivalent length concept. As can be understood from the brief development which follows, lengths of pipe equivalent to a valve or fitting require an assumption of some constant friction factor even though it is known that f varies with flow rate.

Let L_e = equivalent length of straight pipe. Then

$$F = K \frac{u^2}{2g_c} = \frac{fL_e}{D} \frac{u^2}{2g_c}$$

or
$$L_e = \frac{KD}{f} \qquad (11)$$

Thus L_e varies with flow rate. Values of K, on the other hand, are relatively constant above Reynolds numbers of 1000 where most practical problems occur.

Values of K, the resistance coefficient, for a wide variety of valves and fittings are given in Tables 4.2, 4.3, 4.4 and Figs. 4.4 and 4.5 (pages 129 to 133). These data are based on the indicated references. Much experimental work remains to be done before more reliable values can be reported. Agreement does not exist between various sources of data, but the values presented here were chosen to yield a consistent set of data for use in design problems.

Single-Phase Flow

Convenient charts for calculating the pressure drop due to friction can be prepared, based on Fig. 4.1 and 4.2. For the designer who must make many such calculations and line-sizing decisions, charts designed for rapid use are essential. As an example, a simplified

chart for a single fluid such as water can be developed at some temperature such as 68°F and for 100 ft of pipe.

$$\Delta p = 0.000217(f)(100)(Q)(62.4)/d^5 \qquad (12)$$

Now f is related to pipe size and Reynolds number which at the constant conditions proposed depends only on pipe size and flow rate. A chart such as Fig. 4.6 (facing page 133) can, therefore, be constructed having parameters of pipe size. An auxiliary correction chart for other water temperatures makes the information complete. Similar charts for gases such as steam (Fig. 4.7, facing page 132) can be constructed for use over lengths of pipe for which the pressure-dependent physical properties can be assumed constant (lengths for which Δp is less than 10% of total pressure).

General charts applicable to all gases or all liquids can also be developed, but they are more complex and confusing to use because of the large number of chart reading operations required. Another approach to the problem will be given here which offers time-saving possibilities as well as less opportunity for chart-reading errors. This time-saving method can be developed by considering Eq. (4):

$$F = \frac{fL}{D} \frac{u^2}{2g_c}$$

The term fL/D is a dimensionless term which for a given length of pipe (arbitrarily set at 100 ft) and inside diameter of pipe depends only on the Reynolds number. It can be thought of as a resistance coefficient, K_p, for pipe similar to that for fittings. Values of K_p per 100 ft of pipe were determined from Figs. 4.1 and 4.2 and plotted in Fig. 4.8, page 134. The variation in K_p with Reynolds number is not great, and this simplifies the trial and error calculations involved in line sizing as described in a following section.

By summation of the values of K for valves and fittings (Tables 4.2 through 4.4 and Figs. 4.4 and 4.5, pages 129 to 133) and straight pipe (Fig. 4.8) in a given system, the pressure drop due to friction can be obtained readily from the product of this summation and the kinetic energy term, $u^2/2g_c$. The kinetic energy can be read rapidly from an alignment chart, such as Fig. 4.9, page 135. These same charts can be used for non-Newtonian fluids, if the generalized Reynolds number is determined as described previously ($D^n u^{2-n} \rho/\gamma$).

Two-Phase Flow (Gas-Liquid)

Mixtures of vapor and liquid are common in process lines, but analysis of mixed flow is difficult. It has been demonstrated that the geometric patterns exhibited by two-phase flow systems are so complex and unstable that rational analysis based on theory is impossible at this time. A general empirical solution has been proposed, but it is tedious to apply. More recently two empirical correlations for horizontal pipes have been developed. These correlations are simpler to use and give reasonable results, considering the complexity of the problem. The correlation by Chenoweth and Martin[5] has been included here because it is readily adaptable to the scheme of calculations already described and gives comparable accuracy for most process calculations (average absolute deviation 20.4%).

The correlation is presented in Fig. 4.10. The liquid volume fraction is used, together with the ratio

$$\frac{(0.01 K_{PG}L + \Sigma K)_L}{(0.01 K_{PL}L + \Sigma K)_G} = \frac{\psi_G \rho_L}{\psi_L \rho_G} \qquad (13)$$

to obtain the ratio $\Delta P_{TP}/\Delta P_L{}^*$ where:

$K_{PG} =$ Resistance coefficient from Fig. 4.8 based on a fictitious Reynolds number $Re^*_G = \dfrac{6.3W}{d\mu_{cp_G}}$

      where $\ W =$ total flow of gas plus liquid in lb/hr

            $d =$ ID of pipe in inches

           $\mu_{cp_G} =$ viscosity of gas only in cp

$K_{PL} =$ Resistance coefficient from Fig. 4.8 based on fictitious Reynolds number $Re^*_L = \dfrac{6.3W}{d\mu_{cp_L}}$

      where $\mu_{cp_L} =$ viscosity of liquid only in cp

$K =$ summation of resistance coefficient of all fittings

$\rho_L =$ liquid density

$\rho_G =$ gas density

$\Delta P_{TP} =$ predicted two-phase pressure drop

$\Delta P_L{}^* = \rho_L(0.01 K_{PL}L + \Sigma K)\dfrac{u_t{}^2}{2g_c}$, psf

      where: $u_t =$ total velocity of gas plus liquid based on density of liquid only

         $= \dfrac{\rho_L}{144}(0.01 K_{PL}L + \Sigma K)J_t$, psi   (13A)

      where $J_t =$ velocity head for total fluid based on liquid density only

This is a fictitious "all-liquid" pressure drop. From the ratio $\Delta P_{TP}/\Delta P_L{}^*$ and the calculated value $\Delta P_L{}^*$, ΔP_{TP} can be determined. The use of Fig. 4.10, page 136, in conjunction with the other figures for rapid solution of gas-liquid flow problems is demonstrated in the examples.

Few data have been obtained for vertical lines and until such data are obtained it is suggested that this same correlation be used for approximating losses due to friction. As always, of course, static head changes in vertical lines must be calculated too. High velocities (above 15 ft/sec) are recommended in vertical lines to prevent irregular or pulsating flow.

This correlation is based on no change in phase of the flowing fluids. If phase changes occur (flashing), a step-wise solution must be employed. The system should be analyzed in small increments over which average conditions are assumed to prevail. Based on the calculated pressure drop, the conditions in the next section can be estimated by flash equilibrium calculations. Such studies should be made by a competent process engineer.

Pressure Drop in Condensate Lines

In power plants valuable condensates are returned under their own pressure to the boiler system. Condensates in process plants, however, are often collected in areas some distance from the steam-generating plant. Under these conditions it often becomes more economical to collect the condensate at the point of use and pump it back to the steam plant. In such instances the friction losses are calculated in the manner already described for single fluids.

If condensate is returned under its own pressure or passed through reducing valves to lower pressure steam lines, the flow of two phases must be considered since the condensate will flash. Allen[1] has proposed a method for determining pressure drop in flashing condensate systems. He has particularly emphasized the need for careful calculations of flow through pressure reducing valves. Such calculations are preferably done in cooperation with an instrument engineer.

Two-Phase Flow: Solid-Liquid (Slurries)

Much more work is necessary on fundamentals of slurry flow before exact expressions can be presented. Many slurries such as clay in water behave as non-Newtonian fluids. As discussed previously, exact calculations for these fluids require laboratory data and a sound background in fluid dynamics. However, approximate methods for calculating friction losses have been presented for aqueous slurries in turbulent flow. For convenience these methods can be classified according to two general types of slurries, homogeneous and heterogeneous.

HOMOGENEOUS SLURRIES

Slurries such as clay and fine ash under 30 μ in size in water behave as homogeneous fluids, and friction losses can be estimated as described for single-phase fluids, except that the Reynolds number should be approximated by using the viscosity of the suspending medium.[4, 17, 20]

HETEROGENEOUS SLURRIES

Heterogeneous slurries are composed of larger particles of solid (above 30 μ), so that even in turbulent flow the composition of solids is not uniform over the cross section (for example, sand in water). Durand and Condolios[7] proposed the following equation for slurries that are composed of graded particles and flowing at velocities above 3 ft/sec.

$$\frac{h_m \frac{\rho_m}{\rho_w} - h_w}{h_w} = 121C\left[\frac{gD\left(\frac{\rho_s}{\rho_w} - 1\right)}{u^2} \cdot \frac{V}{gD_p\left(\frac{\rho_s}{\rho_w} - 1\right)^{1/2}}\right] \quad (14)$$

where C = concentration of solids by volume
D_p = particle diameter in feet
D = inside diameter of pipe in feet
g = acceleration due to gravity, 32.17 ft/sec²
h_m = head loss due to friction for slurry in $\frac{\text{ft-lb}_f}{\text{lb}_m \text{ of slurry}}$
h_w = head loss that would be obtained for pure water alone $\frac{\text{ft-lb}_f}{\text{lb}_m \text{ of water}}$
ρ_m = density of slurry, lb/cu ft
ρ_w = density of water, lb/cu ft
ρ_s = density of solid, lb/cu ft
V = free falling velocity of the particles in water, ft/sec
u = velocity of flow of mixture, ft/sec

This equation has been applied with reasonable accuracy for closely sized particles. If the slurry is composed of mixed sizes, an equivalent particle diameter to be substituted for D_p in Eq. (14) has been suggested, based on the following equation:[19]

$$D_e = \frac{\Sigma w}{\Sigma\left(\frac{w}{D_p}\right)} \quad (15)$$

where D_e = equivalent diameter
w = fraction by weight of particles of diameter D_p

In designing piping systems for heterogeneous slurries it is also necessary to estimate the velocity below which appreciable settling of the particles occurs. This velocity is termed the minimum velocity and may be roughly

estimated for aqueous slurries from the following equation[20]

$$u_m^{1.225} = 7.22\, gD_p'\left(\frac{D\rho_m}{\mu}\right)^{0.775}\frac{\rho_s - \rho_w}{\rho_w} \quad (16)$$

where u_m = minimum velocity
μ = viscosity of medium (water), cp
D_p' = particle diameter such that 85% of particles by weight are smaller than D_p', ft

Until further work is done, these equations can be used as an approximation in the design of hydraulic systems. Actual laboratory tests are preferable as design aids, especially for slurries for which little experience has been accumulated.

Two-Phase Flow: Solid-Gas

The movement of solids in air or other gas streams is of importance in pneumatic conveying systems and in fluid catalyst systems. Because of the complexities of the variables involved, the design of conveying systems is empirical and depends largely on experience with the particular solid-gas system. Generalized correlations of restricted applicability have been developed, but the designer is urged to obtain expert advice from equipment manufacturers or other designers with experience in this field. Vogt and White[22] have presented an empirical correlation for granular solids and a good review of the literature.

Friction Losses for Non-Circular Ducts

The customary procedure for the solution of friction loss problems in non-circular ducts has been to use the relationships already discussed but to substitute an equivalent diameter D_e for the circular tube's inside diameter. By analogy with a circular duct the equivalent diameter is defined as four times the cross-sectional area of flow divided by the wetted perimeter. This ratio is called the hydraulic radius and for annuli it becomes $D_2 - D_1$, the difference between the inside diameter of the larger tube and the outside diameter of the smaller.

$$D_e = 4\left(\frac{\pi D_2^2/4 - \pi D_1^2/4}{\pi D_2 - \pi D_1}\right) = D_2 - D_1 \quad (17)$$

This customary procedure has not been thoroughly substantiated for all cases. A recent study[23] on annuli has indicated that the friction factor based on the hydraulic radius just defined must be corrected for accurate predictions, as shown in Fig. 4.11, page 136. In lieu of similar data for other cross sections, a friction factor based on an equivalent diameter must be determined without such correction.

PROCESS LINE SIZING TECHNIQUES

The sizing of process lines can be divided into two categories: (1) lines which do not contain pumping equipment, (2) lines which contain pumps or compressors. Lines in the first group are sized on the basis of available pressure drop, while those in the second must be based on an economical pipe size (the smaller the pipe, the less the pipe costs but the higher the pumping costs, and thus an optimum size must exist). Paradoxically, however, it is uneconomical to make economic studies on each process line for a proposed plant. The designer, therefore, must have some means of determining which lines justify careful cost analysis.

The problem is as follows: Based on previous experience defined by allowable velocities or allowable pressure drops as given in Table 4.5, page 136, the designer can select a conservative line size which will definitely fulfill the requirements. But will the next smaller size be more economical? The only manner in which this can be determined definitely is by a careful economic analysis. If the possible saving, however, is not far from the cost in manhours required to make the study, it is better to select the conservative size without further use of valuable time. If, on the other hand, the possible savings are substantial, then further detailed study is indicated. The data given in Tables 4.6 through 4.9, pages 137–138, afford a rapid way to estimate piping costs, which will aid in making such decisions.

The actual determination of a line size based on an allowable pressure drop involves a trial and error calculation. Because of the nature of the relationship between pipe size and friction losses, a line size must be assumed and the corresponding pressure drop calculated. The labor involved, however, has been greatly exaggerated by many writers. By using charts such as Figs. 4.4 to 4.9, pages 133–135, only a few seconds are required per calculation.

Flow in Sewers

Sewer design calculations are most conveniently made by using the old but reasonably accurate empirical formula suggested by Manning. It is applicable to sewers flowing either partly full or full.

$$V = \frac{1.486}{N} R^{2/3} S^{1/2} \qquad (18)$$

where V = velocity of flow, ft/sec
N = coefficient of roughness (see Table 4.10, page 139)
R = hydraulic radius, ft; $A/$(wetted perimeter), where A is the cross-sectional area of flow
S = head loss per foot of length or required slope of sewer

S is the head loss per foot of length and, since sewers usually flow under atmospheric pressure (gravity flow), this value also represents the slope of the sewer necessary to induce the desired velocity. Values of N are known to vary with depth in a partly filled sewer, but reliable values are not yet known. The design values given in Table 4.10, page 139, are used both for sewers flowing full and partly full.

The relationships between sewer capacity, velocity, and depth can be calculated as ratios for a fixed pipe diameter and slope. These are plotted as Fig. 4.12, page 139, and are determined from Eq. (18) and the trigonometric relations between depth and cross-sectional area of flow. Velocity and depth for sewage flowing below capacity in a given sewer can be determined from Fig. 4.11 if the velocity at full capacity is known (flowing full). It can be seen by reference to Fig. 4.12 that the velocity in a sewer at a fixed slope flowing more than half-full is equal to or greater than the velocity when flowing full.

Sewer design depends on selection of a slope such that the design capacity will flow at a velocity high enough to transport the solids in the sewage (self-cleaning), but low enough to avoid damage to the sewer by erosion. Based on accumulated evidence in operating systems, minimum and maximum design velocities for sewers flowing full are given in Table 4.11, page 140. The usual design range of 3 to 5 ft/sec has been selected in preparing a convenient table based on Manning's formula from which the required slope for each flow rate can be determined (Table 4.12, page 140).

Since recommended velocities are normally given for sewers flowing full, it is convenient to place calculations for sewers flowing partly full on an equal self-cleaning basis. Fair and Geyer[8] state that the degree of self-cleaning for sewers flowing full is the same as that when partly full, if the tractive force on sewer deposits remains the same. They develop the following relationship:

$$\frac{v_s}{V} = \left(\frac{r}{R}\right)^{1/6} \qquad (19)*$$

where v_s = self-cleaning velocity in partly full sewer equivalent to that in full sewer, ft/sec
V = design velocity in full sewer, ft/sec
r = hydraulic radius of flowing portion of partly full sewer
R = hydraulic radius of full sewer $D/4$

By employing Eqs. (18) and (19) and the geometry associated with partly filled sewer flow, Fig. 4.13 (page 140) was developed. It provides a rapid way to estimate the hydraulic factors that produce equal self-cleaning characteristics at all depths of flow. Sewer design calculations are presented in the examples.

*The form given here assumes no change in roughness with depth.

ANALYSIS OF COMPLEX PIPING NETWORKS

Water and steam distribution systems in process plants are often complex and require careful analysis to insure that every section of the plant will be amply supplied. A number of methods of analysis have been developed by hydraulic engineers, including electric analyzers. Of these the Hardy-Cross Method is most applicable to process plant problems, because it provides a logical basis for trial and error solution of the flow rates in various portions of the system.

The friction loss equation can be converted to a simpler form as follows:

$$F = \frac{fLu^2}{D2g_c}$$

$$u = \frac{Q}{(7.48)\,(60)\left(\dfrac{\pi D^2}{4}\right)} = 0.00283\,\frac{Q}{D^2}$$

where Q = flow rate in gpm
D = inside diameter in feet

$$F = \frac{fL(0.00283)^2 Q^2}{D^5 2g_c} \simeq \frac{fLQ^2}{d_i^5 g_c} = RQ^2 \qquad (20)$$

where d_i = inside diameter in inches

$$R = \frac{fL}{d_i^5\,g_c}$$

$$F = \frac{\text{ft-lb}_f}{\text{lb}_m}$$

For a single line and in the turbulent region the value of R can be calculated and assumed to be relatively constant over a moderate range of flow rates.

A similar equation for steam can be written in the form

$$\Delta p = R'W^2 = \text{psi pressure drop}$$

where W = steam flow in lb/hr and

$$R' = \frac{(L)(7.41 \times 10^{-3})}{d}, \qquad (21)$$

based on Gutermuth's equation. R' can be approximated for each section of the system and assumed to be constant over a moderate range of flow rates.

Referring to Fig. 4.18, page 151, a preliminary flow distribution in terms of Q_1, Q_2, and Q_3 is assumed. Q_A, Q_B, and Q_S are known. For the first assumption an equal velocity basis may be used as a criterion. Using the assumed flow rates, values of R are then determined for each section of the system.

At steady-state conditions the algebraic summation of all friction losses in ft-lb$_f$/lb$_m$ should be equal to zero.

That is to say, when proceeding around the loop in a clockwise direction, starting at and returning to point A, the sum of the head losses will be zero.

$$R_1Q_1^2 + R_2Q_2^2 - R_3Q_3^2 = 0 \qquad (22)$$

The negative sign on the third term results because the flow is in the opposite direction. It is convenient to call the clockwise direction positive and the counterclockwise direction negative.

Initially assumed flow rates, Q_1', Q_2', Q_3', are used in Eq. 21. Calling the amount that these assumptions are in error q, the following expression can be written:

$$R_1(Q_1' + q)^2 + R_2(Q_2' + q)^2 - R_3(Q_3' - q)^2 = 0 \qquad (23)$$

which upon expansion is

$$R_1Q_1'^2 + 2R_1Q_1'q + R_1q^2 + R_2Q_2'^2 + 2R_2Q_2'q \\ + R_2q^2 - R_3Q_3'^2 + 2R_3Q_3'q - R_3q^2 = 0$$

Neglecting the q^2 terms which should be small, we obtain the following equation for q:

$$q = -\frac{R_1Q_1'^2 + R_2Q_2'^2 - R_3Q_3'^2}{2(R_1Q_1' + R_2Q_2' + R_3Q_3')} \qquad (24)$$

or for the general use

$$q = -\frac{\Sigma_a RQ'^2}{\Sigma\, 2RQ'} \qquad (25)$$

where $\Sigma_a RQ'^2$ = the algebraic sum of RQ'^2 terms
$\Sigma\, 2RQ'$ = arithmetic sum of $2RQ'$ terms

Equation 25 reduces the trial-and-error calculations, so the probable flow rate and direction in each section of the system can be determined rapidly. The values of R can be checked, based on the calculated flow rates, if these calculated values differ markedly from the original assumptions. Example calculations are given at the end of this chapter to demonstrate the entire procedure.

PLUMBING

Building Water Supply Systems

The sizing of building water supply mains and branches is facilitated by the experience-based rules and data presented in the American Standard National Plumbing Code (ASA A40.8). Just as in any system involving many users, inadequate supply to certain users can be avoided by sizing the distribution system so that the pressure drops from the main to all fixtures are the same.

Data for estimating the maximum required water flow based on so-called fixture-units can be obtained from Table 4.13 and Fig. 4.14, pages 141 and 142. Table 4.14, page 143, gives minimum sizes of fixture supply piping which are useful in sizing lines to single fixtures. Other useful data are given in Table 4.15, page 143.

Building Drainage and Vent Systems

Design data similar to that given for building water supply are given in Tables 4.16, 4.17, 4.18, and 4.19 (pages 144–146). The "National Plumbing Code" (ASA A40.8) should be referred to for design details.

INDEX OF WORKING TABLES AND FIGURES FOR
PRESSURE DROP AND LINE SIZING CALCULATIONS

TABLE 4.2
RESISTANCE COEFFICIENTS FOR VALVES
(All valves wide open)

Valve type	Screwed	Flanged
Angle[1]		
Check		
Ball[2]	70 (all sizes)	70 (all sizes)
Lift (or piston)[2]	12 (all sizes)	12 (all sizes)
Swing[1]		2 (all sizes)
Gate[1]		
Globe[1]		
Plug (all sizes)[2]	1.0	0.5

(Varies greatly with design. Consult manufacturer
for more accurate data if required.)

[1] Based on *Pipe Friction Manual,* 3rd ed., copyright 1961, Hydraulic Institute, 122 East 42nd St., New York 17, N.Y. Charts reproduced by permission.

[2] Average of manufacturer's data.

TABLE 4.3
RESISTANCE COEFFICIENTS FOR STANDARD FITTINGS[1]

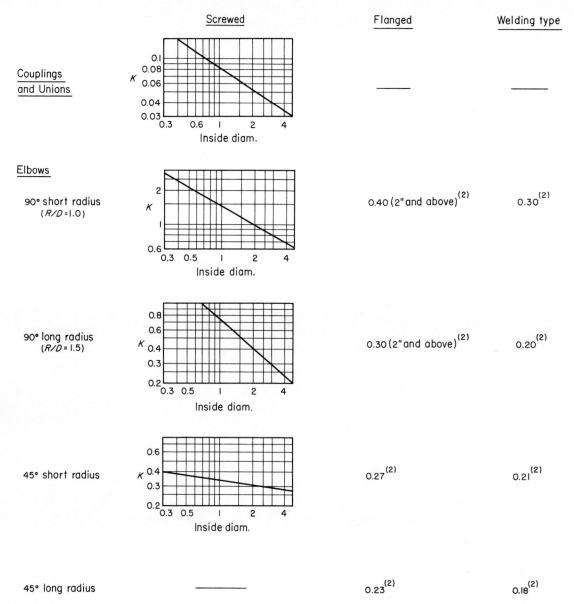

	Screwed	Flanged	Welding type
Couplings and Unions		———	———
Elbows			
90° short radius (R/D = 1.0)		0.40 (2" and above)[2]	0.30[2]
90° long radius (R/D = 1.5)		0.30 (2" and above)[2]	0.20[2]
45° short radius		0.27[2]	0.21[2]
45° long radius	———	0.23[2]	0.18[2]

[1] Based on *Pipe Friction Manual,* 3rd ed., copyright 1961, Hydraulic Institute, 122 East 42nd St., New York 17, N.Y.   Charts reproduced by permission.

[2] Because of scatter of experimental data there seems to be no justification for reporting variations of K with diameter.   The most consistent data are those of A. Hoffman, *Transactions of Munich Hydraulic Institute,* Bulletin No. 3, ASME 1945 authorized translation.   These data, which show variations of K with roughness and $R/D,$ were used in compiling the suggested values for flanged and welding-type fittings.   Flanged fittings are cast and are rougher than corresponding welding-type fittings.

TABLE 4.3 (*Concluded*)

RESISTANCE COEFFICIENTS FOR STANDARD FITTINGS*

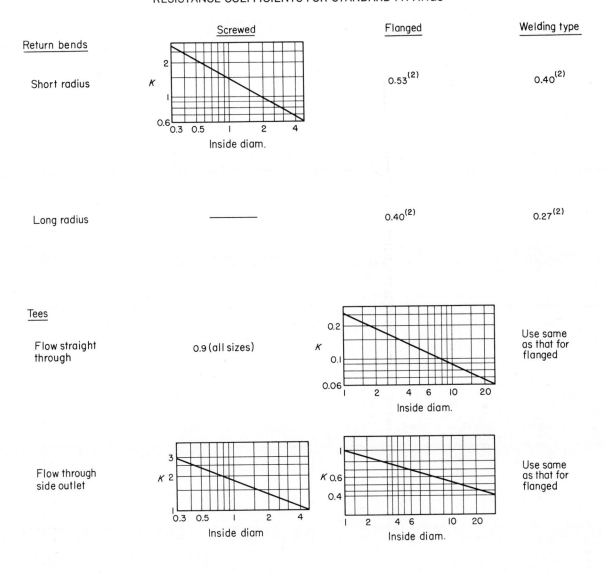

* See notes (1) and (2) on previous page.

TABLE 4.4
RESISTANCE COEFFICIENTS FOR ENTRANCES AND EXITS

Type			Exit from[1] tank	Entrance to tank
Bell−mouth (well−rounded)	Exit →	← Entrance	0.05	1.0
Sharp−edged slightly−rounded	Exit →	← Entrance	0.5 0.23	1.0 1.0
Inward projecting pipe	Exit →	← Entrance	1.0	1.0

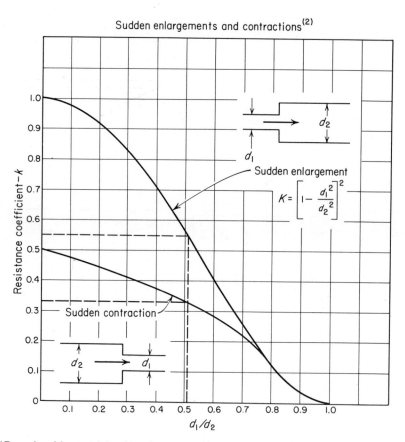

Sudden enlargements and contractions[2]

$$K = \left[1 - \frac{d_1^2}{d_2^2} \right]^2$$

Sudden enlargement

Sudden contraction

Resistance coefficient−k

d_1/d_2

[1] Reproduced by permission from *Pipe Friction Manual,* 3rd ed., copyright 1961, Hydraulic Institute, 122 East 42nd St., New York 17, N.Y.

[2] Reproduced by permission from Crane Co., Technical Bulletin 410, Chicago (1957).

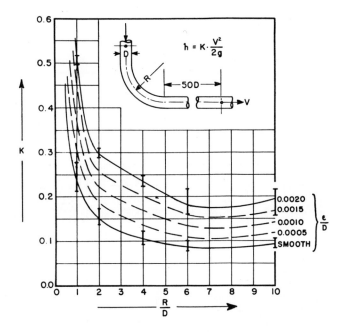

FIG. 4.4. Resistance coefficients for 90° bends of uniform diameter.
(Reproduced by permission from *Pipe Friction Manual,* 3rd ed., copyright
1961, Hydraulic Institute, 122 East 42nd St., New York 17, N.Y.)

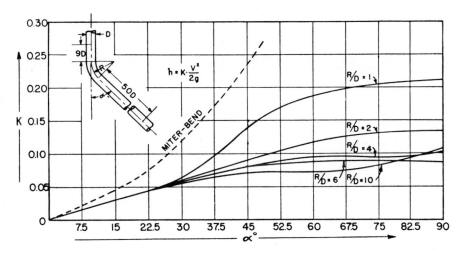

FIG. 4.5. Resistance coefficients for bends of uniform diameter and smooth surface at Reynolds No.
2.25×10^5. (Reproduced by permission from *Pipe Friction Manual,* 3rd ed., copyright 1961, Hydraulic
Institute, 122 East 42nd St., New York 17, N.Y.)

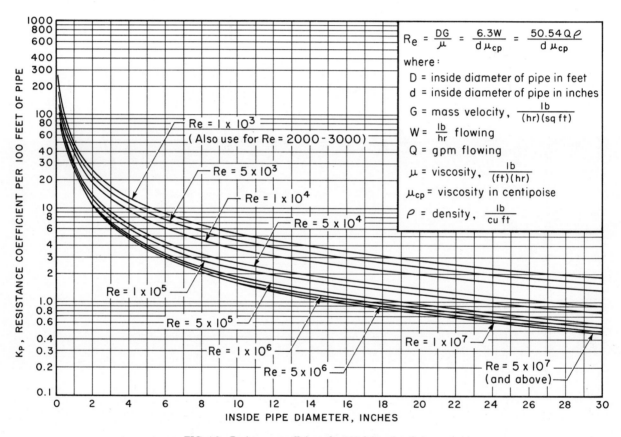

FIG. 4.8.   Resistance coefficients for 100-ft lengths of pipe.

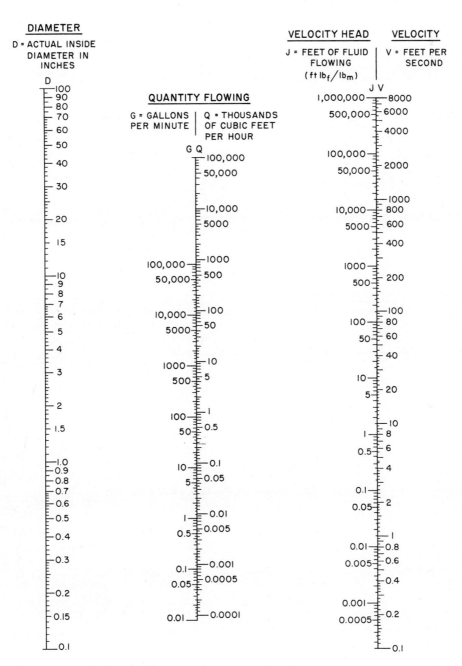

FIG. 4.9. Velocity and velocity-head chart.

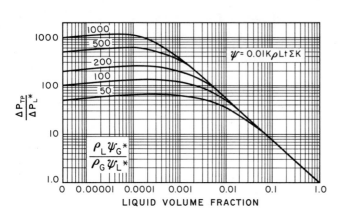

FIG. 4.10. Pressure drop ratio for gas-liquid flow. (Reproduced by permission of J. M. Chenoweth and M. W. Martin, C. F. Braun Co. Paper presented at ASME Petroleum Mechanical Engineering Conference, New Orleans, La., Sept. 25, 1955.)

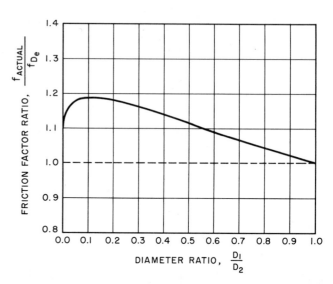

FIG. 4.11. Correction factor for equivalent diameter. [Reproduced by permission from J. E. Walker, G. A. Wahn, and R. R. Rothfus, *AIChE Journal*, **3**, 484 (1957).]

TABLE 4.5

SUGGESTED DESIGN BASIS FOR PROCESS LINES

(Tube Turns Division of Chemetron Corp.)

	Reasonable velocity (Tube Turns)	Allowable pressure drop due to friction
Pump discharge	* $d/2 + 4$ ft/sec	2 psi/100 ft
Pump suction	One-third the above	0.5 ft-lb$_f$/lb$_m$ per 100 ft
Steam or vapor	* d in 1000 ft/min	0.5 per cent of line pressure per 100 ft
Gravity flow of liquids	—	0.2 ft-lb$_f$/lb$_m$ per 100 ft
Water lines	5–7 ft/sec	

* d = inside diameter of pipe in in.

Reproduced from: H. F. Rase and M. H. Barrow, *Project Engineering of Process Plants,* John Wiley and Sons, New York, 1957.

TABLE 4.6
PIPE AND FITTING PRICES*

Base price: Carbon Steel ASTM 106 Grade A or equal $0.105 per pound

Base weights: (Based on carbon steel pipe)

	Weight (lb/ft)		
Size, in.	Schedule 10	Standard weight	Extra strong
2	2.64	3.65	5.02
3	4.33	7.57	10.25
4	5.61	10.79	14.99
6	9.29	18.97	28.55
8	13.40	28.55	43.40
10	18.70	40.48	54.70
12	24.20	53.60	65.40
14	36.70	55.00	72.10
16	42.10	63.00	82.80
18	47.40	71.00	93.50
20	52.70	78.60	104.10
24	63.40	94.60	125.50

Fittings	L_e/D, equivalent length to diameter ratio	
90° elbow, long and short radius	1.3	
Tees, full size outlet	2.1	
Tees, reducing outlet	2.6	(12″ and below)
	2.1	(14″ and above)
180° bends, long radius	3.0	
Reducers, concentric	0.50	
Reducers, eccentric	0.60	
Lap-joint stub ends	0.80	

Material factors for pipe and fittings	Multiply by
Carbon steel	1
4–6 per cent chrome, ½ per cent Moly	3.25
Stainless type 304	10.2
Stainless type 316	16.07

* Reprinted by permission: H. F. Rase, *Petroleum Refiner, 32,* No. 8, 141–144 (1953).

TABLE 4.7
FLANGE PRICES*

Base Prices
Basis: 150 lb ASA RF Welding neck ASTM A-181
(approximately 40¢/lb)

Size	Cost, $
2	3.5
3	5
4	6.5
6	9
8	14
10	20
12	35

Rating Factors	Multiply by	
Rating, lb	Raised face	Ring type joint
150	1.0	1.5
300	1.22	3.5
600	4.3†	5.0‡
900	5.5†	6.5‡
1500	9.0†	9.5‡

Flange Type Factors	Multiply by
Welding neck	1.0
Slip-on	0.65

Material Factors	Multiply by
Carbon steel ASTM A-181	1.0
4–6 per cent chrome, ½ per cent Moly (ASTM 182 F5)	3.50§
Stainless type 316	5.0 §

* Reprinted by permission: H. F. Rase, *Petroleum Refiner, 32,* No. 8, 141–144 (1953).

† Use 4.0 for sizes 2, 3, and 4 in. only.

‡ Use 4.5 for sizes 2 in. and 3 in. and 5 for 4 in.

§ For 150 lb and 300 lb welding neck only.

TABLE 4.8
FABRICATION AND ERECTION COSTS

Fabrication Costs	Multiply carbon steel costs of pipe, fittings, and flanges by
Material	
Carbon steel	1.0
All nickel and alloy steels (18-8, Monel)	2.0
5 per cent chrome (no nickel) and higher steels	2.5

Erection Costs

Approximately equal to cost of fabricated carbon steel pipe; i.e., material cost of carbon steel pipe, fittings, and flanges plus fabrication cost. This also applies for alloy pipe installations if no field welding is required. If field welding is required, the following multipliers are suggested for application to the cost of an equivalent carbon steel system to allow for special fit-up and field stress relieving.

Stainless steels	1.5
4–6 per cent chrome	1.2

* Reprinted by permission: H. F. Rase, *Petroleum Refiner, 32,* No. 8, 141–144 (1953).

TABLE 4.9
VALVE PRICES*

Size	Base price (adjusted) cost steel 150 lb O.S. and Y gate valve
2	70
3	85
4	120
6	205
8	290
10	400
12	530

Rating Factors Rating, lb	Multiply by	
	Raised face	Ring type joint
(125)	(See Cast Iron below)	
150	1.0	. . .
300	1.6	1.7
400	2.4	2.5
600	3.4	3.6
900	4.3	4.6
1500	6.7	7.1

Material Factors	Multiply by
Materials	
5 per cent chrome ½ per cent moly	1.25
Monel	2.1
Stainless steel, type 316	1.7 (150 lb only)
Cast iron (125 lb)	0.45

Valve Type Factor	Multiply by
Valve type	
Gate	1.0
Globe	1.2
Check	1.0 for class 150
	0.70 for all others

* Reprinted by permission: H. F. Rase, *Petroleum Refiner, 32,* No. 8, 141–144 (1953).

ILLUSTRATION OF USE OF TABLES 4.6 THROUGH 4.9—ESTIMATION OF A GIVEN LINE.* Determine the erected and delivered cost of a 6-in. Schedule 40, 150 lb ASA rating (RF), carbon steel pump discharge line which is 325 ft long and is composed of 12 flanges, 16 weld ells, 1 tee, 1 check valve, and 2 gate valves.

(a) Pipe and fittings (See Table 4.6.)

$$\frac{L_e}{D}, \text{equivalent length, ft}$$

325 ft pipe		325
16 ells	(16)(1.3)(6) =	125
	(1)(2.1)(6) =	13
	Total:	463 ft

Total weight: $(463)\left(\frac{lb}{ft}\right) = (463)(18.97) = 8780$

Cost: $(8780)(0.105) = \$922$

(b) Flanges (See Table 4.7.)
$(12)(9) = \$108$

(c) Fabrication (See Table 4.8.)
Total cost flanges and pipe = 922 + 108 = \$1030
Fabrication cost = (1.0)(1030) = \$1030

(d) Erection costs (See Table 4.8.)
Cost = 1030 + 1030 = \$2060

(e) Valve costs (See Table 4.9.)
2 gate valves (2)(205)   = \$410
1 check valve (1)(205)(1.0) = 205
                             \$615

Total cost of system delivered and erected

$$(a) + (b) + (c) + (d) + (e) = \$4735$$

If this line were constructed of Schedule 40, 4–6 per cent chrome-moly with 300 lb RTJ chrome-moly flanges, the cost would be calculated by using the appropriate factors.

(a)	(922)(3.25)	= \$ 3,000
(b)	(108)(3.5)(3.5)	= 1,322
(c)	(1030)(2.5)	= 2,580
(d)	1030 + 1030	= 2,060
(e)	Gates: (2)(205)(1.7)(1.25) =	872
	Check: (1)(205)(1.7)(1.25) =	436
	Total	\$10,270

* Reprinted by permission—H. F. Rase, *Petroleum Refiner, 32*, No. 8, 141–144 (1953).

TABLE 4.10

DESIGN VALUES OF ROUGHNESS COEFFICIENT N FOR SEWER LINES*

Conduit Material	N Design Value
Tile pipe, vitrified (glazed)	0.014
unglazed	0.015
Concrete pipe	0.015
Cast-iron pipe, coated	0.013
Brick sewers, glazed	0.013
unglazed	0.015
Steel pipe, welded	0.013
riveted	0.017
Concrete-lined channels	0.016

* Reproduced by permission: G. M. Fair and J. C. Geyer, *Water Supply and Waste-Water Disposal,* John Wiley and Sons, New York, 1954.

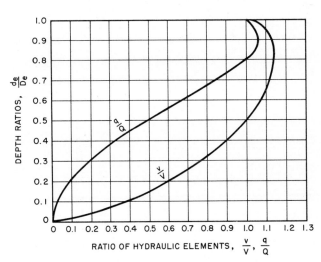

FIG. 4.12. Flow rate and velocity ratios for sewers flowing partially full for fixed diameter and slope.

d_e, q, v = depth, capacity and velocity, respectively, for partially filled sewer.

D_e, Q, V = depth, capacity, and velocity, respectively, for same sewer running full.

(Reproduced from G. M. Fair and J. C. Geyer, *Water Supply and Waste-Water Disposal,* John Wiley and Sons, New York, 1954.)

TABLE 4.11
RECOMMENDED SEWER DESIGN VELOCITIES
Basis: Flowing full

	Velocity, ft/sec	
	Minimum	Maximum
Sanitary Sewers	2 to 2.5	8 for concrete* 12 for tile
Storm Sewers	2.5 to 3	ditto
Usual Design Value for Process Plants	3	5

* Data of Fair and Geyer[8]

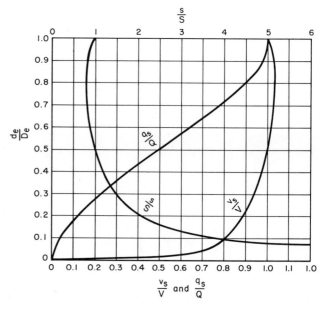

FIG. 4.13. Sewer hydraulic factors which produce equal self-cleaning at all depths.

D_e, Q, V, S = depth, capacity, velocity, and slope for sewer flowing full.
d_e, q_s, v_s, s = depth, capacity, velocity, and slope for same sewer flowing partially full at rate and slope so as to produce equal self-cleaning.

(Reproduced from G. M. Fair and J. C. Geyer, *Water Supply and Waste-Water Disposal*, John Wiley and Sons, New York, 1954.)

TABLE 4.12
DESIGN DATA FOR SEWERS FLOWING FULL
(Based on Manning's Formula)

	Minimum Velocity—3'/sec		Maximum Velocity—5'/sec	
Pipe Size	Gallons Per Minute	Slope or Head Loss Per Foot ft/ft	Gallons Per Minute	Slope or Head Loss Per Foot ft/ft
4	120	.017	200	.044
6	260	.01	450	.027
8	480	.0076	800	.02
10	750	.006	1250	.015
12	1100	.0048	1800	.012
15	1675	.0036	2800	.009
16	1900	.0033	3200	.0086
18	2450	.0029	4100	.0077
20	3000	.0026	5000	.0067
24	4200	.002	7600	.006

Note: Sewers less than 4" are not ordinarily used in plants.

TABLE 4.13

DEMAND WEIGHT OF FIXTURES IN FIXTURE UNITS (*a*)

Fixture or Group (*b*)	Occupancy	Type of Supply Control	Weight in Fixture Units (*c*)
Water closet	Public	Flush valve	10
Water closet	Public	Flush tank	5
Pedestal urinal	Public	Flush valve	10
Stall or wall urinal	Public	Flush valve	5
Stall or wall urinal	Public	Flush tank	3
Lavatory	Public	Faucet	2
Bathtub	Public	Faucet	4
Shower head	Public	Mixing valve	4
Service sink	Office, etc.	Faucet	3
Kitchen sink	Hotel or restaurant	Faucet	4
Water closet	Private	Flush valve	6
Water closet	Private	Flush tank	3
Lavatory	Private	Faucet	1
Bathtub	Private	Faucet	2
Shower head	Private	Mixing valve	2
Bathroom group	Private	Flush valve for closet	8
Bathroom group	Private	Flush tank for closet	6
Separate shower	Private	Mixing valve	2
Kitchen sink	Private	Faucet	2
Laundry trays (1–3)	Private	Faucet	3
Combination fixture	Private	Faucet	3

(*a*) For supply outlets likely to impose continuous demands, estimate continuous supply separately and add to total demands for fixtures.

(*b*) For fixtures not listed, weights may be assumed by comparing the fixture to a listed one using water in similar quantities and at similar rates.

(*c*) The given weights are for total demand. For fixtures with both hot and cold water supplies, the weights for maximum separate demands may be taken as three-fourths the listed demand for supply.

Reproduced by permission: "National Plumbing Code" (ASA A40.8-1955), published by American Society of Mechanical Engineers, 345 East 47th St., New York 17, N.Y.

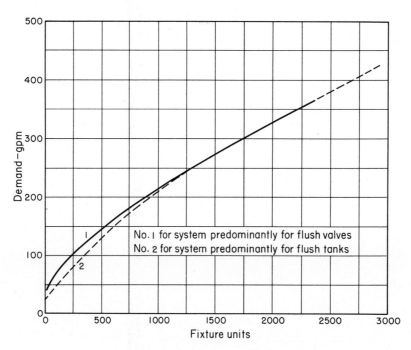

Chart I. Estimate curves for demand load

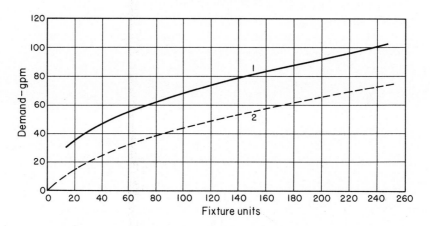

Chart 2. Enlarged scale demand load

FIG. 4.14. (See also Table 4.13). Conversion of fixture units to flow rate. [Reproduced by permission from "National Plumbing Code" (ASA 40.8—1955), American Society of Mechanical Engineers, 345 East 47th St., New York 17, N.Y.]

TABLE 4.14*
MINIMUM SIZE OF FIXTURE SUPPLY PIPE

Type of Fixture or Device	Pipe Size (inches)	Type of Fixture or Device	Pipe Size (inches)
Bath tubs	½	Shower (single head)	½
Combination sink and tray	½	Sinks (service, slop)	½
Drinking fountain	⅜	Sinks flushing rim	¾
Dishwasher (domestic)	½	Urinal (flush tank)	½
Kitchen sink, residential	½	Urinal (direct flush valve)	¾
Kitchen sink, commercial	¾	Water closet (tank type)	⅜
Lavatory	⅜	Water closet (flush valve type)	1
Laundry tray, 1, 2, or 3		Hose bibbs	½
compartments	½	Wall hydrant	½

For fixtures not listed, the minimum supply branch may be made the same as for a comparable fixture.

* Reproduced by permission: "National Plumbing Code" (ASA A40.8-1955), published by American Society of Mechanical Engineers, 345 East 47th St., New York 17, N.Y.

TABLE 4.15*
RATE OF FLOW AND REQUIRED PRESSURE DURING
FLOW FOR DIFFERENT FIXTURES

Fixture	Flow Pressure (a) psi	Flow Rate gpm
Ordinary basin faucet	8	3.0
Self-closing basin faucet	12	2.5
Sink faucet—⅜ in.	10	4.5
Sink faucet—½ in.	5	4.5
Bathtub faucet	5	6.0
Laundry tub cock—½ in.	5	5.0
Shower	12	5.0
Ball-cock for closet	15	3.0
Flush valve for closet	10–20	15–40 (b)
Flush valve for urinal	15	15.0
Garden hose, 50 ft and sill cock	30	5.0

(a) Flow pressure is the pressure in the pipe at entrance to the particular fixture considered.

(b) Wide range due to variation in design and type of flush-valve closets.

* Reproduced by permission: "National Plumbing Code" (ASA A40.8-1955), published by American Society of Mechanical Engineers, 345 East 47th St., New York 17, N.Y.

TABLE 4.16
FIXTURE UNITS PER FIXTURE OR GROUP*

Fixture Type	Fixture-Unit Value as Load Factors	Minimum-Size of Trap[2] Inches
1 bathroom group consisting of water closet, lavatory and bathtub or shower stall	Tank water closet 6 Flush-valve water closet 8	
Bathtub[1] (with or without overhead shower)	2	1½
Bidet	3	2
Combination sink-and-tray	3	Nominal 1½
Combination sink-and-tray with food-disposal unit	3 4	1½ Separate traps 1½
Dental unit or cuspidor	1	1¼
Dental lavatory	1	1¼
Drinking fountain	½	1
Dishwasher[2] domestic	2	1½
Floor drains[3]	1	2
Kitchen sink, domestic	2	1½
Kitchen sink, domestic, with food waste grinder	3	1½
Lavatory[4]	1	Small P.O. 1¼
Lavatory[4]	2	Large P.O. 1½
Lavatory, barber, beauty parlor	2	1½
Lavatory, surgeon's	2	1½
Laundry tray (1 or 2 compartments)	2	1½
Shower stall, domestic	2	2
Showers (group) per head[2]	3	
Sinks		
Surgeon's	3	1½
Flushing rim (with valve)	8	3
Service (Trap standard)	3	3
Service (P trap)	2	2
Pot, scullery, etc.[2]	4	1½
Urinal, pedestal, syphon jet, blowout	8	Nominal 3
Urinal, wall lip	4	1½
Urinal stall, washout	4	2
Urinal trough[2] (each 2-ft section)	2	1½
Wash sink[2] (circular or multiple), each set of faucets	2	Nominal 1½
Water closet, tank-operated	4	Nominal 3
Water closet, valve-operated	8	3

[1] A shower head over a bathtub does not increase the fixture value.

[2] For fixtures not listed or for devices with intermittent flows use the following data:

Fixture drain or trap size	Fixture-unit value
1¼ inches and smaller	1
1½ inches	2
2 inches	3
2½ inches	4
3 inches	5
4 inches	6

For a continuous or semicontinuous flow into a drainage system, such as from a pump, pump ejector, air-conditioning equipment, or similar device, two fixture units shall be allowed for each gallon-per-minute of flow.

[3] Size of floor drain shall be determined by the area of surface water to be drained.

[4] Lavatories with 1¼ or 1½ inch trap have the same load value; larger P.O. plugs have greater flow rate.

* Reproduced by permission: "National Plumbing Code" (ASA A40.8-1955), published by American Society of Mechanical Engineers, 345 East 47th St., New York 17, N.Y.

TABLE 4.17*
BUILDING DRAIN AND SEWER SIZE FROM FIXTURE UNITS
(See Table 4.16)

Diameter of Pipe	Maximum Number of Fixture Units That May Be Connected to Any Portion[1] of the Building Drain or the Building Sewer			
	Fall per Foot			
	$\frac{1}{16}$-Inch	$\frac{1}{8}$-Inch	$\frac{1}{4}$-Inch	$\frac{1}{2}$-Inch
Inches				
2			21	26
2½			24	31
3		20[2]	27[2]	36[2]
4		180	216	250
5		390	480	575
6		700	840	1,000
8	1,400	1,600	1,920	2,300
10	2,500	2,900	3,500	4,200
12	3,900	4,600	5,600	6,700
15	7,000	8,300	10,000	12,000

[1] Includes branches of the building drain.

[2] Not over two water closets.

* Reproduced by permission: "National Plumbing Code" (ASA A40.8-1955), published by American Society of Mechanical Engineers, 345 East 47th St., New York 17, N.Y.

TABLE 4.18*
HORIZONTAL DRAINAGE BRANCH AND WASTE STACK SIZES FROM FIXTURE UNITS[4]
(See Table 4.16)

Diameter of Pipe	Maximum Number of Fixture Units That May Be Connected To:			
	Any Horizontal[1] Fixture Branch	One Stack of 3 Stories in Height or 3 Intervals	More Than 3 Stories in Height	
			Total for Stack	Total at One Story or Branch Interval
Inches				
1¼	1	2	2	1
1½	3	4	8	2
2	6	10	24	6
2½	12	20	42	9
3	20[2]	30[3]	60[3]	16[2]
4	160	240	500	90
5	360	540	1,100	200
6	620	960	1,900	350
8	1,400	2,200	3,600	600
10	2,500	3,800	5,600	1,000
12	3,900	6,000	8,400	1,500
15	7,000	. . .	. . .	. . .

[1] Does not include branches of the building drain.

[2] Not over two water closets.

[3] Not over six water closets.

[4] Any structing with building drain should have at least one stock-vent carried full size through roof not less than 3″ in diameter or the size of building whichever is the lesser.

* Reproduced by permission: "National Plumbing Code" (ASA A40.8-1955), published by American Society of Mechanical Engineers, 345 East 47th St., New York 17, N.Y.

TABLE 4.19
SIZE AND LENGTH OF VENTS FOR BUILDING DRAINAGE SYSTEM*

Use of Soil or Waste Stack	Fixture Units Con-nected	Diameter of Vent Required (Inches) (a)								
		1¼	1½	2	2½	3	4	5	6	8
		Maximum Length of Vent (Feet) (b)								
Inches										
1¼	2	30								
1½	8	50	150							
1½	10	30	100							
2	12	30	75	200						
2	20	26	50	150						
2½	42	. . .	30	100	300					
3	10	. . .	30	100	200	600				
3	30	. . .	. . .	60	200	500				
3	60	. . .	. . .	50	80	400				
4	100	. . .	. . .	35	100	260	1000			
4	200	. . .	. . .	30	90	250	900			
4	500	. . .	. . .	20	70	180	700			
5	200	. . .	. . .	. . .	35	80	350	1000		
5	500	. . .	. . .	. . .	30	70	300	900		
5	1100	. . .	. . .	. . .	20	50	200	700		
6	350	. . .	. . .	. . .	25	50	200	400	1300	
6	620	. . .	. . .	. . .	15	30	125	300	1100	
6	960	. . .	. . .	. . .	. . .	24	100	250	1000	
6	1900	. . .	. . .	. . .	. . .	20	70	200	700	
8	600	. . .	. . .	. . .	. . .	. . .	50	150	500	1300
8	1400	. . .	. . .	. . .	. . .	. . .	40	100	400	1200
8	2200	. . .	. . .	. . .	. . .	. . .	30	80	350	1100
8	3600	. . .	. . .	. . .	. . .	. . .	25	60	250	800
10	1000	. . .	. . .	. . .	. . .	. . .	. . .	75	125	1000
10	2500	. . .	. . .	. . .	. . .	. . .	. . .	50	100	500
10	3800	. . .	. . .	. . .	. . .	. . .	. . .	30	80	350
10	5600	. . .	. . .	. . .	. . .	. . .	. . .	25	60	250

(a) Not less than 1¼ inches or one-half of diameter of drain to which connected.

(b) 20% of total length may be installed horizontally.

* Reproduced by permission: "National Plumbing Code," (ASA A40.8) published by American Society of Mechanical Engineers, 345 East 47th St., New York 17, N.Y.

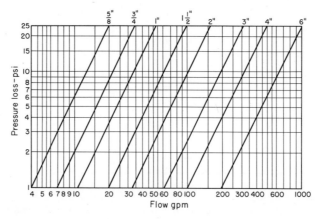

FIG. 4.15.   Pressure losses in disc-type water meters.   [Reproduced by permission from "National Plumbing Code" (ASA 40.8—1955), American Society of Mechanical Engineers, 345 East 47th St., New York 17, N.Y.]

EXAMPLE PROBLEMS

A number of example problems are solved in this section so that the principles already discussed and use of the design charts may be demonstrated. Every step in the calculations is included to avoid any possible ambiguity.

Example 1. Evaluation of Static Head. The static head is defined as the vertical distance in feet, expressed as ft-lb$_f$/lb$_m$, between the point of reference and some intermediate point or terminal point in the piping system. If the system terminates in a free level of liquid, that level would be the terminal point. If it terminates in a free discharge, the point of discharge would be the terminal point. These facts are illustrated in Fig. 4.16.

Note that in Cases A and B point 1 is below point 2. This means that the pressure at 2 will be less than the pressure at 1 by an amount equivalent to the static head *S*, plus, of course, friction losses between 1 and 2. In case C, however, the pressure at point 2 will be increased by

the amount *S* over that of point 1, less the friction losses. These facts are illustrated further in Example 2.

Example 2. Calculation of Pressure at Points in System. Figure 4.17 shows the layout of a system in which a light naphtha is pumped from tank *V*-1 to tower *T*-1 by pump *P*-1. The suction pressure and discharge pressure at the pump, points 1 and 2, respectively, are required. The following data apply:

Fluid: light naphtha, 86.2° API at 60°F, *K* = 11.8.*
Flow rate: 475 gpm measured at 60°F.
Pumping temperature: 106°F.
Piping: Schedule 40.
    Fittings are standard welding-type.
    Valves are flanged-type.

Calculation of Properties of Fluid at Pumping Temperature.
Specific gravity at 60/60°F = 0.650 (Fig. A.3, Appendix)
Specific gravity at 106°/60°F = 0.625 (Fig. A.1, Appendix)
Density at 106°F: (0.625)(62.4) = 39.0 lb/cu ft
Viscosity at 106°F: 0.17 centistokes (Fig. A.6*c*, Appendix) or (0.17)(0.625) = 0.106 centipoise
Flow at 106°F: (475)(0.65/0.625) = 494 gpm

Pressure at Point 1. Apply Eq. 3 between upstream point *V* and downstream point 1, or reason according to tabulation on page 117.

$$\frac{P_V - P_1}{\rho} = \frac{u_1^2 - u_V^2}{2g_c} + (Z_1 - Z_V)\frac{g}{g_c} + F$$

*K is the characterization constant for the naphtha. It is used in evaluating properties, as shown in the Appendix.

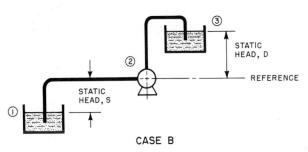

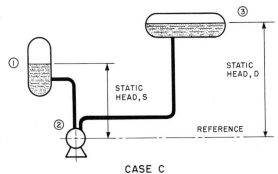

CASE A

CASE B

CASE C

FIG. 4.16. Static heads, Example 1.

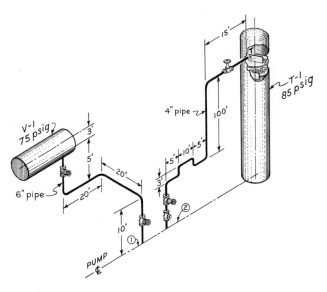

FIG. 4.17. Piping layout for Example 2.

or $\quad \dfrac{P_1}{\rho} = \dfrac{P_V}{\rho} + \dfrac{u_V^2}{2g_c} - \dfrac{u_1^2}{2g_c} + (Z_V - Z_1)\dfrac{g}{g_c} - F$

or $\quad P_1 = \left(\dfrac{P_1}{\rho}\right)\left(\dfrac{\rho}{144}\right) = \dfrac{(336.8)(39)}{(144)} = 91.2 \text{ psia}$

or $\qquad 91.2 - 14.7 = 76.5, \text{ psig.}$

Pressure at Point 2. Apply Eq. 3 between upstream point 2 and downstream point T, or reason according to tabulation on page 117.

$$\dfrac{P_2 - P_T}{\rho} = \dfrac{u_T^2 - u_2^2}{2g_c} + (Z_T - Z_2)\dfrac{g}{g_c} + F$$

or $\quad \dfrac{P_2}{\rho} = \dfrac{u_T^2 - u_2^2}{2g_c} + (Z_T - Z_2)\dfrac{g}{g_c} + F + \dfrac{P_T}{\rho}$

Head Term	Head in ft-lb$_f$/lb$_m$

A. Pressure head at starting

point V-1, $\dfrac{P_V}{\rho}$

75 psig + 14.7 psi = 89.7 psia

$\dfrac{(89.7)(144)}{39} =$ · · · · · · · · 321

B. Velocity head term, $\dfrac{u_V^2}{2g_c} - \dfrac{u_1^2}{2g_c}$

For 6-in. Schedule 40 pipe,
ID = 6.065 in. (Table 3.2)
Velocity at V-1 is negligible
because tank is large.
Velocity head at Point 1 is
0.5 ft-lb$_f$/lb$_m$ from Fig. 4.9.

Difference $\dfrac{u_V^2}{2g_c} - \dfrac{u_1^2}{2g_c} = 0 - 0.5 =$ · · · −0.5

(Or from tabulation on page 117, velocity head is higher at point 1 than at starting point, V. Therefore, pressure head decreases.)

C. Static head term, $(Z_V - Z_1)\dfrac{g}{g_c}$

$\dfrac{g}{g_c} = 1$

$(3 + 5 + 10) - 0 =$ · · · · · · · +18

(Downstream point 1 is lower than starting point V-1; therefore, pressure head increases.)

D. Friction term, $-F$ (see Fig. 4.8)

$\text{Re} = \dfrac{50.54 Q \rho}{d\mu_{\text{cp}}} = \dfrac{(50.54)(494)(39)}{(6.065)(0.106)} = 1,520,000$

Pipe and Fittings	Number or Length	K	Total
Straight Pipe	5 + 20 + 20 + 10 = 55 ft	3.2 per 100 ft (Fig. 4.8)	1.76*
Gate valves	2	0.1 (Table 4.2)	0.20
Exit from tank	1	0.5 (Table 4.4)	0.50
Ells	3	0.3 (Table 4.3)	0.90
		Total K	3.36

$*(3.2)\left(\dfrac{55}{100}\right) = 1.76.$

(Decrease in direction of flow)
Total friction loss = −(3.36) (velocity head from Fig 4.9) − (3.36)(0.5) = · · · −1.68

E. $\dfrac{P_1}{\rho}$, pressure head at point 1 =

$321 - 0.5 + 18 - 1.68 =$ · · · · +336.8

Head Term	Head in ft-lb$_f$/lb$_m$

A. Pressure head at starting

point T-1, $\dfrac{P_T}{\rho}$

$\dfrac{(85 + 14.7)(144)}{39} =$ · · · · · · 368

B. Velocity head term, $\dfrac{u_T^2}{2g_c} - \dfrac{u_2^2}{2g_c}$

For 4-in. Schedule 40 pipe,
ID = 4.026 in. (Table 3.2)
Velocity at T-1 is negligible
inside tower.
Velocity head at Point 2 is
2.1 ft-lb$_f$/lb$_m$ (Fig. 4.9)

$\dfrac{u_T^2}{2g_c} - \dfrac{u_2^2}{2g_c} =$ · · · · · · · −2.1

(Higher at point 2 than at starting point T. Therefore, pressure decreases.)

C. Static head term, $(Z_T - Z_2)\dfrac{g}{g_c}$

$10 + 100 =$ · · · · · · +110

(Lower at point 2 than at starting point T.)

D. Friction term, (see Fig. 4.8)

$Re \dfrac{50.54 Q \rho}{d\mu c_p} = \dfrac{(50.54)(494)(39)}{(4.026)(0.106)} = 2,300,000$

Pipe and Fittings	Number or Length	K	Total
Straight pipe	10 + 5 + 3 + 10 + 3 + 5 + 100 + 15 = 151	5.1 per 100 ft (Fig. 4.8)	7.7*
Check valves (swing)	1	2 (Table 4.2)	2.0
Gate valves	2	0.15 (Table 4.2)	0.3
Ells	7	0.3 (Table 4.3)	2.1
Entrance to tank	1	1.0 (Table 4.4)	1.0
		Total	13.1

$*(151)\left(\dfrac{5.1}{100}\right) = 7.7.$

Increase in direction of flow
(13.1)(velocity head from Fig. 4.9)
(13.1)(2.1) = · · · · · · +27.7

E. $\dfrac{P_2}{\rho}$, pressure head at Point 2 =

$$368 + 2.1 + 110 + 27.7 = \qquad +507.8$$

or $\quad P_2 = \dfrac{P_2}{\rho} \dfrac{\rho}{144} = \dfrac{(507.8)(39)}{144} = 137.5$ psia

or $\quad 137.5 - 14.7 = 122.8$ psia

Pressure Drop Calculations for Other Fluids

The following examples illustrate calculations of pressure drop due to friction for other fluids—gases, non-Newtonian, and two-phase. Static head, velocity head changes, and losses through fittings are calculated as in Example 1 and are thus not illustrated in this group.

Resistance coefficients for fittings given in Tables 4.2 to 4.4 are based on data for gases and liquids but can be used to estimate losses for two-phase flow through valves and fittings, if average densities are used in determining volume and velocity of flow.

Example 3. Calculation of Pressure Drop in Short Gas Lines. In a 2-in. Schedule 40 pipe, 400 lb/hr of a mixture of 80.2% methane, 10.3% ethane, 6.5% propane, 2.0% iso-butane, and 1.0% *n*-butane is flowing at 205°F and 1655 psig. Calculate the pressure drop due to friction in 100 ft of pipe.

Density of gas at 205°F and 1655 psig: 4.35 lb/cu ft (See Example A.2 in Appendix.)
Mean viscosity 0.018 cp (See Example A.5 in Appendix.)
From Fig. 4.8:

$$\text{Re} = \frac{(6.3)(W)}{(d)(\mu c_p)} = \frac{(6.3)(8700)}{(2.067)(0.0163)} = 1,630,000$$

ID 2-in. Schedule 40 pipe = 2.067 in. (Table 3.2)
$K_p = 11$ (from Fig. 4.8)
Flow rate: (8700 lb/hr)/(4.35 lb/cu ft) = 2000 cfh
Velocity head: 10 ft-lb$_f$/lb$_m$ (Fig. 4.9)
Head loss per 100 ft = (11)(10) = 110 ft-lb$_f$/lb$_m$

or $\dfrac{(110)(\rho)}{144} = \dfrac{(110)(4.35)}{144} = 3.32$ psi/100 ft

Example 4. Repeat Example 2 for a Line 12,000 ft Long. For this long line the assumption of average density is not valid and Weymonth's formula (Eq. 7) must be used.

Use inlet average *z* of 0.9 and average molecular weight, $M = 20.6$ (see calculation A.2 in Appendix).

$$p_1{}^2 - p_2{}^2 = \frac{16fLW^2zRT}{g_cM\pi^2D^5(144)^2} \quad \text{(Equation 7)}$$

where $p =$ psia
$\quad$ Re = 1,630,000
$\quad f = 0.0195$ (Figs. 4.1 and 4.2)
$\quad p_1 = 1670$ psia
$\quad (1670)^2 - (p_2{}^2) =$

$$\frac{(16)(0.0195)(12,000)\left(\dfrac{8700}{3600}\right)^2(0.9)(1544)(205+460)}{(32.17)(20.6)(3.14)^2\left(\dfrac{2.067}{12}\right)^5(144)^2}$$

$$p_2 = (2.79 \times 10^6 - 1.00 \times 10^6)^{1/2} = 1340\ \text{psia}$$

$$\cdots$$

$$\therefore \Delta p = 1670 - 1340 = 330\ \text{psi}/12{,}000\ \text{ft*}$$

Example 5. Non-Newtonian Fluid. In a 3-in. Schedule 40 pipe, 4% paper pulp is flowing at the rate of 300 gpm. Calculate the pressure drop due to friction for 100 ft of pipe. It is known that this fluid behaves in a non-Newtonian manner. The density of the pulp-water mixture is 64.0 lb/cu ft.

$$\text{Re} = N_{Re} = \frac{D^n u^{2-n}\rho}{\gamma} \quad \text{(Equation 8)}$$

From Table 4.1:
$$n = 0.575$$
$$\gamma = 6.13$$
From Fig. 4.9:
$$u = 13.5\ \text{ft/sec}$$
$$\text{velocity head} = 2.8\ \text{ft-lb}_f/\text{lb}_m$$

From Table 3.2: $D = \dfrac{3.068}{12} = 0.256$ ft

$$\text{Re} = \frac{(0.256)^{0.575}(13.5)^{1.425}(64)}{6.13} = 200$$

(If the Re would have been over 2100, Fig. 4.3 and Eq. 9 would have been used.)

$$K_\rho = \frac{76800}{(\text{Re})(D_i)}$$

$$K_p = \frac{76800}{(200)(3.067)} = 125 \qquad \text{(see notes on Fig. 4.8)}$$

$$F = (125)(2.8) = 350\ \text{ft-lb}_f/\text{lb}_m$$

or $\quad = \dfrac{(350)(64)}{144} = 155$ psi/100 ft

Example 6. Two-Phase (Gas-Liquid) Flow. Calculate the pressure drop due to friction for a mixture of gas and liquid flowing in a 100 ft length of 2-in. Schedule 40 pipe under the following conditions (see page 122):
Flow rate: 3800 lb/hr

Liquid volume fraction: 0.2 $\quad \dfrac{\text{volumes-liquid}}{\text{volume of mixture}}$

Liquid viscosity: 0.2 cp
Gas viscosity: 0.02 cp
Liquid density: 40 lb/cu ft
Gas density: 3 lb/cu ft

---

* The method of Example 2 would have given a high value of 398 psi. Greater accuracy can be obtained by repeating Example 4, using an average value of *z* based on a pressure of $\dfrac{1670 + 1340}{2}$.

Fictitious Reynolds number based on liquid viscosity

$$\mathrm{Re}_L^* = \frac{6.3\,W}{d\mu_{cp_L}} = \frac{(6.3)(3800)}{(2.067)(0.2)} = 56{,}400$$

$$\mathrm{Re}_G^* = \frac{6.3\,W}{d\mu_{cp_G}} = \frac{(6.3)(3800)}{(2.067)(0.02)} = 564{,}000$$

$K_{P_L} = 13$ from Fig. 4.8
$K_{P_G} = 11$ from Fig. 4.8

Volume of flow based on liquid density

$$\frac{3800}{40} = 95 \text{ cfh or } 0.095 \text{ thousands of cfh}$$

From Fig. 4.9 $J_t = 0.023$ ft-lb$_f$/lb$_m$

$$\Delta p_L^* = \frac{\rho_L}{144}(0.01 K_{PL} L + \Sigma K)J_t \qquad \text{(Equation 13A)}$$

$$\Delta p_L^* = \frac{40}{144}[(0.01)(100) + 0](0.023) = 0.083 \text{ psi}$$

$$\frac{\psi_G \rho_L}{\psi_L \rho_G} = \frac{(0.01)(11)(100)(40)}{(0.01)(13)(100)(3)} = 11.2$$

$$\frac{\Delta p_{TP}}{\Delta p_L^*} = 4.05 \qquad \text{(Fig. 4.10)}$$

$$\Delta p_{TP} = (0.083)(4.05) = 0.34 \text{ psi/100 ft}$$

This is the estimated two-phase pressure drop.

Example 7. Sizing Lines. The requirement is to pipe 38,745 lb/hr of 56.4° API crude oil at 60°F (characterization factor $K = 11.0$) to the suction of a pump. Size this line, using Schedule 40 pipe.

From Appendix A:

  Specific gravity = 0.753   (Table A.3)
  Density = (0.753)(62.4) = 47.0 lb/cu ft
  Viscosity = 0.43 centistokes   (Fig. A.6b)
  or (0.43)(0.753) = 0.324 centipoises

GPM:

  7.48 gallons/cu ft

$$\frac{(38745)(7.48)}{(47)(60)} = 103 \text{ gpm}$$

From Table 4.5, page 136, the recommended allowable pressure drop due to friction for pump suction lines is 0.5 ft-lb$_f$/lb$_m$ for 100 ft. Hence, it is necessary to determine a line size that will produce a pressure drop in this range. Assume a line size and then calculate the pressure drop or head loss due to friction. This process can be made easy by using Fig. 4.9, page 135, in conjunction with the suggested velocities in Table 4.5 for determining a reasonable preliminary line size. Thus the velocity should be in the range of

$$\frac{1}{3}\left(\frac{d}{2} + 4\right) \text{ ft/sec} = \frac{d}{6} + 1.3.$$

For 4-in. Schedule 40 line:

$$\frac{d}{6} + 1.3 = \frac{4.026}{6} + 1.3 = 1.97 \text{ ft/sec}$$

From Fig. 4.9 velocity = 2.6 ft/sec
Close; therefore, try 4-in. Schedule 40:
$J = 0.11$ ft-lb$_f$/lb$_m$ (Fig. 4.9)

$$\mathrm{Re} = \frac{(50.54)(103)(47)}{(4.026)(0.324)} = 188{,}000$$

$K_P = 5.6$
$F = (5.6)(0.11) = 0.615$ ft-lb$_f$/lb$_m$ for 100 ft

This is close enough to the design basis of 0.5. To go from 4 in. to 6 in. would increase cost while further reducing pressure drop. When the reduction in pressure drop must be compared with the increased cost of the pipe and fittings, a rough approximation of the costs can be obtained, as shown in the example accompanying Tables 4.6, 4.7, 4.8, and 4.9, pages 137–138.

Example 8. Sewer Design. A sewer is to be designed to handle a maximum rate of 480 gpm and a minimum of 150 gpm of industrial wastes. Determine the size and slope required. (See page 125.)

(a) Using a design basis of 3 ft/sec, select a sewer size for a sewer flowing full at maximum flow rate. From Table 4.12: 8-in. sewer of slope of 0.0076 ft/ft is selected.

(b) Now determine slope required for self-cleaning at minimum flow equal to that for a sewer flowing full at 3 ft/sec.

$$\frac{q_s}{Q} = \frac{150}{480} = 0.31$$

From Fig. 4.13 at $\dfrac{q_s}{Q} = 0.31$, $\dfrac{d_e}{D_e} = 0.36$

and at $\dfrac{d_e}{D_e} = 0.36$, $\dfrac{s}{S} = 1.3$

∴ slope required for equal self cleaning = $(1.3)(0.0076) = 0.01$ ft/ft

At $\dfrac{d_e}{D_e} = 0.36$, $\dfrac{v_s}{V} = 0.96$ or

$$v_s = (3)(0.96) = 2.9 \text{ ft/sec}$$

(c) Check velocity at new slope for maximum flow when full, using Fig. 4.12.

At $\dfrac{q_s}{Q} = 0.31$, $\dfrac{d_e}{D_e} = 0.39$

and at $\dfrac{d_e}{D_e} = 0.39$, $\dfrac{v}{V} = 0.89$

or $V = \dfrac{v_s}{0.89} = \dfrac{2.9}{0.89} = 3.26$ ft/sec

This does not exceed 5 ft/sec maximum. Hence final design is 8 in. sewer at 0.01 ft/ft slope (¼ in. per ft).

Example 9. Complex Piping Network. Size the lines in the illustrated loop (Fig. 4.18) for distributing water to major users in a plant. Minimum pressure required at A and B is 60 psig. (See page 126.)

Solution. The arrows indicate the initial assumed direction of flow.

For first assumption and design basis let

$$Q_1 = 1000 \text{ gpm}$$
$$Q_3 = 2000 \text{ gpm}$$
$$Q_2 = 0$$

Then size lines using 5 to 7 ft/sec as suggested in Table 4.5. From Fig. 4.9 use 10 in. for line 1, 12 in. for line 3, and 10 in. for line 2 to complete loop and allow for future expansion.

These preliminary sizes seem logical since the larger line is in the section of the loop that has a potential user at C. Based on these preliminary decisions, check the system for flow and pressures at points A and B.

From Eq. 20: $R = \dfrac{fL}{d^5 g_c}$

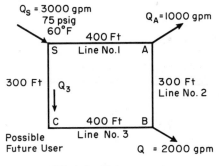

FIG. 4.18. Piping network.

Line No.	Line Size	ID Std. Wt.	Assumed Flow	Re	$f†$	R	RQ^2	$2RQ$
1	10″	10.02″	1000	273,500*	0.0163	2×10^{-6}‡	2§	4×10^{-3}
2	10″	10.02″	…	…	…	…	…	…
3	12″	12.0″	2000	455,000	0.0155	1.35×10^{-6}	5.4	5.40×10^{-3}

$$* Re = \frac{(50.54)(1000)(62.4)}{(10.02)(1.15)} = 273,500 \qquad ‡ \frac{(0.0163)(400)}{(10.02)^5(32.17)} = 2 \times 10^{-6}$$

$$†f \text{ from Fig. 4.1 and 4.2} \qquad § (2 \times 10^{-6})(1 \times 10^3)^2 = 2$$

$$\Sigma_a RQ'^2 = 2 + 0 - 5.4 = -3.4$$
$$\Sigma 2RQ' = 0.004 + 0.0054 = 0.0094$$

$$q = -\frac{-3.4}{0.0094} = 362$$

Hence, second assumption
$$Q_1 = 1000 + 362 = 1362$$
$$Q_2 = 362$$
$$Q_3 = 2000 - 362 = 1638$$

Line No.	Assumed Flow	R^*	RQ^2	$2RQ$
1	1362	1.9×10^{-6}	3.53	5.18×10^{-3}
2	362	1.65×10^{-6}	0.21	1.20×10^{-3}
3	1638	1.37×10^{-6}	3.68	4.48×10^{-3}

* R based on new flow rate

$$\Sigma_a RQ'^2 = 3.53 + 0.21 - 3.62 = 0.06$$
$$\Sigma 2RQ' = 0.00518 + 0.00120 + 0.00448$$
$$= 0.01086$$

$$q = \frac{-6.06}{0.01086} = -5.5 \text{ call } 6$$

Next assumption:
$$Q_1 = 1362 - 6 = 1356$$
$$Q_2 = 362 - 6 = 356$$
$$Q_3 = 1638 + 6 = 1644$$

For more accuracy, further trials can be made until the

q error is zero (R does not change rapidly with small changes in flow rate). The values just obtained, however, are satisfactory for most purposes.

Pressure at A = (pressure at S) − (loss in head expressed as psi)

$$75 - R_1 Q_1^2 \frac{\rho}{144} = 75 - (1.9 \times 10^{-6})(1356)^2 \frac{62.4}{144}$$

$$= 75 - 1.5 = 73.5 \text{ psig}$$

Pressure at $B = 73.5 - R_2 Q_2^2 \dfrac{\rho}{144} = 73.5 - 0.1 =$

73.4 psig. Hence pressure at A and B above minimum allowable. Smaller lines could be selected, especially between A and B. Final design depends on economic factors and anticipated expansion in service.

REFERENCES

1. Allen, W. F., Jr., *Trans. ASME* **73**, 257 (1951).
2. Alves, G. E., D. F. Boucher, and R. L. Pigford, *Chem. Eng. Progr.* **48**, 385 (1952).
3. Bertuzzi, A. F., M. R. Tek, and F. H. Poettmann, *J. Petroleum Technol.* (1956).
4. Caldwell, D. H., and H. E. Babbitt, *Trans. Am. Inst. Chem. Engrs.* **33**, 237 (1941).

5. Chenoweth, J. M., and M. W. Martin, "A Pressure Drop Correlation for Turbulent Two-Phase Flow of Gas-Liquid Mixtures in Horizontal Pipes," Presented at ASME Petroleum Mechanical Engineering Conference, New Orleans (1955).

6. Crane Co. "Flow of Fluids" Tech. Paper No. 410 (1957).

6A. Dodge, D. W., and A. B. Metzner, *AIChE Journal* **5,** 189 (1959).

7. Durand, R., and E. Condolios, *Compte Rendudes deuxieme Journees de l'Hydralique,* June 1952.

8. Fair, G. M., and J. C. Geyer, *Water Supply and Waste-Water Disposal,* John Wiley and Sons, New York, 1954.

9. Hoffman, A., *Trans. Munich Hydraulic Inst.,* Bulletin 3, translated by ASME (1945).

10. Hydraulic Institute, *Pipe Friction Manual,* New York (1954).

11. Kittredg, C. P., and D. S. Rowley, *Trans. ASME* **79,** 1759 (1957).

12. Lockhart, R. W., and R. C. Martinelli, *Chem. Eng. Progr.* **45,** 39 (1949).

13. McKnown, J. S., *Trans. Am. Soc. Civil Engrs.* **119,** 1103 (1954).

14. Metzner, A. B., *Advances in Chemical Engineering,* 77–153, Academic Press, New York, 1956.

15. Metzner, A. B., and N. C. Reed, *AIChE Journal* **1,** 434 (1955).

16. Moody, L. F., *Trans. ASME* **66,** 671 (1944).

17. Newitt, D. M., J. F. Richardson, M. Abbott, and R. B. Turtle, *Trans. Inst. Chem. Engrs.* (London) **33,** 93 (1955).

18. Pigott, R. J. S., *Trans. ASME* **79,** 1767 (1957).

19. Smith, R. A., *Trans. Inst. Chem. Engrs.* (London) **33,** 85 (1955).

20. Spells, K. E., *Trans. Inst. Chem. Engrs.* (London) **33,** 79 (1955).

21. Tube Turns Corp., *Piping Engineering,* Part 3.01, Louisville, Kentucky, 1951.

22. Vogt, E. G., and R. R. White, *Ind. Eng. Chem.* **40,** 1731 (1948).

23. Walker, J. E., G. A. Whan, and R. R. Rothfus, *AIChE Journal* **3,** 484 (1957).

5

Piping specifications

Example Specifications—page 154

Piping specifications and piping drawings supply all the data needed for the accurate procurement and proper erection of a piping system. This chapter is designed as a guide in preparing the specifications—an indispensable member of this combination. The guide is arranged according to the order in which specifications should be prepared. By following the steps as outlined and using the suggested forms shown, it should be possible to prepare a good set of piping specifications with minimum effort.

DETERMINING TYPE AND NUMBER OF SPECIFICATIONS

The first step in preparing specifications is to list the fluids being handled and ranges of operating conditions for each fluid. Study of this list will reveal certain logical categories into which the specifications can be divided on the basis of operating pressure, temperature, and materials of construction. Great care and wisdom should be used to select the number of such categories. An excessive number will make design construction and maintenance difficult and costly, while too few will result in excessive first cost. A single specification designed to be suitable for all piping in a plant must be designed for the most severe conditions. In such a case, all other piping will be overdesigned.

It is convenient at this stage to prepare a list of fluids (Table 5.1) to be used in conjunction with major piping materials for developing an outline of the specifications (Table 5.2). By studying each material and the corresponding fluids it is possible to develop logical groupings of fluids into a minimum number of specifications.

The reasoning used in categorizing the example fluids is shown in Table 5.2. Although other materials are also important the flanges, valves, and piping are the keys to the division of specifications. It is advantageous to consider the more severe services first. Flange ratings are determined from Table 3.33 and piping thicknesses calculated as shown on page 43.

PREPARING DETAILED SPECIFICATIONS

After an outline of the specifications has been approved, detailed specifications can be prepared. These include general notes outlining the desired piping practices. The general notes are then followed by the detailed piping specifications in which all pipe, fittings, valves, and accessories for each specification are described. Since valves vary in design it has become common practice to use a manufacturer's name in the valve description and then include an "or equal" phrase to indicate willingness to consider competitive valves. The tables in Chapter 2 will aid in selecting many of the materials, but familiarity with manufacturer's literature and all new piping developments is essential in writing good specifications.

The specifications which follow are based on the example already considered. Manufacturer's names are, of course, omitted. This example can serve as a guide in preparing specifications. Editorial comments in brackets serve to point out major issues which must be considered. This same specification can also be prepared in an abbreviated tabular form as shown in Table 5.3 for Specification "M" only. Tabulations are easier to use since all the data appear on one page. The ex-

TABLE 5.1
SUMMARY OF FLUIDS FOR EXAMPLE PIPING SPECIFICATION

Fluid designation	Service and fluid	Operating pressure, psig	temperature, °F
HC	Hydrocarbon #1* (heater/crossovers)	250	800 Maximum
HB	Hydrocarbon #2*	250	800
HA	Hydrocarbon #3*	250	750 Maximum
H	Hydrocarbon #4*	100	600
S	Steam	275	410
SE	Steam, exhaust	40	290
SC	Steam, condensate	15	240
A	Air, utility	100	Ambient
AC	Air, instrument, dried	50	Ambient
F	Fuel, natural gas, dry	50	Ambient
C	Caustic	50	Ambient
IN	Instrument piping	Same as line services	
WF	Water, fire protection	100	Ambient
WC	Water, process cooling	75	85
W	Water, utility	75	Ambient
WS	Water, sanitary	40	Ambient

* Allow 0.10″ corrosion allowance for these services.
Reproduced from: H. F. Rase and M. H. Barrow, *Project Engineering of Process Plants,* John Wiley and Sons, New York, 1957.

ample specifications which follow should not be construed as final recommendations for the type of services indicated. It is certainly possible that more economical specifications can be written. Those included in this example are given to illustrate methods for developing and presenting specifications.

Typical Piping Design Specifications

[Editorial comments are given within brackets.]

1.000. GENERAL

1.100. *Basis of Design.* The design, fabrication, and erection of all piping and accessories shall conform to practices specified in the Code for Pressure Piping ASA B31.1, or latest issue thereof, and to the drawings and the following specifications. All references to the code herein shall be understood to refer to the above ASA Code. [Definite reference to the code eliminates lengthy descriptions of design and fabrication methods.]

1.200. *Definitions.* The terminology used throughout this specification and on the drawings will be in general that conforming to current trade practices.

Anchor. Point where piping is fixed is called the anchor point. The device attached to pipe or the complete structure if it serves the single purpose of fixing the pipe is called an anchor.

Bleed. A small valve provided for drawing off liquids.

Blind. A plate sized to be inserted in a flange to isolate a portion of a system.

Block Valve. A valve furnished to shut off a separate system. The term is loosely used and may be understood to mean any valve which is used for shutoff service rather than for throttling.

B.O.P. Bottom of outside of pipe. Used for pipe support location.

Control Valve. Any one of a number of different types of valves, remotely operated from some type of instrument.

Directions. The Plant North direction shown on plot plans and other drawings will serve for orientation of equipment.

Drip-Leg. A vertical section of pipe located in horizontal piping to deflect and "catch" condensate.

F.W. Field Weld.

Gradient. The successive drop in elevation of piping to insure gravity flow and drainage.

Guide. Device controlling the direction of piping movement.

Hanger. A rod and clamp, a chain, or a spring device used for supporting pipe is called a hanger.

Line. A pipe run from one point to another. The Line Number is a number and symbol appearing on the piping drawings which identifies the pipe according to size, process fluid, general location, and specification.

P.E. Indicates plain-end. Used to differentiate between threaded-end pipe and mill cut. A procurement term.

P.S. The designation P.S. is used to indicate Pipe Support, which support may be one of the above anchors, guides, or shoes or several combined to form the support.

Random Pipe. Pipe furnished in random, not cut, lengths.*

* Carbon steel pipe is never bought in *exact* lengths. The vendor offers his pipe at a price per hundred ft. When pipe is shipped and when received, it is measured. Vendor submits invoices according to the length shipped.

Random lengths of pipe are usually 18 to 20 ft long. Double random is double this length, and is available in certain sizes and is requested when 40-ft sections would be advantageous for straight runs of pipe.

Shoe. Device welded to or clamped to a pipe which provides a bearing.

S.P. Sample point. A small valve with necessary piping to draw samples.

Spec. The line specification.

Spool. A short piece of pipe provided with flanges. (The term is also used to indicate pieces of fabricated pipe, but a piece of fabricated pipe does not necessarily always have flanges on each end. Lists of spool numbers are called spool-sheets. These are often used as notification to a fabricator that he may proceed with fabrication, i.e., receipt of the spool sheet releases all the listed spools.)

Spool Number. A number identifying a piece of fabricated pipe.

S.T. Steam trap. Usually schematically indicated, but un-derstood to mean the trap and necessary valving and piping for by-passing in emergencies.

Strainer. A perforated metal sheet temporarily placed in a flanged joint to protect equipment from foreign matter in initial operation. Permanent strainers shall also be furnished where indicated.

1.300. *Line Numbering System.* The following numbering sys-tem shall be used as line identification on piping appearing on flow sheets and piping drawings. The complete line number shall consist of the following components, as illustrated: 4-HA-101-M. The first numeral is the nominal line size. The HA designates the service; the 100, the area number; and the M designates the piping specification applicable.

Line numbers shall change: (1) when size changes; (2) when

TABLE 5.2
SELECTING THE NUMBER AND TYPE OF SPECIFICATIONS

Specification	Service	Flanges	Pipe	Valves	Reasons for Category
M	SE, SC, A, H, F, C	150 lb ASA	2″ and below: Sched. 80 ASTM A53 Gr A or ASTM A83 3″ to 10″: Sched. 40 ASTM A53 Gr A	Cast steel 150 lb flanged valves and forged steel screwed valves	These services fall into 150 lb ASA class and require all steel materials, no cast iron (see Code). ASTM A53 pipe is satisfactory for services below 900°F. A83 is indicated as an alternate in case A53 is not available in small sizes.
N	S, HA, HB, HC	300 lb ASA	2″ and below: Sched. 80 ASTM A106 Gr A 3″ to 10″: Sched. 40 ASTM A106 Gr A 2″ and larger: 0.375″ wall (0.375″ wall commonly available in these large sizes)	Cast steel 300 lb flanged valves and forged steel screwed valves	Another "all steel" spec. requiring 300 lb flange rating. ASTM A106 Gr A is used here as a safety factor since operating temperatures of 800° are rather close to the limit of 900°F. for ASTM A53.
P	WF, W, WC, WS	150 lb ASA	Same as in N for above ground. Use cast iron, below ground and ASTM A120 galvanized for WS (drinking water)	Cast iron flanged valves and brass screwed valves	Cast iron can be used for valves in water service. In itself this economy justifies a separate specification for water. In addition, underground pipe is cast iron, and galvanized pipe must also be specified. 150 lb ASA steel flanges are used so that piping may be welded.
Q	AC	125 lb ASA Cast iron	Same as above ground pipe for N	Same as in P	These lines will be small and the cheaper cast iron screwed flanges will be used. Also only steel pipe is used for air. These two facts along with other minor details justify a separate spec.
Z	IN	Same as line service	Same as line service-Copper tubing for air transmission	Same as line service	Instrument lines require special notes and the air transmission lines are tubing thus justifying a separate spec.

TABLE 5.3
TABULAR-TYPE PIPING SPECIFICATION

(Example shows one specification only. Other specifications would be tabulated in line.)

Spec.	Service	ASA Rating	Pipe	Fittings 1½" and smaller	Fittings 2" and above	Flanges	Bolting*	Gaskets
M	A 100 lb air C 50 lb caustic F 50 lb fuel H 150 lb hydrocarbon SC 15 lb condensate SE 40 lb exhaust	230 psi at 100°F 150 psi at 500°F 100 psi at 750°F (maximum)	ASTM A53 Grade A 2" and smaller: schedule 80 (may be ASTM A106 or A83 if A53 is unavailable) 3" through 10": Schedule 40	Screwed, 3000 lb CWP forged carbon steel Unions: 2000 lb CWP forged carbon steel, ground joint, integral steel seats	Buttwelding type, seamless carbon steel, ASTM A106 and A234 Grade A Flanged fittings (do not ordinarily use): 150 lb ASA $\frac{1}{16}$" raised face	150 lb ASA welding neck $\frac{1}{16}$" raised face ASTM A181, Class I	Service H: Spec. 1 Services A, C, F, SC, and SE: Spec. 2	Asbestos filled double-jacketed corrugated iron, except for services A and SC use sheet asbestos ring gaskets

* Bolting Specs.

Spec. 1: ASTM A193, Grade B7, Class 7 fit for bolt; Class 2B fit for nuts, with two (2) heavy series hexagonal nuts each. Nuts shall conform to ASTM A194, Class 2H, oil quenched, not forged.

Spec. 2: ASTM A307, Class 2A fit for bolts and 2B fit for nuts.

Spec. 3: Bolting for cast-iron pipe shall conform to ASA A21.11 requirements.

TABLE 5.3 (*Concluded*)

(Example shows one specification only. Other specifications would be tabulated in line.)

Screwed Valves (1½" and under)				Flanged Valves (2" and above)			
Cock (plug)	Gate	Globe	Check	Cock	Gate	Globe	Check
Lubricated ¼": 2000-lb WOG forged carbon steel, ½" to 1½": 300 lb WP cast carbon steel, wrench operated	Services C, F, H and SE, 1½" and smaller: 600 lb SWP forged carbon steel, inside screw stem, union bonnet, 11–13% chrome stainless steel trim For services A and SC, 1½" and smaller: 150 lb SWP brass, rising stem, inside screw	Services C, F, H and SE, 1½" and smaller: 600 lb SWP forged carbon steel, inside screw stem union bonnet, 11–13% chrome stainless steel trim Services A and SC, 1½" and smaller: 250 lb SWP brass rising stem, union bonnet	Services C, F, H and SE, 1½" and smaller: 600 lb SWP carbon steel, horizontal, piston type, 11–13% chrome stainless steel trim Services A and SC, 1½" and smaller: 250 lb. SWP brass, horizontal piston type, nickel alloy seat and disc	Lubricated 2" to 6": 150 lb ASA cast carbon steel, wrench operated	Services A, C, F, H and SE, 2" and larger: 150 lb ASA cast carbon steel, $\frac{1}{16}$" RF, OS & Y 11–13% Chrome stainless steel trim Service SC, 2" and larger: 125 lb ASA Cast iron OS&Y, brass trim and stem	Services A, C, F, H and SE, 2" to 6": (use gate-valves above 6") 150 lb ASA cast carbon steel, $\frac{1}{16}$" RF, OS&Y 11–13% chrome stainless steel trim Service SC, 2" and larger: use gate valves	Services A, C, F, H and SE, 2" and larger: 150 lb ASA cast carbon steel, $\frac{1}{16}$" RF, swing-type 11–13% chrome stainless steel trim Service SC, 2" and larger: 125 lb ASA cast iron, swing-type, brass trim

specification changes; (3) when branches leave original line; (4) wherever an additional line number would simplify the engineering and drafting. [Line numbering affords an efficient means for locating lines and referring each line to a specific specification. Thus long design notes are not required on the drawings.]

1.400. *Line Sizes.* Piping shall be sized in accordance with best engineering practice for pressure drops consistent with proper operation of equipment. In general the following limitation shall apply:

1.401. Sizes ⅜ in., 1¼ in., 2½ in., 3½ in., 4½ in., 5 in., and 7 in. shall not be used. [Sometimes used in naval architecture.]

1.402. Drains, sample, and vent lines shall be ¾ in. minimum unless otherwise noted. Drains for hot piping shall be not less than 1½-in. nominal size.

1.403. Except for instrument transmission lines, piping run on support shall not be smaller than 2 in. Smaller sizes may be used for short runs less than one support space in length in special cases.

1.404. Underground lines beyond process area limits shall be 3 in. or larger. Sanitary water may be 2 in. minimum. Certain utility lines may be smaller, but must be considered as special cases and approval secured. [This is done to insure adequate future capacity without costly excavation for replacement.]

1.500. *General Notes on Piping Design*

1.501. *Aboveground Lines.* All process and utility lines within the process area and all yard lines outside from the process area except water lines, sewer lines, and certain designated pump and compressor lines shall be run aboveground on concrete sleepers, or overhead on structural steel supports. [Underground process lines cause excessive maintenance problems.]

1.502. *Arrangement of Aboveground Lines.* Aboveground lines running in the same plane shall be arranged so that the bottoms of the lines, exclusive of insulation, are at the same elevation. Sufficient height must be allowed to clear the insulation of insulated lines, and this height shall govern the elevation of uninsulated lines as well. [Simplifies design of supports.]

1.503. *Elevation of Aboveground Lines.* When piping is run in parallel groups, one elevation shall be selected for lines running to Plant North and South and another elevation for lines running East and West. [Plant North is an arbitrary direction set on Plot Plans for orientation purposes in all work.] If piping is run in "banks" or layers on pipe supports, the clear dimension between bottom of pipe in one layer and top-of-pipe in adjacent layer shall be three times the nominal diameter of the largest pipe in either group, plus twice the extension of a flange beyond the outside diameter of the largest pipe in either adjacent layer. Lines shall ordinarily change elevation when they change direction, except at the discretion of the designer in special cases and for aboveground lines not running in a bank of piping. The dimension of clearance shall be the same as that given above. The largest pipe size in the bank shall always govern for clearances. [Any other arrangement would result in chaos.]

1.504. *Elevation of Underground Lines.* Changes in direction of all underground lines, except sewers and drains, shall be accompanied by a change in elevation, except in the case of lines 24 in. and over. All underground piping shall be arranged as far as possible for free draining to some low point. Access shall be provided at drainage points.

1.505. *Location of Valves, Use of Chain Operators.* Diaphragm control valves, motor-operated valves, special manual control valves, and lubricated plug cocks shall be located so that they may be readily accessible from platforms, when elevated, or at grade, when possible.

The same will apply, in general, for block valves, check valves, etc., unless adjustment of the valve is not required during operation, in which case extension stems or chain operators shall be used. Seven feet three inches (7'3") is the maximum distance the center of valve wheel may be located above the operating level without the use of a chain operator or extension. Chains shall extend to within three feet (3') of the platforms, from which the valve will be operated.

Use of chain-operated valves, however, shall be kept to a minimum. Valves requiring extensions, stems, or chain operators shall be noted on drawings. Reach-rods or extension stems, field fabricated, shall be provided for valves 1½ in. and below rather than chain wheels. [All valves must be readily operable.]

1.506. *Overhead Clearance.* Overhead lines running from vessels or other elevated equipment to steel pipe supports or to concrete sleepers shall have a minimum clearance of 25'0" above plant roadways, and railroad (rails). Nine feet (9'0") shall be the minimum piping clearance above walkways and platforms and above grade in the immediate process area. [Facilitates movement of machinery and equipment.]

1.507. *Bent Pipe.* Pipe shall not be bent to a mean radius of less than five (5) times the nominal pipe size. Exceptions, when necessary, shall be accomplished by means of short-radius welded fittings. Such special fittings shall be noted on arrangement drawings. A minimum straight run equivalent to two or more pipe diameters should be allowed between two adjacent pipe bends wherever possible. Carbon-steel pipe 4 in. and below may be cold bent on bending machines. Alloy pipe or other pipe subject to cold working stresses shall be considered as special and shall be bent according to specified procedures.

1.508. *Angle Valves.* The use of angle valves shall be avoided.

1.509. *Drains, Vents, and Test Openings.* Drains shall be provided at low points and vents at high points. The drain and/or vent shall consist of a 6000-lb forged-steel pipe coupling welded into the line, a short schedule 160 nipple, and a ¾-in. gate valve in accordance with the line specification. In hot process lines the drain or vent shall not be less than 1½-in. nominal pipe size. For these connections a hexagon-shaped threaded steel plug* shall be made-up into the valve. All such plugs shall extend at least 3 in. beyond threaded portion.

Test connections or threaded connections for future use in piping or vessels shall be fabricated the same as drains and vents, except valve may be omitted and connection plugged. Shipping plugs shall be removed from such connections and replaced with hexagon steel plugs as noted above. Plugs for insulated equipment shall be fabricated to extend beyond insulation.

---

*The plug consists of a piece of carbon-steel hexagon barstock threaded with ordinary pipe threads on one end. So-called threaded bull-plugs often crush during tightening, damaging the pipe coupling or the threads.

1.510. *Service Connections and Utilities.* Three-quarter-inch (¾-in.) hose (manufactured, heavy) connections for steam, air, and water shall be provided at convenient points in the process area for general utility purposes. Utility lines at vertical vessels are to be run adjacent to vessels and outside the insulation but not outside the tower platforms.

1.511. *Connections for Pressure Gages and Pressure Instruments.* Connections for pressure instruments, to indicate or record pump discharge pressures, shall be located in the discharge piping between the pump flange, near the pump, and the first valve. Other points shall be noted on drawings specifically. All such pressure-indicating or pressure-recording points shall be accessible. Test points may be accessible by ladder. Bleed-off valving for pressure instrumentation on hot lines shall be provided with a double block (two valves) and bleed valve located between the two blocks. Instrument bleed sizes need not follow drain and vent sizing, but any bleed piping for hot services shall be not less than ½-in. nominal size and shall be arranged so that a rod may be driven up the valve and fitting. Bleed piping for hot fluids which may ignite on exposure to air should be supplied with a simple pipe coil cooler with water jacket. Bleeds from overhead lines and from gage glasses shall be piped to grade or nearest open drain, except that the end of such bleed piping shall be visible from the primary or secondary bleed valve.

Valving arrangements for instrument connections will be shown on instrument connection detail drawings.

1.512. *Thermo-wells, Test Wells, Orifice Flanges, Miscellaneous Instruments.* All temperature instrument points and orifice flange locations shall be noted on piping drawings. The necessary fittings and arrangement shall be shown on instrument schedule and instrument installation drawings. All such installations shall be field-fabricated, unless otherwise noted. Orifice flanges shall be located in the exact position shown on drawings. Thermo-wells should be accessible when practical.

1.513. *Steam Traps.* Typical and special steam trap assemblies shall be shown in detail on "Steam-Trap Schedule Drawing" and identified by the trap number. The steam-trap location will be shown on piping drawings by symbol and number.

1.514. *Expansion Loops.* If stresses produced by expansion or contraction of piping cannot be reduced to the allowable limits of stress as defined by the Code, by changes in direction or elevation, expansion loops shall be provided. Such loops, for piping 6 in. and above, shall be fabricated from welding fittings. Cold spring shall not be used.

All expansion loops shall in general be in a vertical plane, rather than horizontal. Loops may be hung downward from supports, and completely fabricated from welding fittings, i.e., two 90° weld ells—one 180° weld fitting, or four 90° long-radius weld ells. Expansion loops shall be designed for a maximum stress of only 75% of that allowed by the code under the service.

1.515. *Piping at Pumps, Compressors, and Turbines*

1.5151. *Pumps.* Check valves shall be provided in discharge lines of all centrifugal pumps upstream of first block valve. [Prevents backing-up of fluid through pump when pump is not operating.]

1.51511. Temporary strainers shall be fabricated from machine-perforated sheet metal not less than 1/16 in. thick for piping up to 8-in. nominal size and not less than ⅛ in. thick for all sizes above 8 in.

1.51512. Pumps and piping shall be so arranged that minimum clearance between projections of adjacent pumps or machines shall be 2'6". The minimum operating aisle, which shall be at the driver end, shall be 5'0". Adequate clearance shall be provided for the withdrawal of reciprocating parts or parts which require horizontal withdrawals.

1.51513. Suction and discharge piping shall be arranged in accordance with the latest revision of the "Standards of the Hydraulic Institute," 122 East 42nd St., New York, N.Y. Water pumps shall be installed with flooded suctions wherever possible. Otherwise, foot valves and ejector or injector systems shall be provided.

1.51514. Hot-piping connections into equipment such as pumps and turbines shall be arranged and anchored to eliminate all possible piping stress being carried into machine.

1.51515. Piping connections to equipment shall not be run horizontally over or across equipment. Equipment must be left clear for removal by crane or other maintenance equipment. See 1.603.

1.51516. Lubricating system, gland-seal, and small water-cooling piping shall be shown in detail on drawings to avoid possible connection errors.

1.5152. *Compressors.* Compressor piping shall be arranged to avoid excessive or cyclic vibration. See 1.51516 for lubricating oil and water systems. [Compressor manufacturers furnish piping recommendations.]

1.5153. *Turbines.* Mechanical expansion joints shall be provided in all exhaust-steam and live-steam lines connecting to equipment. See 1.51516 for lubricating oil and water systems.

1.516. *Steam, Exhaust and Condensate Lines.* All secondary or saturated-steam piping and exhaust-steam piping shall be run with a gradient. Drip-legs with steam traps shall be provided in this piping at all junction points and other points considered necessary.

Steam traps shall be provided in all vertical expansion loops. The discharge from all steam traps shall be run to the nearest condensate return header, except as noted on drawings.

Main steam distribution headers shall be run with a gradient, and drip-legs shall be provided for condensate removal.

Steam connections from steam headers to equipment drivers shall be provided with a gate valve located in a horizontal run at the header, and a globe valve adjacent to the driver. This valving shall be in addition to any control valves. Exhaust steam connections at the driver shall be provided with a gate valve adjacent to the driver. No block valve shall be provided at exhaust steam headers. ½-in. minimum nominal pipe size condensate blow-off connections shall be provided in both steam and exhaust connections at equipment and run to the nearest open box drain. Such lines shall not be submerged in liquid or run to bottom of the drain box.

1.600. *Fabrication and Fittings*

1.601. *Shop- and Field-Fabricated Pipe.* All pipe 4 in. and above shall be shop-fabricated, utilizing welding fittings. [On foreign projects it is often economical to shop-fabricate (weld) pipe down to 2-in. size to save U.S. expatriate labor.] All 3-in. piping shall be field fabricated utilizing welding fittings. All 2-in. piping (2½ in. when used) and smaller shall be field-fabricated, utilizing screwed fittings. In certain cases for 2-in. piping only, welded joints may be more practical than screwed. Where welding is to be utilized it shall be clearly noted (for 2-in. piping only) on drawings. All galvanized piping shall be run screwed. No welding shall be performed on galvanized piping.

1.602. *Changes in Direction*

1.6021. Changes in direction of all screwed piping shall be made with pipe bends where possible (see 1.507); otherwise, use screwed fittings, or as noted above in 1.601.

1.6022. Changes in direction of welded piping shall ordinarily be made with seamless buttwelding elbows. Bent pipe may be used, depending on the specification, the service, or the application. See 1.507. Pipe bends in piping 6 in. and above shall be considered as special cases and must be authorized.

1.603. *Flanged Fittings and Spools.* Flanged fittings shall be kept to an absolute minimum. Short flanged spools (see 1.200) for piping 4 in. and above shall be provided at each piece of equipment for removal purposes. [Makes removal of equipment possible with only minor work on piping.] At pumps and turbines the spool may be eliminated, if valving or expansion joints provide removable pieces of piping. See 1.51515.

1.604. *Reducers.* Changes in size of screwed lines shall be effected with forged-steel (only) screwed reducing fittings, except at screwed control valves where swaged nipples (seamless only) may be used. For flanged or welded lines, seamless buttwelding reducers shall be used. The use of reducing screwed tees shall be limited to side outlet reducing only.

1.605. *Branch Connectors.* Branch connections in shop-fabricated piping may be made without the use of buttwelding fittings when properly reinforced according to code requirements. Buttwelding fittings shall be used for branch connections in field fabricated welded lines. Flanges shall be provided for the reduced connection near the branch.

Manifold connections may be made by means of 6000-lb forged-steel couplings welded into the pipe. All such connections must conform to code practice. Large-size (above screwed) piping may be welded directly into headers for manifolds. Thread-end nipples welded into pipe bosses shall not be used. [The exposed threads on nipples are easily damaged.]

1.606. *Maximum Length (Shipping).* No shop-fabricated integral piece of pipe shall exceed a maximum length of 40'0'', nor a width or height exceeding the limits of a nine-foot (9') rectangle, regardless of whether this limitation is indicated on the drawings. Field welds shall be understood to be required whenever the pieces exceed these limits. [These are typical. Dimensions are governed by shipping regulations.]

1.607. *Connection Flanges and Blind Flanges.* Flanges shall be supplied in piping outside the process area limits only at selected points and shall be omitted or kept to a minimum. Flanges in steam-piping systems particularly shall be kept to a minimum. All piping run as headers shall be terminated by a blind flange.

1.608. *Buttwelding Elbows.* Buttwelding 90° elbows shall be long-radius type [creates less turbulence]. Exceptions, where necessary, shall be noted on arrangement drawings.

1.700. *Testing*

1.701. *Field Testing.* After erection but before application of insulation, all lines shall be given a hydrostatic test as follows:

1.7011. Each section of piping shall be tested hydrostatically at a pressure equal to two (2) times the lowest primary service pressure rating of the fittings, valves, and flanges in the line, but in no case less than one and one-half times the actual normal working pressure. [In certain hydrocarbon and chemical processes all water must be removed from any part of the process system before operation. Such processes cannot be tested with water since the complete removal of the water from the system could not be effected without disassembly, thus nullifying the test. Some other acceptable process fluid must be used for testing. Air may not be used since it will also leave water or lubrication oil as a contaminant. Such requirements must be developed early so that the fluid will be available.]

1.7012. The test pressure must not be greater than one and one-half times the maximum allowable working pressure for the pipe or as determined in accordance with testing requirements in the Code.

1.7013. The test pressure must not be greater than the maximum allowable pressure for the weakest piece of equipment installed in the line and included in the hydrostatic test. In general, major equipment such as vessels, exchangers, pumps, and compressors shall be isolated from pipeline hydrostatic test. When necessary for practicability, exchangers and vessels may be included with the connected piping, provided the piping test pressures are within the allowable cold pressure limits of the equipment.

1.7014. The lowest hydrostatic test pressure shall be 100 psig.

1.7015. The following equipment shall be excluded from the general hydrostatic test, but shall be tested with compressed air (and soap suds) at pressure equal to the allowable working pressure of the equipment and lines or equal to the maximum compressed air pressure, whichever is the lower.

1.70151. Instrument Air Lines.

1.70152. Airlines to Air Motor-Operated Valves.

1.70153. Pressure Parts of Instruments in Gas or Vapor Service. All other pressure parts of instruments, however, shall be subjected to the general hydrostatic test, except when test pressure exceeds normal working pressure, in which case the instrument shall be isolated.

1.70154. Plant Air Lines.

1.70155. Plant Fuel Gas Lines.

1.7016. Relief valves shall be excluded from the general hydrostatic tests. Blinds shall be installed between the relief valve inlet or outlet and the section of pipe being tested.

1.7017. Where the test is to be made by using water or other fluid, all air shall be vented from the lines as fluid is admitted.

1.7018. The piping and equipment being subjected to the hydrostatic test shall be maintained under pressure for a sufficient length of time to permit thorough inspection for leaks and defects.

1.7019. All underground pipe anchors for bell and spigot elbows and tees shall be completed previous to testing, to avoid the possibility of blowing out.

1.7020. *Shop Testing.* All flanged, shop-fabricated, carbon steel piping for foreign shipment shall be hydrostatically tested before shipment to twice the pressure corresponding to the ASA flange rating of the piece being tested. All alloy or heat-treated shop-fabricated piping shall be tested in the same manner, whether for foreign or domestic use.

Carbon steel pipe fabricated without flanges will be subject to field test only after erection, and the shop test will be waived.

All shop tests shall be performed in the presence of and be witnessed by the purchaser's inspector or delegated representative.

[The following are the several categorized specifications.]

1.800. *Pipe Specifications*

1.801. *Specification M.* This specification shall apply to the following services:

A	100-lb Air	F	50-lb Fuel
C	50-lb Caustic	H	150-lb Hydrocarbon
SC	15-lb Condensate	SE	40-lb Exhaust

Rating: 230 psi at 100°F
150 psi at 500°F
100 psi at 750°F (maximum)

1.8011. *Pipe.* Seamless carbon steel, random lengths, P.E. for sizes 2 in. and below, bevelled for welding 3 in. and above, conforming to ASTM Spec. A53 Grade A, latest revision. [A53 may be used below 900°F and is cheaper than A106.]

2 in. and smaller. Schedule 80. [May be ASTM A106 or A83 if A53 is unavailable.]

3 in. through 10 in. Schedule 40.

1.8012. *Flanges.* Forged carbon steel, 150-lb ASA welding neck, 1/16-in. raised face conforming to ASTM Spec. A181, latest revision, Class I, 0.35% maximum carbon. Note: Bore of all weld neck flanges to be the same as the ID of pipe with which used.

1.8013. *Fittings.* 1½ in. and smaller. Screwed type, 3000-lb CWP forged carbon steel.

2 in. and larger. Buttwelding type, seamless carbon steel conforming to ASTM Spec. A234 and A106, latest revisions, Grade A. Inside diameter and wall thickness to be same as pipe with which used.

2 in. and larger. Flanged type, 150-lb ASA standard cast carbon steel, 1/16-in. raised face.

Cast Steel Flanged Fittings. Do not ordinarily use.

1.8014. *Unions.* 1½ in. and smaller. 2000-lb CWP forged carbon steel screwed, ground joint, integral steel seats.

2 in. and larger. Use flanges as specified above.

1.8015. *Bushings and Plugs.* 1½ in. and smaller. Steel hexagon bushings, and steel hexagon barstock plugs. Use reducers where reduction is more than two sizes.

1.8016. *Bolting.* Alloy steel bolt studs, see 1.901.

1.8017. *Gaskets.* Asbestos-filled double-jacketed corrugated iron, except for services A and SC, sheet asbestos ring gaskets shall be used.

1.8018. *Valves*

1.80181. *Gate Valves.* 1½ in. and smaller. 600-lb SWP forged carbon steel screwed, inside screw stem, union bonnet, 11–13% chrome stainless steel trim (V35). Note: For services A and SC, see under 1.80381.

2 in. and larger. 150-lb ASA standard cast carbon steel, flanged, OS&Y, 11–13% chrome stainless steel trim; 1/16-in. raised face (V1). Note: For service SC, see 1.80381.

1.80182. *Globe Valves.* 1½ in. and smaller. 600-lb SWP forged carbon steel screwed, inside screw stem, union bonnet, 11–13% chrome stainless steel trim (V34). Note: For sevices A and SC, see 1.80382.

2 in. and larger. 150-lb ASA standard cast carbon steel, flanged, OS&Y, 1/16-in. raised face, 11–13% chrome stainless steel trim (V2). Note: For service SC, see 1.80382. [Use Gate Valves for sizes above 6 in.]

1.80183. *Check Valves.* 1½ in. and smaller. 600-lb SWP forged carbon steel screwed, horizontal, piston type, 11–13% chrome stainless steel trim (V19). Note: For services A and SC, see 1.80383.

2 in. and larger. 150-lb ASA standard cast carbon steel, flanged, 1/16-in. raised face, swing type, 11–13% chrome stainless steel trim (V3). Note: For service SC, see 1.80383.

1.80184. *Plug Valves.* ¼-in. size. 2000-lb WOG forged carbon steel, screwed, lubricated plug cock (V110).

1½ in. and smaller. 300-lb WP cast carbon steel, screwed, wrench operated, lubricated plug cock (V58).

2 in. to 6 in. sizes. 150-lb ASA standard cast carbon steel, flanged, 1/16-in. raised face, wrench operated, lubricated plug cock (V54).

1.802. *Specification N.* This specification shall apply to services S, HA, HB, HC 300-lb process oil and vapor, 275-lb steam:

Rating: 500 psi at 100°F
375 psi at 500°F
300 psi at 750°F

1.8021. *Pipe.* Seamless carbon steel, random lengths, P.E. for sizes 2 in. and below, bevelled for welding 3 in. and above, conforming to ASTM Spec. A106 Grade A, latest revision.

2 in. and below. Schedule 80.

3 in. to 10 in. Schedule 40.

12 in. and larger. 0.375-in. wall.

1.8022. *Flanges.* Forged carbon steel, 300-lb ASA welding neck, 1/16-in. raised face, conforming to ASTM Spec. A181, latest revision. Class I, 0.35% maximum carbon, except for HC, use RTJ Flange Facing.

Note: Bore of all weld neck flanges to be same as the ID of pipe with which used.

1.8023. *Fittings.*

1½ in. and smaller. Screwed type, see 1.8013.

2 in. and larger. Buttwelding type, see 1.8013.

Note: Cast Steel Flanged Fittings; do not ordinarily use.

1.8024. *Unions.* 1½ in. and smaller. 2000-lb CWP forged carbon steel screwed, see 1.8014.

2 in and larger. Use flanges as specified above.

1.8025. *Bushings and Plugs.* 1½ in. and smaller. See 1.8015.

1.8026. *Bolting.* Alloy steel bolt studs, see 1.901.

1.8027. *Gaskets.* Asbestos filled corrugated iron, except for service HC which shall be 90 Brinell soft, carbon steel, octagonal ring.

1.8028. *Valves*

1.80281. *Gate Valves.* 1½ in. and smaller. 600-lb SWP forged carbon steel screwed, inside screw stem, union bonnet, 11–13% chrome stainless steel trim (V35).

2 in. and larger. 300-lb ASA standard cast carbon steel,

flanged, OS&Y, 11–13% chrome stainless steel trim, ¹⁄₁₆-in. raised face (V4).

1.80282. *Globe Valves.* 1½ in. and smaller. 600-lb SWP forged carbon steel screwed, inside screw stem, union bonnet, 11–13% chrome stainless steel trim (V34).

2 in. and larger. 300-lb ASA standard cast carbon steel, flanged, OS&Y, ¹⁄₁₆-in. raised face, 11–13% chrome stainless steel trim (V5).

Note: Use gate valves for sizes above 6 in.

1.80283. *Check Valves.* 1½ in. and smaller. 600-lb SWP forged carbon steel screwed, horizontal, piston type, 11–13% chrome stainless steel trim (V19).

2 in. and larger. 300-lb. ASA standard cast carbon steel, flanged, ¹⁄₁₆-in. raised face, swing type, 11–13% chrome stainless steel trim (V6).

1.80284. *Plug Valves.* ¼-in. size. 2000-lb WOG, forged carbon steel, screwed, lubricated plug cock (V110).

1½ in. and smaller. 300-lb WP cast carbon steel, screwed, wrench operated, lubricated plug cock (V58).

2-in. to 4-in. sizes. 300-lb ASA standard cast carbon steel, flanged, ¹⁄₁₆-in. raised face, wrench operated, lubricated plug cock (V59).

1.803. *Specification P.* This specification shall apply to services W, WC, WS, WF:

WF 100-lb Fire Water   WC 75-lb Cooling Water
W   75-lb Service Water   WS 45-lb Sanitary Water
        Rating: 125 psi at 150°F

Note: All pipe and fittings for WS service shall be galvanized.

1.8031. *Pipe*

1.80311. *Pipe Above Grade.* Seamless carbon steel, random lengths, P.E. for sizes 2 in. and below, bevelled for welding 3 in. and above, conforming to ASTM Spec. A53, Grade A, latest revision, for above grade use only. For WS use ASTM A120 galvanized.

2 in. and smaller. Schedule 80, except for WS which may be Schedule 40.

3 in. thru 10 in.   Schedule 40.
12 in. thru 20 in.   0.375-in. wall pipe.
24 in. and above.   0.250-in. wall, seamless or weld.

1.80312. *Pipe Below Grade.* 2 in. and smaller. Same as above, except galvanized.

3 in. and above. Cast-iron pipe, mechanical joint in accordance with ASA A21.6 or 8, Class 22.

1.8032. *Flanges.* 150-lb ASA standard forged carbon steel, ¹⁄₁₆-in. welding neck, raised face, except to be plain face where connected to plain-faced cast-iron valves, fittings, and equipment. Flanges shall conform to ASTM Spec. A181, latest revision, Class I, 0.35% maximum carbon. For steel pipe only.

Note: Bore of all weld neck flanges to be same as the ID of pipe with which used.

1.8033. *Fittings*

1.80331. *Fittings Above Grade.* 1½ in. and smaller. 300-lb standard malleable iron, screwed. Galvanized for WS up to 2 in.

2 in. and larger. Buttwelding type seamless carbon steel, conforming to ASTM Spec. A234 and A106, latest revision, Grade A, at same ID and thickness as pipe with which used.

2 in. and larger. Flanged type. Class 125-lb ASA standard cast iron.

1.80332. *Fittings to Size Required Below Grade.* Class 150 mechanical joint fittings having the same laying dimensions as ASA Class 125 standard flanged fittings.

1.8034. *Unions.* 1½ in. and smaller. 300-lb standard malleable iron, screwed, ground joint, brass to iron seat.

2 in. and larger. Use flanges as specified above, or mechanical joints. On cast-iron pipe in accordance with ASA A21.11.

1.8035. *Bushings and Plugs.* Hexagon steel bushings and hexagon steel plugs. Use reducers where reduction is more than two sizes. Note: All changes from cast iron to steel shall be effected by a flanged cast-iron stub piece connected to a mechanical joint.

1.8036. *Bolting.* Standard carbon steel square head machine bolts with semi-finished steel nuts for steel piping. Bolting shall be supplied with cast-iron piping in accordance with ASA A21.11.

1.8037. *Gaskets.* ¹⁄₁₆-in. thick compressed asbestos. For rapid-faced joints. Use ring gaskets for raised-faced joints and full face gaskets for plain-faced joints. Plain composition rubber gaskets shall be supplied with cast-iron pipe.

1.8038. *Valves*

1.80381. *Gate Valves.* 1½ in. and smaller. 150-lb SWP brass, screwed, rising stem, inside screw (V47).

2 in. and larger. 125-lb ASA standard cast iron, flanged, OS&Y, brass trim; and stem (V26).

2 in. and larger. 125-lb ASA standard cast iron, flanged, OS&Y (fire service double-disk, nonrising stem, brass trim, only) (V117).

1.80382. *Globe Valves.* 1½ in. and smaller. 250-lb SWP brass, screwed, rising stem, union bonnet (V51).

2 in. and larger. Use gate valves.

1.80383. *Check Valves.* 1½ in. and smaller. 250-lb SWP brass screwed, horizontal piston type, Ni-alloy seat and disk (V101).

2 in. and larger. 125-lb ASA standard cast iron, flanged, swing type, brass trim (V28).

1.80384. *Fire Hydrants.* Two hose connection, plant type hydrants (VF).

1.804. *Specification Q.* This specification shall apply to service AC only:

50-lb Air for Instruments, dried.
        Rating, 125 PSI at 150°F

1.8041. *Pipe.* Seamless carbon steel conforming to ASTM Spec. A53, Grade A, latest revision, galvanized. ASTM A83 acceptable for 2 in. and smaller.

2 in. and smaller. Schedule 80.
3 in. thru 10 in. Schedule 40.

1.8042. *Flanges.* Cast iron, 125-lb ASA screwed, flat face, galvanized.

1.8043. *Fittings.* 2 in. and smaller. Screwed type. 300-lb standard malleable iron, galvanized screwed.

3 in. to 6 in. and larger. Flanged type. 125-lb ASA standard cast iron. Note: Use either screwed or flanged as required.

Note: The use of cast-iron flanged fittings shall be kept to a minimum.

1.8044. *Unions.* 2 in. and smaller. 2000-lb CWP forged carbon steel screwed, ground, joint, integral steel seats.

3 in. and larger. Flanges are permissible as specified above.

1.8045. *Bushings and Plugs.* 1½ in. and smaller. Steel hexagon bushings, and steel hexagon barstock plugs. Use reducers where reduction is more than two sizes.

1.8046. *Bolting.* Carbon steel machine bolts, see 1.901.

1.8047. *Gaskets.* ¹⁄₁₆-in. compressed asbestos gasket.

1.8048. *Valves.* See 1.8038 for Valve specifications.

1.805. *Specification Z.* This specification shall apply to instrument services only:

Instrument Services Rating, see Line Services.

1.8051. *Pipe.* Instrument lead lines, or lines which carry fluids to and from instrument elements shall be the same specification or better, than the line service from which instrument line originates. Schedule 160 nipples shall be used up to first block valve where any danger of physical damage to piping is possible. Extensive runs of small sized lines shall be avoided, but where necessary structural support must be provided. Instrument piping of process fluids exceeding 25 ft must be authorized prior to installation.

1.8052. *Tubing.* Copper tubing shall be used for *air* transmission lines. All tubing shall be ¼-in. OD 0.032-in wall ASTM, B68, nonarsenical cold drawn, vacuum annealed.

1.8053. *Fittings.* Flared-type fittings shall be used for all copper tubing lines.

1.8054. *Valves.* Valving used in instrument lead piping shall be of the same specification as the originating line specification.

1.900. *Miscellaneous Specifications*

1.901. *Alloy-Steel Bolting.* Alloy-steel bolting shall be provided for services S, H, HA, HB, and HC and shall conform to ASTM A193, Grade B7, Class 7 fit for bolt, Class 2B fit for nuts, with two (2) heavy series hexagonal nuts each. Nuts shall conform to ASTM Spec. A194, latest revision, Class 2H, oil quenched, hot forged. All other bolting shall conform to ASTM A307 with Class 2A fit for bolts and 2B fit for nuts, excepting bolting furnished by pipe vendor for *W* services, which shall meet ASA A21.11 requirements.

1.902. *Cleaning and Painting*

1.9021. After erection all lines shall be cleaned internally by circulation to remove all dirt and other foreign matter before the plant begins operation. Care shall be taken to thoroughly clean air lines to air actuated diaphragm control valves before they are connected, and after connection up to last fitting before instrument or valve.

1.9022. All underground (buried) shop-fabricated steel pipe 3-in. thru 10-in. size shall have coating on the outside only and wrapping applied as follows:

    (*a*) Clean by sandblasting.

    (*b*) Apply one coat of bitumastic primer.

    (*c*) Apply first coat of bitumastic enamel.

    (*d*) Apply second coat of bitumastic enamel.

    (*e*) Spirally wrap one layer of 15-lb asbestos felt.

    (*f*) Spirally wrap one layer of 60-lb Kraft paper.

1.9023. All underground (buried) steel pipe shall have inside and outside cleaned and wrapping applied as follows:

Specifications Inside (for 12 in. and above)

    (*a*) Clean by sandblasting.

    (*b*) Apply one coat bitumastic primer.

    (*c*) Apply one heavy coat bitumastic enamel.

Specifications for coating and wrapping outside of pipe shall be same as for 3-in. thru 10-in. sizes as specified above, 1.9022.

1.903. *Fireproofing.* Pipe supports carrying main steam headers or hydrocarbons through areas subject to fire hazards shall be fireproofed to the point of the cross load-bearing members, or as noted on drawings.

1.904. *Steam Traps.* Inverted bucket-type steam traps shall be provided for all services except those handling small quantities of steam for which case impulse-type traps may be used. All bucket-type traps shall have forged or cast steel bodies with allowable working pressure of 600 psi minimum. All other traps shall be steel and have service rating at least two times line pressure.

(End of Example Specification)

6

Piping design and drafting procedures

Piping design-drafting consumes the major portion of the total design-drafting effort for process plants. It also requires the contributions of many branches of engineering, for piping is the connecting link in a process and thus is found in every part of the plant associated with almost every major piece of equipment and structure. This close interrelationship between piping design and other phases of plant design demands competent designers with knowledge of major characteristics of process equipment, principles of fluid flow, stress effects on various pipe configurations, safety in design, costs of materials, and piping erection procedures.

In this chapter the techniques employed in piping design-drafting are described. The following chapter then deals with the details of piping layout and arrangement in which these techniques are used.

PROCESS PLANT DRAWINGS

The tangible products of an engineering design group are drawings, material lists, and other data for procurement. The role of piping design is better understood if we realize the functions of the many types of drawings required for design and erection of a process plant and how the piping designer participates in preparing and using these drawings. The following discussions are presented in the usual order in which the drawings are prepared.

Flow Diagram

The flow diagram is the first drawing produced in a plant-design project and is the central source of informa-

tion for all design groups. It depicts the process as conceived by the chemical engineers and is a representation of what is called process design. The equipment needed to make the desired product has been specified by these engineers and the flow through piping from one piece of equipment to another has been indicated schematically.

A typical flow diagram is given in Fig. 6.1. It is a schematic representation of the process and shows all the equipment and associated piping by symbolic representation. The goal of the flow diagram is clarity, and the placement of the equipment on it is not necessarily related to its ultimate placement in the plant. Note on Fig. 6.1 that each piece of equipment and connecting pipe bears an item number. The numbering of equipment and piping aids greatly in all phases of design and construction. When the plant is large, it is often convenient to use a numbering system that will indicate the area in which the equipment or piping is located.

Numbering process lines has proved a valuable aid to the piping designer and seems to be one of the best means of conveying information about each pipeline to the designer. The kinds of valves, fittings, and instruments to be installed in the line can be communicated by means of symbols (Tables 6.1 and 6.2) which can be keyed to the piping specifications by means of a letter designation. The pressure and temperature ratings, materials of construction, and mode of installation of valves and piping are thoroughly described in these specifications (see Chapter 5). A typical line-numbering system might employ a number such as 3″06403E which would have the following interpretation:

3″	0	6	403	E
Line Size	Fluid Class	Area Number	Line Number	Piping Specification

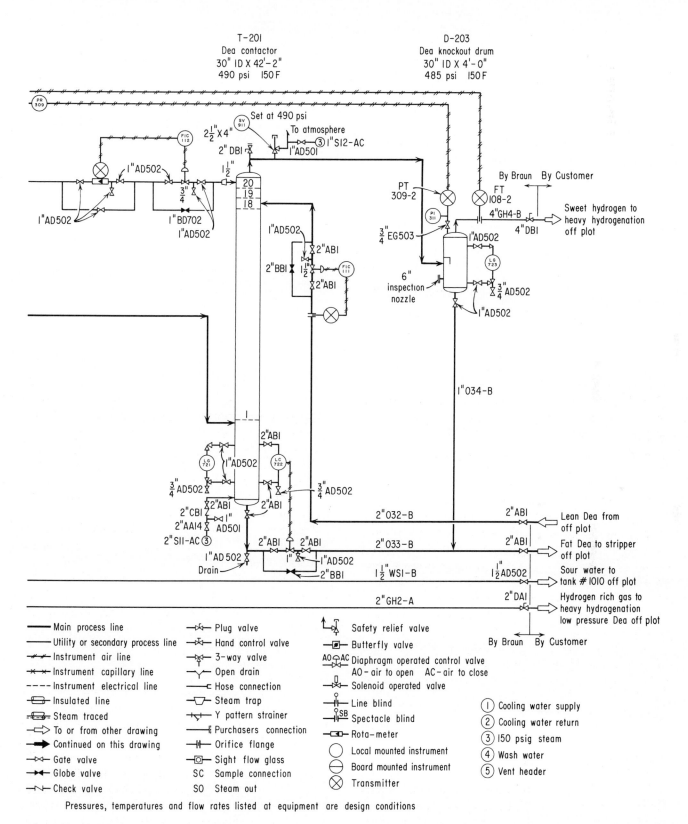

FIG. 6.1. Flow diagram. This is part of a large flow diagram, shown only to indicate method of presentation. (Reproduced by permission of C. F. Braun & Co., Engineers and Constructors.)

TABLE 6.1

TYPICAL VALVE AND PIPE SYMBOLS FOR FLOW DIAGRAMS

ITEM	SYMBOL		
Valves & Fitting		**Line Designations**	
Angle Valve		Main Process	
Blind Flange		Secondary	
Check Valve		Air	
Figure 8 Flange		Condensate	
Flange		Sewer	
Gate Valve		Steam	
Globe Valve		Steam Traced	
Hose Connection		Water	
Plug Cock			
Reducer			
Strainer			

Note: Usually no attempt is made to distinguish flanged and screwed fittings on flow diagrams. The piping specification as noted by the line number designation provides this information.

Table reproduced from H. F. Rase and M. H. Barrow, *Project Engineering for Process Plants,* John Wiley and Sons, New York, 1957.

Thus the line number designates a particular line and gives its size, area location, the applicable piping specification, and the type of fluid flowing in it.

In addition to the flow diagram and piping specification a list of "Piping Process Data" for each line appearing on the flow diagram is prepared. Operating temperatures and pressures are included in this list so that the piping designer may allow for expansion stresses when necessary in his arrangements and provide for insulation where required for heat conservation or personnel protection. On certain small projects these data may be placed by the line number on the flow diagram. Usually, however, this unnecessarily clutters the drawing and a separate list is preferred by many designers.

The piping designer refers to the flow diagram many times throughout the design project. He will find it the common meeting ground for all design groups and will use it in checking his own work to be certain that process requirements have been met.

Plot Plans (See also page 181)

The plot plans show the location of all major equipment on the plant site. Foundation centerlines are given as well as the outline and location of structures and buildings, paved areas, roadways, railway spurs, walkways, and maintenance ways. Such an important document represents the combined decisions of many people involved in the project, but the piping designer responsible for piping layout plays a major role. It is he who must advise on space needs for equipment and piping, and his recommendations are useful guides in setting the final layout and plot plan. Since the piping designer must fit the piping to the arrangement shown on the plot plan, it is poor economy to prepare these arrangements without his help.

Included on the plot plan are lists of all major equipment, drawings (title and number), and reference sources for plant elevation data. The plot plan, like the flow diagram, is a key reference for all design groups throughout the project, and continues in use during plant construction when the flow diagram is less important.

Vessel Drawings

Vessel drawings proceed through several stages. Early in the project, vessel outline sketches are prepared by the vessel designers (Fig. 6.2). These are preliminary drawings which are suitable for initial arrangements of vessels and piping and for placing orders with fabricators. Completed vessel drawings are dimensioned schematics issued to the vessel fabricator for use in preparing his shop detail drawings. They must be checked by the piping designer for proper nozzle arrangement; and, whenever possible, nozzles are rotated in accordance with the recommendations of the piping designer to better suit piping needs.

Foundation Drawings

These drawings show dimensions and outlines of foundations, location and size of reinforcing steel, and bending details for steel. The piping designer must work closely with the foundation designer since underground water piping and sewers require large piping which can interfere with the foundations. Thus underground piping, underground electrical conduit, and foundations are developed simultaneously. Often all three are ultimately shown on the same drawings or associated drawings, so that underground work may proceed simultaneously.

Structural Drawings

Structural steel, reinforced concrete, building framing, and supports for piping and equipment are shown on the structural steel drawings. These drawings are used by the fabricator to prepare his shop drawings and erection drawings, which are schematic arrangements showing each separate piece of structural steel in its installed position and designated with a piece number which will also appear on the steel itself.

Here again the piping designer becomes involved. He must develop the heights, widths, and locations of pipe supports, and he must check all structural steel to

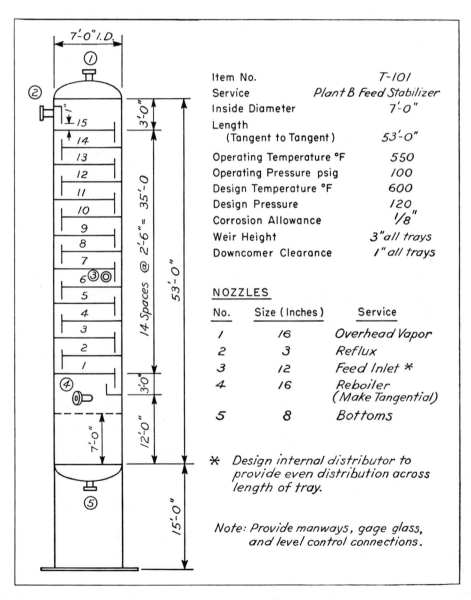

FIG. 6.2. Process vessel sketch. (Reproduced from H. F. Rase and M. H. Barrow, *Project Engineering of Process Plants,* John Wiley and Sons, New York, 1957.)

TABLE 6.2

INSTRUMENTATION FLOW PLAN SYMBOLS

Basic Instrumentation Symbols

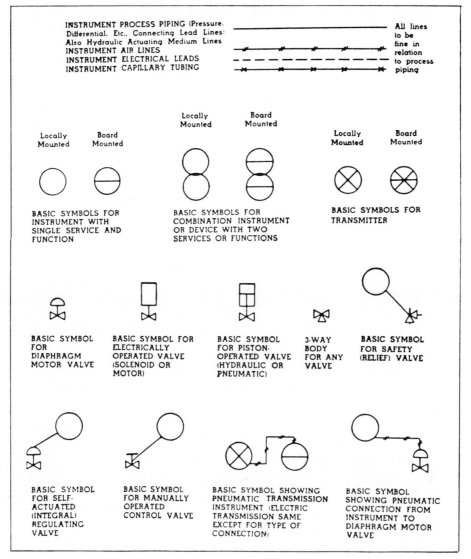

Reprinted by permission of Instrument Society of America.

be certain there are no interferences between piping and structures. This checking must be done on the original structural-steel drawings since the structural-steel fabricator's shop drawings are submitted for checking only when the steel fabricator requests permission to substitute sizes of structural members other than specified. In such cases the check is only partial and the piping designer seldom becomes involved.

Piping Drawings

Process plant piping drawings describe the arrangement of piping and may be drawn as plans and elevations (sections) or in three dimensions (isometric). In either case, there must be some type of plan drawing to illustrate the location of major piping runs and pipe racks. These drawings, together with the piping specifications and pipe process data lists, are used by the construction crews in erecting the piping; and by the fabricator in preparing his shop detail drawings (spool sheets) for piping that is to be shop-fabricated.

When pipe is shop-fabricated each piece is marked with an identifying number (spool number) which is matched in the field with numbers appearing on the piping drawings (Figs. 6.3 and 6.4).* Pipe fabricators furnish prints of their shop drawings called spool sheets

* One advantage of isometrics is seen by the location of spool numbers. With the plan and elevation method, numbers shown on the plan must not be repeated on the elevation, and it is necessary for the person doing the shop drawings to refer constantly to both plan and elevations.

TABLE 6.2 (*Continued*)
INSTRUMENTATION FLOW PLAN SYMBOLS
Typical Symbols for Flow

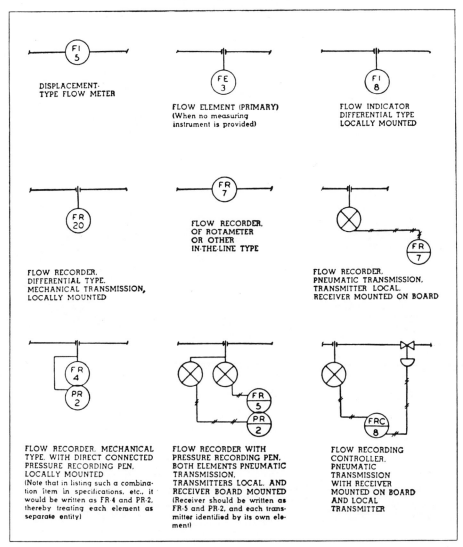

Reprinted by permission of Instrument Society of America.

(see Fig. 8.1), but these are not usually checked, since this would constitute double checking of the piping. However, critical spool sheets are checked by the piping designer. These include spools requiring heat treatment, alloy spools, or spools for overseas jobs where errors, though the fabricator's responsibility, would cause excessive delays. Spool sheets are sometimes used by the fabricator to invoice each piece of fabricated pipe. In these instances, the sheets are checked by the piping designer for material and labor only.

Piping is subject to change during the development of a plant design, and it is imperative that these changes be brought immediately to the attention of the fabricator, because if changes are made without the knowledge of the fabricator, he cannot be held responsible. Of course, errors caused by the fabricator must be cor-

rected by him or corrected in the field and backcharged to him.

Electrical Drawings

Electrical drawings give the location of all conduit runs, switchgear, and lighting fixtures, and also give tabulations of cables and circuits. These drawings are schematic and usually are not dimensioned, except to locate large-size conduit and major electrical equipment. Since electrical conduit is more easily arranged and generally much smaller than piping, it is common practice to erect most of the piping before installing electrical conduit and to rearrange or adjust conduit runs when conflicts occur. However, when the electrical installation is extensive and large conduit (3 in. and

TABLE 6.2 (*Continued*)
INSTRUMENTATION FLOW PLAN SYMBOLS
Typical Symbols for Temperature

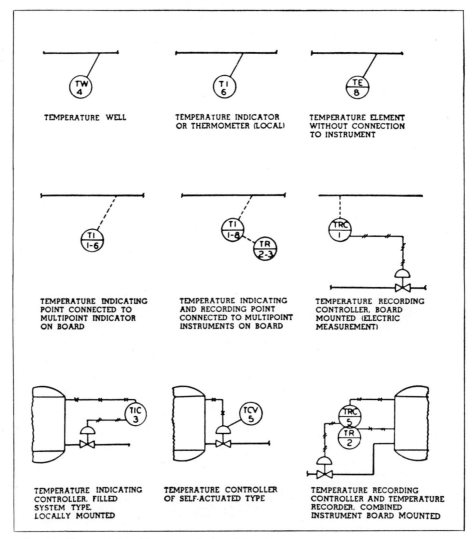

Reprinted by permission of Instrument Society of America.

above) is utilized, the arrangements must be planned with more care, since rerouting large conduit to clear process piping can be a difficult procedure.

Instrumentation Drawings

Instrumentation drawings show the location of all instruments and raceways, give details of connections and control panels, and present space allowances for pneumatic or other control elements. Instrument location lists, similar to pipe process-data lists, are also prepared. They indicate instrument element locations and establish a check list for control valve sizes. These data are most important to the piping designer.

Architectural Drawings

Architectural drawings give architectural details of such structures as control houses, shops, office buildings, laboratories, and process buildings. Process buildings house process equipment which invariably must be connected by piping, which in turn affects the space needs in the building and design of the structural supports.

Manufacturer's Drawings

Manufacturer's drawings are prepared by the manufacturer of process plant equipment such as vessels, heat exchangers, pumps, and instruments. They are often

TABLE 6.2 *(Continued)*
INSTRUMENTATION FLOW PLAN SYMBOLS
Typical Symbols for Level

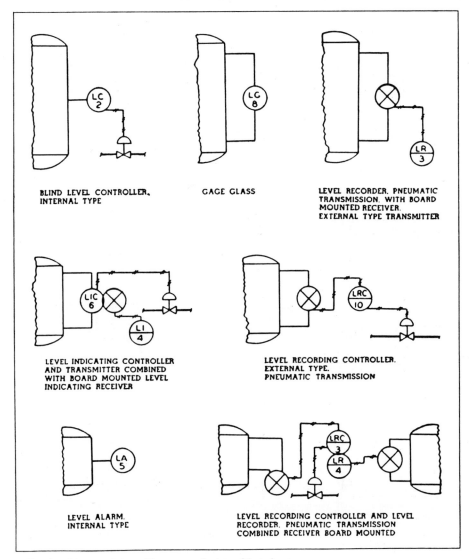

Reprinted by permission of Instrument Society of America.

called *vendor prints* and are most vital to the piping designer. Once received, the piping designer can proceed with final design and layout, for these prints certify the size and location of all piping connections in addition to giving exact dimensions of the equipment.

PIPING-DESIGN DRAFTING PROCEDURES

The piping designer can begin detailed design when he receives the flow diagrams, plot plans, piping specifications, pipe process-data list, instrument location list, vessel outline drawings, and vendor's prints. One of three procedures is used to develop detailed design.

These procedures are "Plan-and-Section," "Three-Dimension," and "Model."

The Plan-and-Section method is the conventional or classical drawing procedure of orthogonal projections. All piping is drawn in plan at various elevations. Sections are taken through the piping and elevations are drawn. (See Figs. 6.3 and 6.4,* pages 175 and 176.)

The Three-Dimension method utilizes what is known as "isometric" or three-dimensional drawing. Although some plans and elevations (sections) are drawn, most of the piping is drawn on small sheets in three dimensions. (See Fig. 6.4.)

* Table 6.3 at that end of this chapter gives standard graphical symbols used in piping drawings.

172

Piping design and drafting procedures / 6

TABLE 6.2 (*Continued*)
INSTRUMENTATION FLOW PLAN SYMBOLS
Typical Symbols for Pressure

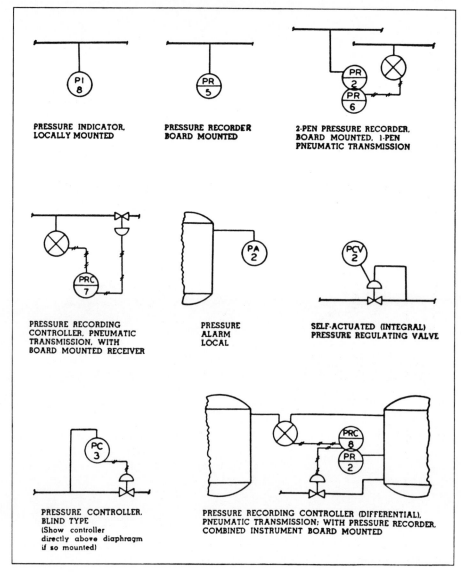

Reprinted by permission of Instrument Society of America.

The "Model" method employs scale models together with a limited number of drawings. A plot plan of some kind is necessary. After equipment is located on the model, isometric sheets may be prepared directly from the model as piping is installed on the model.

The following steps are employed in piping design when using each of these methods:

I. PLAN-AND-SECTION METHOD

A. Layout. Major pipe-ways are determined. Studies are made of various hot lines, alloy piping, and headers in order to fix the arrangement of all large piping. The drawing scale and number of plan drawings are decided.

B. Plan Drafting. All piping in any area is drawn in plan on one or several drawings at as many elevations or planes as are required for clarity.

C. Section Drafting. Simultaneously with plan drafting, other personnel are assigned to draw "sections" or elevations along certain section lines in the area.

D. Drawing Checking. This is usually done in two steps, at times simultaneously. One check is made to be certain that all lines shown on the flowsheets are on the piping drawings, and that the lines shown are designated by the proper specification number. The other check is to verify that process requirements are met, valves are in an operable location, all mechanical clearances are satisfactory,

TABLE 6.2 *(Continued)*
INSTRUMENTATION FLOW PLAN SYMBOLS
Typical Miscellaneous Symbols

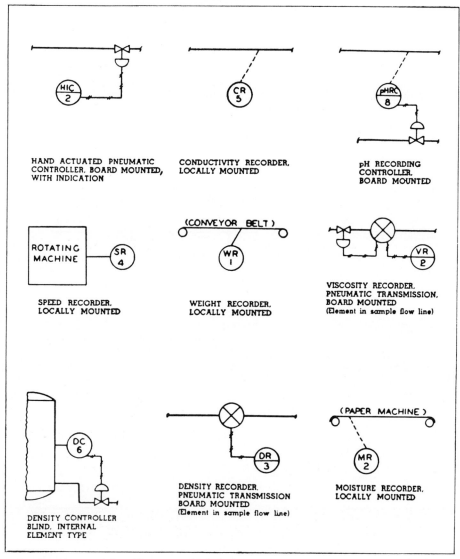

Reprinted by permission of Instrument Society of America.

intermediate dimensions upon addition check with over-all dimension, specifications have been properly applied, and insulation is located correctly.

E. Material Listing. After the drawings are checked, the piping material is listed for procurement. When shop-fabricated piping is used, all the fabricated pieces are listed as "spool" numbers, but material is not listed as this is done by the fabricator as part of his shop work. All other piping materials and accessories necessary for the permanent plant must be listed.

II. THREE-DIMENSION METHOD

A and B. Layout and Plan Drafting. This is the same as plan-and-section method, with the exception that the detail is very limited. All major lines are shown, but secondary piping is not shown in much detail.

C. Three-Dimension Drafting. The major difference in method occurs where section drawings would be started. An isometric take-off man, an experienced piping designer, makes three-dimensional free-hand sketches of each line shown in the plans.* These sketches are then passed on to less-experienced personnel to complete as finished isometric drawings. Usually only one or two lines are shown on each isometric drawing sheet.

* The word *isometric* is loosely used for simplification. Several three-dimensional methods are used.

TABLE 6.2 *(Concluded)*
INSTRUMENTATION FLOW PLAN SYMBOLS
Typical Symbols for Combined Instruments

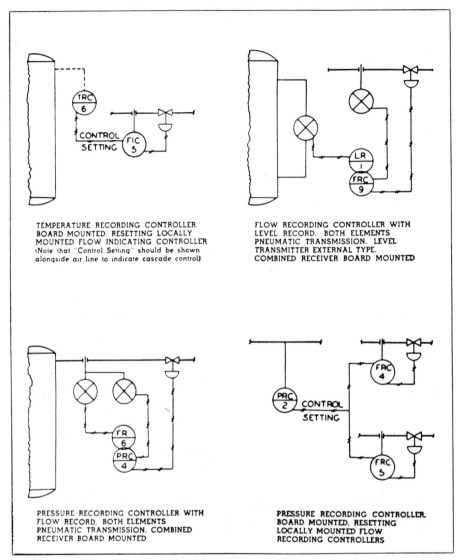

TEMPERATURE RECORDING CONTROLLER.
BOARD MOUNTED. RESETTING LOCALLY
MOUNTED FLOW INDICATING CONTROLLER
(Note that "Control Setting" should be shown
alongside air line to indicate cascade control)

FLOW RECORDING CONTROLLER WITH
LEVEL RECORD. BOTH ELEMENTS
PNEUMATIC TRANSMISSION. LEVEL
TRANSMITTER EXTERNAL TYPE.
COMBINED RECEIVER BOARD MOUNTED

PRESSURE RECORDING CONTROLLER WITH
FLOW RECORD. BOTH ELEMENTS
PNEUMATIC TRANSMISSION. COMBINED
RECEIVER BOARD MOUNTED

**PRESSURE RECORDING CONTROLLER.
BOARD MOUNTED. RESETTING
LOCALLY MOUNTED FLOW
RECORDING CONTROLLERS**

Reprinted by permission of Instrument Society of America.

Some firms use large sheets, drawing a major part of a particular area of the plant in isometric views. It is usually more practical to use smaller sheets thus spreading drafting over more personnel, which results in faster overall completion of the drafting steps.

D. Drawing Checking. After step *C*, the procedures vary with each organization. It is difficult to divide the checking among a number of men. Generally, checking is more difficult when using the isometric method than with the plan-and-section method, but much depends upon the particular type of installation and the personnel.

F. Materials Listing. Material listing is much sim-

plified when small isometric sheets are used, as all materials required for the particular piping can be listed directly on the sheet showing that piping.

III. MODEL METHOD

A. Layout. Scale models are prepared for all major pieces of equipment, preferably by using the same scale as that to be used in the piping drawings. Foamed polystyrene and wood are popular materials for this purpose. A cross-sectioned sheet, the same scaled size as the plot plan, is prepared and placed on a firm base. The equipment locations may then be studied to determine the most acceptable arrangement. Most of the layout sketching

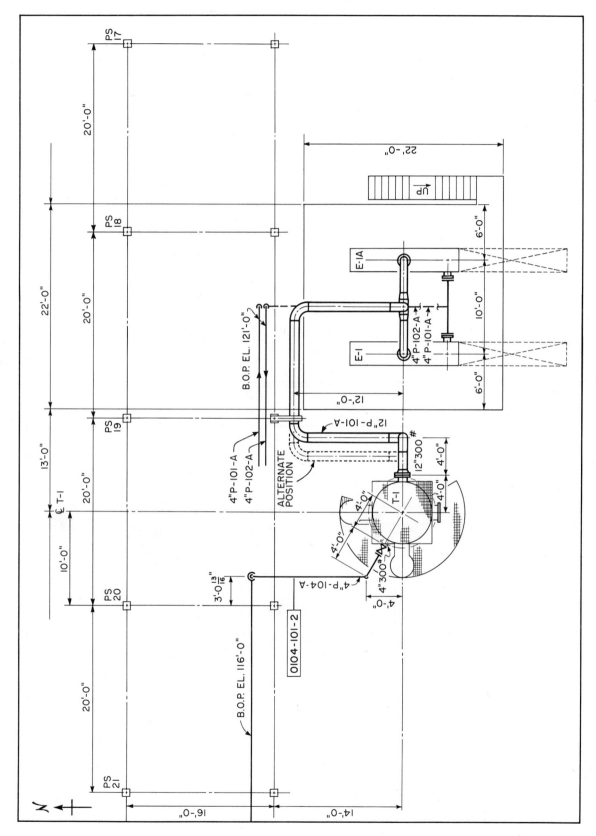

FIG. 6.3. Piping drawing—Plan.

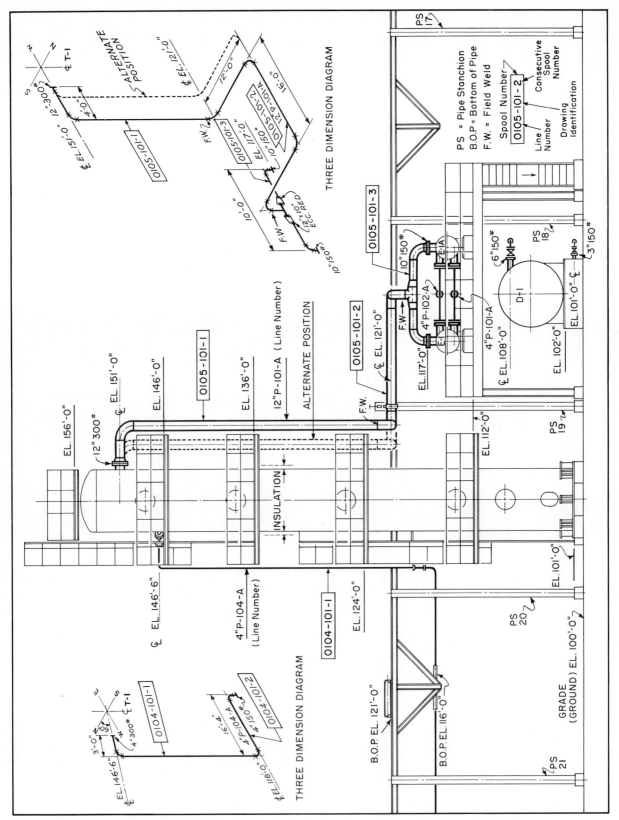

FIG. 6.4. Piping drawing—Elevation and three dimension.

is thus eliminated. However, some schematic sketching is necessary to determine the number of pipe runs in the various areas so that pipe rack widths can be selected.

B. Plan Drafting. The detail of this step is much reduced but all piping must be located and dimensioned.* Since the equipment models are available, it is possible to use less-experienced men instead of layout personnel in this work. While the major plan drawing is proceeding, more detailed work is also continuing on the model. Collaboration between personnel in both functions is accomplished by preparing the major plan drawing in close proximity to the actual model. The drawing is the dimensional certification of the equipment located on the model.

C. Section Drafting. Section drafting is not ordinarily used with a model, except for certain preliminary sketches. With the plan dimensional drawing and access to the model, isometric sketches may be prepared. The free-hand sketch or take-off step II.*C* is not required.

D. Checking. Dimensional and flow-sheet data are needed for checking as in the other two methods, but if the model is completed in great detail, the rigorous check of the flow diagram may not be necessary. In general, the checking steps will not vary greatly with the other two methods described above, but it must be pointed out that checking varies with the type of plant and the experience of personnel. Mechanical conflicts should become apparent and be eliminated during the construction of the model.

E. Material Listing. Since isometric drafting is used for piping details, the method is the same as for the three-dimension method, provided materials are listed on the isometric sheets.

Comparison of Methods

It would be unrealistic to select any single method of piping design-drafting for all process plant piping installations since each plant is usually unique and the capability of design personnel varies with each design organization. The three methods discussed are used to some extent by nearly all design organizations. Models have not been completely accepted due to the difficulty in installing the method and in estimating actual savings in design time.

The plan-and-section method evolved from academic or standard drafting methods consisting of simply drawing the necessary views or orthographic projections. Undoubtedly this method has the advantage of sim-

plicity in showing simple piping systems such as: (*a*) underground piping, as used for water, sewers, and drains; (*b*) multiple parallel runs of piping, as used for pipe racks; or (*c*) simple plant or pumping station manifolds.

The three-dimension method is extremely useful in showing involved piping such as would be found in a complicated process elevated in a structure. The use of a large number of small sheets permits spreading the work so that detailed drafting may be completed more rapidly.

In order to simplify procedures where a large amount of piping is involved, it is often more economical to fix methods because using separated design methods to suit various types of piping may cause problems in the overall production of drawings. Thus it may be more practical to determine the method on the basis of the size and the type of plant.

By preparing scale models of equipment, the layout of equipment and arrangement of equipment can proceed much faster than by preparing sketches. The advantage of having physical objects to manipulate is well known and needs no lengthy defense. Initial drafting layouts of many process plants are subject to change as the work progresses. With a model many of the problems which can become apparent only as the design is developed are immediately discovered and eliminated. Usually this advantage for piping alone will save the cost of the model because any change in piping affects initial drafting, material listing (and thus procurement), checking, fabricated piping, and, at times, construction work.

The model is useful in the initial layout and thus in the design of structural parts of the plant. By studying the equipment and the model, structural designers can often simplify ladders and platforms, and easily locate miscellaneous supports. The same is true of electrical design for the process plant.

For the engineering service organization, the model simplifies the problem of illustrating various arrangements to a customer, and this is its greatest value. The service engineering organization must usually submit all drawings to its customers for general approval. It is almost impossible to avoid some waste of engineering labor during the interim of (*a*) having the drawings printed, (*b*) transmitting the prints, (*c*) waiting for the customer's study, (*d*) processing the comment prints, (*e*) studying the customer's comments, and (*f*) reassigning personnel to work on the drawings (if work was stopped). While this procedure cannot be avoided, with

* Various colored plastic models of valves, pipe and fittings made to ⅜-in. or other convenient scale are now manufactured. Many model builders use these convenient and economical pieces for laying out piping on the model.

the use of a model the decisions by the customer can usually be made rather rapidly during a conference with the customer, using the model as a focal point of the discussion.

The last use of the model, by the contractor, is during construction. It has often proved practical to ship the model to the job site where it may be studied by the erection personnel in anticipating special erection problems and solving many problems that arise during construction. When construction is complete, the model can be used by the customer in training operating personnel.

Models, unlike drawings, cannot readily be reproduced and are cumbersome to move about. Of course, photographs of the model from different angles can prove helpful as transmittable material. Several organizations are experimenting with the substitution of reproducible dimensioned photographs of models for many piping arrangement drawings. In the meantime, the experimental and open-minded approach generated by interest in models has caused re-examination of the more conventional methods so that these too have been simplified and the number and complexity of drawings greatly reduced. Thus it seems that the decision on the best design-drafting procedure can only be made when based on the factors that impinge on each particular project. It must be remembered that ultimately the competence of the designers will govern the success of a job much more than the technique.

MODERN DRAFTING PRACTICES

Industrial drafting is still the only known simple method of illustrating unique designs by graphical representations in a form which may be easily or inexpensively reproduced. Information given on drawings is used in procurement, manufacture, and erection, and thus must be seen by many persons. Apart from creative design this is the only reason for making drawings; that is, providing an economical medium for a wide distribution of information.

In modern drafting practices all drawings are made with pencil on transparent sheets, which may be paper (vellum), linen (treated cloth), or plastic sheets. Constantly increasing labor costs demand that all excessive time-consuming techniques be eliminated. The detail on drawings is usually the absolute minimum required for conveying the data needed in fabricating and erecting the pipe. Symbols such as those given in Table 6.3, page 179, are used, and there is a minimum attempt at pictorial representation.* Double-line representation of pipe is restricted to large lines (12 in. or above) or lines in critical service requiring great care in placement.

Drawing Scales

A usual scale for piping drawings and models is ⅜ in. = 1 ft. The lowest readable scale is ¼ in. = 1 ft. For congested drawings ⅜ in. may be too small. The best scale is one which provides a clear picture, and arbitrary assignment of fixed scales within an organization for all work is to be avoided.

The primary purpose of a scale for drawings or models is to show relative sizes and location of equipment and piping, and to permit approximate scaling of the drawing during planning and design discussions. Because of this fact, the use of freehand drawing techniques for small details is becoming more common as a substitute for the more tedious mechanical procedures requiring straight-edges, triangles, compasses, and templates.

Dimensioning Methods

Dimensioning methods for plans, elevations, and isometrics are fully illustrated in Figs. 6.3 and 6.4. Note that elevations are greatly simplified by avoiding the use of arrows. Instead, the elevation at grade is arbitrarily set at 100′0″ and elevations at bottom of pipe (BOP), face of flange (FOF), or centerlines of nozzles and equipment are indicated as some value greater than 100′0″. In plans, arrows and dimension lines are used, but they are not broken as is the custom in many other types of drafting. By using solid lines and placing the dimension on the line much time is saved and the information is conveyed just as effectively.

Drawing Reproduction or Printing Methods

A transparent drawing medium is necessary for economical printing of a small number of copies. The printing process employs a paper sensitive to strong light which is mechanically exposed to this light while in contact with the original drawing. The differences in the several processes are primarily in the cost of the light-sensitive materials. Blueprinting is the oldest method and is perhaps the best known. Blueprints show drawn parts white on a blue background while other processes produce the drawn parts colored or black on a white background. Blueprints are less expensive per reproduction than other types of prints, but the printing process is more involved and the machines are more costly. Blueprints are most economical when used for high volume production in which case other types of prints are used for internal reference and check-

* Many companies simplify or modify the ASA symbols given in Table 6.3. The illustrations of piping drawings shown in Fig. 6.3 and 6.4 were prepared to demonstrate several points and were made more pictorial than necessary for piping drawings in order to survive the reduction process necessary for book publication.

ing. It is generally most satisfactory to use the processes which produce a colored or black print on a white background for office use and blueprints for construction or erection, since white background prints may be more easily marked for comments and checking while the dark background of the blueprint survives field use better than other types.

The life of reproductions depends on the process and the type of light-sensitive medium. All have essentially unlimited life under storage conditions or for office use. Some papers tend to fade in continued strong sunlight.

There are many photographic and printing methods which are used to reproduce the original drawing on a transparent sheet. This transparent reproduction can then be used as a drawing and certain desired details added by pencil. This technique is most convenient for frequently used details which require limited additions for each new situation.

Microphotography is being widely used for permanent records and as a means for reducing storage space requirements. Permanent storage of prints can become a real problem through the years. Though most prints do not fade when stored in cabinets, they do consume large amounts of space. Microphotography of originals is gaining increasing acceptance as a means of alleviating this storage problem.

TABLE 6.3
GRAPHICAL PIPING SYMBOLS
(Abstracted by permission: ASA Z32.2.3, published by American Society of Mechanical Engineers. Reproduced from: C. H. Thompson, *Fundamentals of Pipe Drafting,* John Wiley and Sons, New York, 1957.)

NOTE: These symbols, although part of a standard of the American Standards Association, are not necessarily used exactly as shown. Most design corporations are endeavoring to reduce the effort in producing drawings. Simplified symbols are in wide use, and revised standards based on these simplifications are certainly anticipated.

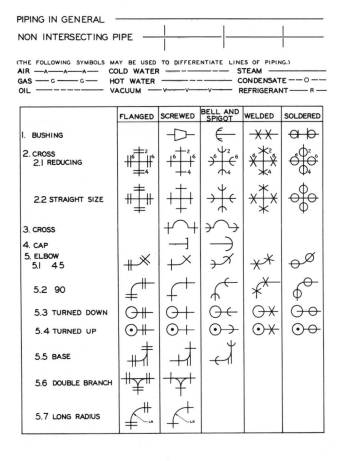

TABLE 6.3 (*Concluded*)
GRAPHICAL PIPING SYMBOLS

	FLANGED	SCREWED	BELL AND SPIGOT	WELDED	SOLDERED
13.4 DOUBLE SWEEP					
13.5 REDUCING					
13.6 SINGLE SWEEP					
13.7 SIDE OUTLET (OUTLET DOWN)					
13.8 SIDE OUTLET (OUTLET UP)					
14. UNION					
15. ANGLE VALVES					
15.1 CHECK					
15.2 GATE (ELEVATION)					
15.3 GATE (PLAN)					
15.4 GLOBE (ELEVATION)					
15.5 GLOBE (PLAN)					
15.6 HOSE ANGLE	SAME AS SYMBOL 23.1				
16. AUTOMATIC VALVES					
16.1 BY-PASS					
16.2 GOVERNORED OPERATED					

	FLANGED	SCREWED	BELL AND SPIGOT	WELDED	SOLDERED
16.3 REDUCING					
17. CHECK VALVE					
17.1 ANGLE	SAME AS SYMBOL 15.1				
17.2 STRAIGHTWAY					
18. COCK					
19. DIAPHRAGM VALVE					
20. FLOAT VALVE					
21. GATE VALVE 21.1*					
21.2 ANGLE GATE	SAME AS SYMBOLS 15.2 & 15.3				
21.3 HOSE GATE	SAME AS SYMBOL 23.2				
21.4 MOTOR OPERATED					
22. GLOBE VALVE 22.1					
22.2 ANGLE GLOBE	SAME AS SYMBOLS 15.4 & 15.5				
22.3 HOSE GLOBE	SAME AS SYMBOL 23.2				
22.4 MOTOR OPERATED					
23. HOSE VALVE 23.1 ANGLE					
23.2 GATE					
23.3 GLOBE					
24. LOCKSHIELD VALVE					
25. QUICK OPENING					
26. SAFETY VALVE					
27. STOP VALVE	SAME AS SYMBOL 21.1				

*ALSO USED FOR GENERAL STOP VALVE SYMBOL WHEN AMPLIFIED BY SPECIFICATION

7

Process piping arrangement

The major effort in piping design for process plants is devoted to developing the many arrangements of piping necessary to convey fluids between equipment in a process unit and between the process unit and storage. The designer must be keenly aware of the principles of both piping and equipment arrangement, for these are two inseparable facets of the overall problem of piping design.

PLANT LAYOUT AND EQUIPMENT ARRANGEMENT

After the flow diagrams and specifications are completed, it is then possible to begin the layout of the plant and the arrangement of equipment. This all-important step in establishing master and unit plot plans for a project should be a cooperative effort between project and process engineers and the structural, vessel, instrument, and piping design groups. The piping designer in particular is concerned with the ultimate layout of the plant; so congested equipment, and thus congested and costly piping, can be avoided by careful planning at this stage. By entering the discussions during the plot-plan stage he can be certain that adequate space is allowed for piping.

Models of equipment are most advantageous at this time. They present all three dimensions simultaneously and thereby facilitate discussions related to space and process requirements.

Master Plot Plan (Fig. 7.1, page 182)

A master plot plan is prepared for a large plant consisting of many process units, offices, shops, laboratories, and warehouses. The plant is divided into blocks surrounded by paved roadways, and the function of each block or area is designated. Safety is an important factor in each decision, for hazardous units must be remote from open flames and sparks (see Table 7.1, page 183, for some rules).

A study of the flow sequence between units and from units to storage will lead to the arrangement using the least piping and materials-handling equipment. Offices, laboratories, and main gate houses should be adjacent to a highway. Warehouses, shops, and packaging and shipping facilities should be accessible to the highway and railroad trackage.

A contour map of the site is used to plan cut and fill, ideal road locations, and plant drainage. If the site is hilly, there may be an advantage to locating product storage tankage in the highest area so that gravity loading may be possible. When the terrain presents a particularly difficult problem some groups find topographical models helpful.

Unit Plot Plans (Fig. 7.2, page 182)

After the master plot plan has been approved, the unit plans for the process units must be prepared. Usually the piping layout designer is asked to suggest arrangements of equipment best suited to the piping conditions. He can make this study with models of the equipment or two-dimensional cutouts by arranging them on a grid representing the area allotted for a given unit. Freehand sketching of major piping which may govern spacing is also helpful at this point.

His suggestions are then discussed with the several specialized designers, the project and process engineers, and the people who must maintain and operate the com-

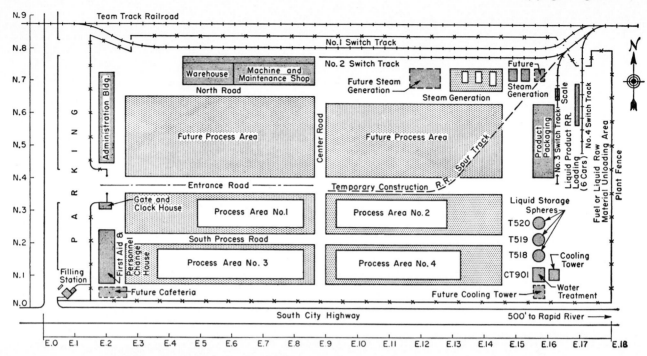

FIG. 7.1. Master plot plan. (Reproduced from H. F. Rase and M. H. Barrow, *Project Engineering of Process Plants,* John Wiley and Sons, New York, 1957.)

pleted plant. When the arrangements are approved, the unit plot plans representing a formal outline of these arrangements are issued for use by all design groups throughout the project.

Estimating Yard Piping Space Requirements

During the plot planning stage estimates of the width of yard pipe supports must be made in order to provide

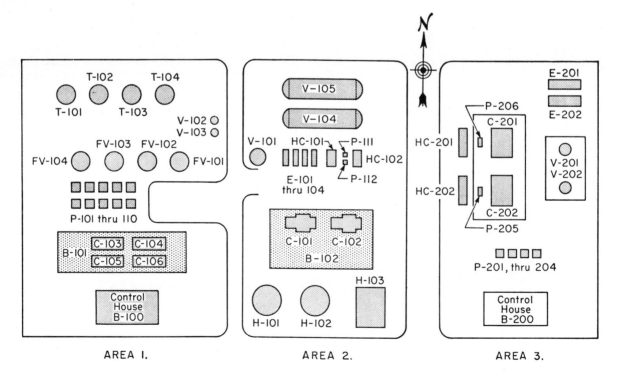

FIG. 7.2. Unit plot plan. Dimensions have been omitted to reduce detail for illustrative purposes. (Reproduced from H. F. Rase and M. H. Barrow, *Project Engineering of Process Plants,* John Wiley and Sons, New York, 1957.)

FIG. 7.3. Elevate mechanical equipment so that servicing can be done in upright position. *Note:* This rule should be applied to large equipment. Small pumps are usually installed on a common shaft centerline with a minimum foundation height of 12 in. A large pump requires a greater foundation mass, and this can be conveniently used to raise the pump to a more serviceable position as shown.

adequate room for these important space consumers. The estimate is best made by laying out lines on the plot plan with the help of the completed engineering flow diagram. The spacings given in Table 7.2, page 184, are used to calculate the width of the support. Space should be allowed for instrument lines and electrical conduit when included on the rack. Allow about 25% overage for expansion.

CRITERIA FOR EQUIPMENT LAYOUT

An experienced piping designer regularly uses certain criteria for equipment layout which are generally applicable. Some of the most common of these are assembled here for ready reference.

Equipment Elevations

It is preferable to place process equipment near grade level when process conditions permit. Substantial savings result because costly elevated structures are eliminated, and operation and maintenance are made easier. For those cases where at-grade placement is possible, it is necessary to recognize the minimum practical elevations dictated by maintenance and operating needs. It will be helpful to consider these factors along with those process requirements and layout factors which make elevated equipment necessary.

MAINTENANCE AND OPERATING REQUIREMENTS

Assuming that there are no special process factors which require the elevation of a piece of equipment, the piping designer may plan the placement of the equipment at grade. He must then decide the minimum permissible elevation above grade. To do this, the operating and maintenance characteristics must be considered.

1. Mechanical Equipment. Servicing or operating mechanical equipment such as pumps, centrifuges, and filters is greatly simplified if the employee can remain in an upright position while performing his duties (see Fig. 7.3).

2. Towers, Drums, and Heat Exchangers. The usual minimum elevations are

Towers   3 to 5 ft skirt height
Drums   3 to 5 ft from drum bottom to grade
Heat exchangers   2.5 to 4 ft from exchanger bottom to grade.

TABLE 7.1

SUGGESTED MINIMUM SPACINGS FOR HAZARDOUS PROCESS UNITS

Description of hazard	Distance to adjacent operating unit. ft
1. Ordinary flammability and low to medium pressures	25–75
2. High flammability and high pressure	100–150
3. Direct fired boilers and furnaces	100–150
4. Blowdown stacks with flare	100–200
5. Loading facilities	100
6. Public roads and railroads	100
7. Cooling tower	100
8. Storage tanks*	75–150

* Minimum distances between storage tanks are given in the National Fire Codes.[10] Wherever possible a distance equivalent to 1 or 1½ times the diameter of the larger tank should be used. Table reproduced from H. F. Rase and M. H. Barrow, *Project Engineering of Process Plants,* John Wiley and Sons, New York, 1957.

TABLE 7.2
SPACING GUIDE FOR YARD PIPING
(Spacing figured for 1-in. between flanges)

Pressure		600								400								
	Size	10	8	6	4	3	2	1½	1	12	10	8	6	4	3	2	1½	1
	1	13⅛	11⅜	10⅛	8½	7¼	6⅜	6¼	5⅝	13⅜	11⅞	10⅝	9⅜	8⅛	7¼	6⅜	6¼	5⅝
	1½	13½	11¾	10½	8⅞	7⅝	6¾	6⅝	6	13¾	12¼	11	9¾	8½	7⅝	6¾	6⅝	6
	2	14	12¼	11	9⅜	8⅛	7¼	7⅛	6½	14¼	12¾	11½	10¼	9	8⅛	7¼	7⅛	6½
	3	14¾	13	11¾	10⅛	8⅞	8	7⅞	7¼	15	13½	12¼	11	9¾	8⅞	8	7⅞	7¼
	4	15½	13¾	12½	10⅞	9⅝	8¾	8⅝	8	15¾	14¼	13	11¾	10½	9⅝	8¾	8⅝	8
150	6	16½	14¾	13½	11⅞	10⅝	9¾	9⅝	9	16¾	15¼	14	12¾	11½	10⅝	9¾	9⅝	9
	8	17¾	16	14¾	13⅛	11⅞	11	10⅞	10¼	18	16½	15¼	14	12¾	11⅞	11	10⅞	10¼
	10	19	17¼	16	14⅜	13⅛	12¼	12⅛	11½	19¼	17¾	16½	15¼	14	13⅛	12¼	12⅛	11½
	12	20½	18¾	17½	15⅞	14⅝	13¾	13⅝	13	20¾	19¼	18	16¾	15½	14⅝	13¾	13⅝	13
	14	21½	19¾	18½	16⅞	15⅝	14¾	14⅝	14	21¾	20¼	19	17¾	16½	15⅝	14¾	14⅝	14
	16	22¾	21	19¾	18⅛	16⅞	16	15⅞	15¼	23	21½	20¼	19	17¾	16⅞	16	15⅞	15¼
	1	13½	11¾	10½	8⅞	7⅝	6¾	6½	5⅞	13¾	12¼	11	9¾	8½	7⅝	6¾	6½	5⅞
	1½	14	12¼	11	9⅜	8¼	7⅜	7⅛	6½	14⅜	12⅞	11⅝	10⅜	9⅛	8¼	7⅜	7⅛	6½
	2	14¼	12½	11¼	9⅝	8⅜	7½	7⅜	6¾	14½	13	11¾	10½	9¼	8⅜	7½	7⅜	6¾
	3	15⅛	13⅜	12⅛	10½	9¼	8⅜	8¼	7⅝	15⅜	13⅞	12⅝	11⅜	10⅛	9¼	8⅜	8¼	7⅝
	4	16	14¼	13	11⅜	10⅛	9¼	9⅛	8½	16¼	14¾	13½	12¼	11	10⅛	9¼	9⅛	8½
300	6	17¼	15½	14¼	12⅝	11⅜	10½	10⅜	9¾	17½	16	14¾	13½	12¼	11⅜	10½	10⅜	9¾
	8	18½	16¾	15½	13⅞	12⅝	11¾	11⅝	11	18¾	17¼	16	14¾	13½	12⅝	11¾	11⅝	11
	10	19¾	18	16¾	15⅛	13⅞	13	12⅞	12¼	20	18½	17¼	16	14¾	13⅞	13	12⅞	12¼
	12	21¼	19½	18¼	16⅝	15⅜	14½	14⅜	13¾	21½	20	18¾	17½	16¼	15⅜	14½	14⅜	13¾
	14	22½	20¾	19½	17⅞	16⅝	15¾	15⅝	15	22¾	21¼	20	18¾	17½	16⅝	15¾	15⅝	15
	16	23¾	22	20¾	19⅛	17⅞	17	16⅞	16¼	24	22½	21¼	20	18¾	17⅞	17	16⅞	16¼
	1	13½	11¾	10½	8⅞	7⅝	6¾	6½	5⅞	13¾	12¼	11	9¾	8½	7⅝	6¾	6½	5⅞
	1½	14	12¼	11	9⅜	8¼	7⅜	7⅛	6½	14⅜	12⅞	11⅝	10⅜	9⅛	8¼	7⅜	7⅛	
	2	14¼	12½	11¼	9⅝	8⅜	7½	7⅜	6¾	14½	13	11¾	10½	9¼	8⅜	7½		
	3	15⅛	13⅜	12⅛	10½	9¼	8⅜	8¼	7⅝	15⅜	13⅞	12⅝	11⅜	10⅛	9¼			
400	4	16	14¼	13	11⅜	10⅛	9¼	9⅛	8½	16¼	14¾	13½	12¼	11				
	6	17¼	15½	14¼	12⅝	11⅜	10½	10⅜	9¾	17½	16	14¾	13½					
	8	18½	16¾	15½	13⅞	12⅝	11¾	11⅝	11	18¾	17¼	16						
	10	19¾	18	16¾	15⅛	13⅞	13	12⅞	12¼	20	18½							
	12	21¼	19½	18¼	16⅝	15⅜	14½	14⅜	13¾	21½								
	1	13½	11¾	10½	8⅞	7⅝	6¾	6½	5⅞									
	1½	14	12¼	11	9⅜	8¼	7⅜	7⅛										
	2	14¼	12½	11¼	9⅝	8⅜	7½											
	3	15⅛	13⅜	12⅛	10½	9¼												
600	4	16⅜	14⅝	13⅜	11¾													
	6	18	16¼	15														
	8	19¼	17½															
	10	21																

300											150											Pressure	
16	14	12	10	8	6	4	3	2	1½	1	16	14	12	10	8	6	4	3	2	1½	1	Size	
15⅞	14⅝	13⅜	11⅞	10⅝	9⅜	8⅛	7¼	6⅜	6¼	5⅝	14⅞	13⅝	12⅝	11⅛	9⅞	8⅝	7⅝	6⅞	6⅛	5⅝	5¼	1	
16¼	15	13¾	12¼	11	9¾	8½	7⅝	6¾	6⅝	6	15¼	14	13	11½	10¼	9	8	7¼	6½	6		1½	
16¾	15½	14¼	12¾	11½	10¼	9	8⅛	7¼	7⅛	6½	15¾	14½	13½	12	10¾	9½	8½	7¾	7			2	
17½	16¼	15	13½	12¼	11	9¾	8⅞	8	7⅞	7¼	16½	15¼	14¼	12¾	11½	10¼	9¼	8½				3	
18¼	17	15¾	14¼	13	11¾	10½	9⅝	8¾	8⅝	8	17¼	16	15	13½	12¼	11	10					4	
19¼	18	16¾	15¼	14	12¾	11½	10⅝	9¾	9⅝	9	18¼	17	16	14½	13¼	12						6	150
20½	19¼	18	16½	15¼	14	12¾	11⅞	11	10⅞	10¼	19½	18¼	17¼	15¾	14½							8	
21¾	20½	19¼	17¾	16½	15¼	14	13⅛	12¼	12⅛	11½	20¾	19½	18½	17								10	
23¼	22	20¾	19¼	18	16¾	15½	14⅝	13¾	13⅝	13	22¼	21	20									12	
24¼	23	21¾	20¼	19	17¾	16½	15⅝	14¾	14⅝	14	23¼	22										14	
25½	24¼	23	21½	20¼	19	17¾	16⅞	16	15⅞	15¼	24½											16	
16¼	15	13¾	12¼	11	9¾	8½	7⅝	6¾	6½	5⅞												1	
16⅞	15⅝	14⅜	12⅞	11⅝	10⅜	9⅛	8¼	7⅜	7⅛													1½	
17	15¾	14½	13	11¾	10½	9¼	8⅜	7½														2	
17⅞	16⅝	15⅜	13⅞	12⅝	11⅜	10⅛	9¼															3	
18¾	17½	16¼	14¾	13½	12¼	11																4	
20	18¾	17½	16	14¾	13½																	6	300
21¼	20	18¾	17¼	16																		8	
22½	21¼	20	18½																			10	
24	22¾	21½																				12	
25¼	24																					14	
26½																						16	
																						1	
																						1½	
																						2	
																						3	
																						4	400
																						6	
																						8	
																						10	
																						12	
																						1	
																						1½	
																						2	
																						3	
																						4	600
																						6	
																						8	
																						10	

Inches in Decimals of a Foot

1/16	3/32	1/8	3/16	1/4	5/16	3/8	1/2	5/8	3/4	7/8
0.0052	0.0078	0.0104	0.0156	0.0208	0.0260	0.0313	0.0417	0.0521	0.0625	0.0729
1	2	3	4	5	6	7	8	9	10	11
0.0833	0.1667	0.2500	0.3333	0.4167	0.5000	0.5833	0.6667	0.7500	0.8333	0.9167

Decimal Equivalents of One Inch

1/64—0.015625	33/64—0.515625
1/32—0.03125	17/32—0.53125
3/64—0.046875	35/64—0.546875
1/16—0.0625	9/16—0.5625
5/64—0.078125	37/64—0.578125
3/32—0.09375	19/32—0.59375
7/64—0.109375	39/64—0.609375
1/8—0.125	5/8—0.625
9/64—0.140625	41/64—0.640625
5/32—0.15625	21/32—0.65625
11/64—0.171875	43/64—0.671875
3/16—0.1875	11/16—0.6875
13/64—0.203125	45/64—0.703125
7/32—0.21875	23/32—0.71875
15/64—0.234375	47/64—0.734375
1/4—0.25	3/4—0.75
17/64—0.265625	49/64—0.765625
9/32—0.28125	25/32—0.78125
19/64—0.296875	51/64—0.796875
5/16—0.3125	13/16—0.8125
21/64—0.328125	53/64—0.828125
11/32—0.34375	27/32—0.84375
23/64—0.359375	55/64—0.859375
3/8—0.375	7/8—0.875
25/64—0.390625	57/64—0.890625
13/32—0.40625	29/32—0.90625
27/64—0.421875	59/64—0.921875
7/16—0.4375	15/16—0.9375
29/64—0.453125	61/64—0.953125
15/32—0.46875	31/32—0.96875
31/64—0.484375	63/64—0.984375
1/2—0.5	1—1.

These are typical distances which are set by permissible clearances for the usual piping from the bottom of such vessels. In order to conveniently remove and install piping, at least 6 in. must be allowed between lowest projection on piping and grade. Do not forget insulation and valves in determining this clearance (see Fig. 7.4). If a control valve must be located in this piping, provide 12 in. below bottom of valve.

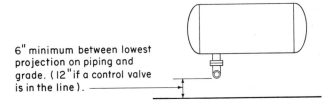

6" minimum between lowest projection on piping and grade. (12"if a control valve is in the line).

FIG. 7.4. Minimum clearances for piping may be set by discharge piping. *Note:* The lowest projection is often a pumpout, steamout, or drain. Adequate clearance below fittings associated with these connections must be provided.

3. Special Equipment. The piping designer should be cautious in establishing elevations for unfamiliar equipment even though the process engineers have not indicated any preference. He should discuss the nature and use of the equipment with them before making a decision. A particular autoclave may be planned with a hand-operated quick-acting dump valve in the bottom outlet. If this is to be the main operating point, the autoclave should be positioned so that the valve can be conveniently reached. If, on the other hand, the valve will rarely be used, a lower elevation of the autoclave may be indicated. The important point is to discuss such problems with the process and/or project engineer. In working with the details of the layout, the piping designer will often discover design problems which may have been overlooked. He can expect nothing but gratitude for this service.

PROCESS REQUIREMENTS

In any plant design project the piping designer will find that the elevation of certain items of equipment have been set by the process engineers. They may be given relative to grade or relative to tangent lines or center lines of some major piece of equipment. These elevations must be rigidly followed, for they are the results of calculations involving the physical and chemical characteristics of the fluids being handled and are the sole responsibility of the process engineer.

As in any design calculation, however, certain judgments must be made in arriving at the final specified condition. Thus the process engineer may wish to discuss his range of possible choices with the piping designer to learn the factors related to layout which, when combined with the process requirements, may enable him to choose the optimum elevation with more confidence.

The usual process reason for elevating equipment above the minimum "at-grade" elevation is the need to provide static head adequate for fluid flow. The examples shown in Fig. 7.5 are typical process-dictated reasons for elevating equipment. There are other reasons based on ease of operation which are characteristic of each type of equipment. If platforms are necessary for larger items of equipment, smaller items should be located on the level most convenient for operating and servicing in association with the major equipment. Remote control of most equipment makes this rule of limited use, but it should not be ignored when applicable.

LAYOUT FACTORS

Equipment may be elevated at times to improve layout and reduce piping. This situation most often arises when platforms already exist for some process reason. In this case, if a line must enter a heat exchanger from the top of a tall reactor and then return to a nearby point, piping can be saved by locating the exchanger in the structure near the outlet. It is not usually advisable, however, to elevate mechanical equipment for such reasons. Maintenance is made too difficult and the structure would have to be designed to withstand vibrations. In all cases, the piping designer must weigh savings in piping costs against possible increases in costs resulting from adding equipment within the elevated structure.

SPECIAL FACTORS

Large settling basins are often built into the ground, using walls of reinforced concrete or sheet piling. In this manner the surrounding undisturbed soil aids in supporting the walls. Pumps are also located below grade at times when suction must be taken from ponds or sumps.

Horizontal Spacing of Equipment

Ease of construction and maintenance and safety are the most important factors in determining the spacing of process equipment.

SAFETY

Operating units in plants are separated not only for convenience of operation and maintenance but also to insure safe conditions. Separating certain equipment within an area may also be dictated by rules of safety. Process engineers must be relied upon to define the nature of hazardous operations and equipment. Once defined,

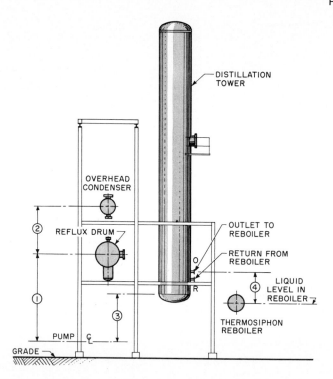

FIG. 7.5. Reasons for elevating equipment to assure adequate static head.

1. Reflux drum—The liquid in this drum is at its boiling point and a certain amount of head above that necessary to overcome friction losses is required for satisfactory pump operation. This head is called net positive suction head (NPSH) and is specified by the pump manufacturer. The process engineer calculates the elevation required to give this head.

2. The product from the overhead condenser must flow by gravity to the reflux drum. The height required to overcome friction, including control-valve pressure drop, is specified by the process engineer.

3. The height of the bottom tangent of the tower to grade is usually set by the NPSH requirements for the pump required for pumping

liquid from the tower. However, note 4 which follows may govern the placement in some cases.

4. The position of a thermosiphon reboiler is governed by the fixed positions of the inlet and return nozzles on the tower. Boiling liquid leaves the tower and liquid plus vapors return. The return fluid is lighter and a siphon effect is created. The process engineer determines the distance "4." It is possible that it can govern the height of the tower. If this occurs, the process engineer should be alerted because he can reduce the distance by increasing the line size. The decision can be made on the basis of a cost comparison of pipe cost versus extra tower height.

Table 7.1 (page 183) will be helpful in establishing preliminary layouts for study by the potential operators of the plant.

CONSTRUCTION

No plot plan is adequate if the problems of construction were not considered in its development. The possibilities for access of mobile cranes and placement of guy derricks must be carefully studied when planning the location and spacing of equipment. Models are most helpful in this task.

MAINTENANCE AND OPERATION

Equipment must be arranged so that access for maintenance and operation is possible without difficulty. Adequate space must be allowed for removing parts for cleaning or servicing. Some of the more common situations related to maintaining and operating equipment

will suffice to suggest the several reasons for space allocations.

Vessels. The spacing of fractionating towers in the same structure is usually set by associated elevated condensing equipment. A center-to-center spacing of 3 to 4 diameters is usually adequate for large vertical vessels without associated equipment in the structure. A space of 4 ft between the shells of horizontal vessels provides the space needed for painting and inspecting these vessels, if piping is not located to obstruct the area.

Heat Exchangers. A typical heat exchanger is shown in Fig. 7.6. The tube bundle must be removed at times for repair and cleaning. To do this the shell is removed and then the floating head. The channel is then removed and the bundle can be pulled out from the shell. Thus space must be provided for swinging the shell cover free of the exchanger and for pulling the bundle (tube bundle length plus 2 to 5 ft).

For elevated exchangers the tube-bundle space must be maintained at ground level so that the bundle can be

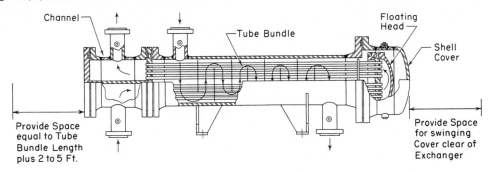

FIG. 7.6. Horizontal spacing at exchangers. (Reproduced by permission of Engineers and Fabricators, Inc., Houston, Texas.)

lowered to that level by means of a mobile crane or a monorail installed in the structure. The platform access space at either end of elevated exchangers need not exceed 4 to 5 ft since the monorail can be extended sufficiently to lower the bundle (see Fig. 7.7).

Exchangers arranged in a row at grade should have 3 ft of clear access between units. The bundle-pulling

area is best oriented toward an adjacent roadway or serviceway. Often a gantry crane is installed for a large number of in-line exchangers (Fig. 7.8). In both the monorail and gantry system the designer must be certain that equipment and piping will not interfere with the hoist or other projections on the lift.

The space at furnaces needed for tube-pulling should

FIG. 7.7. Elevated exchanger and monorail.

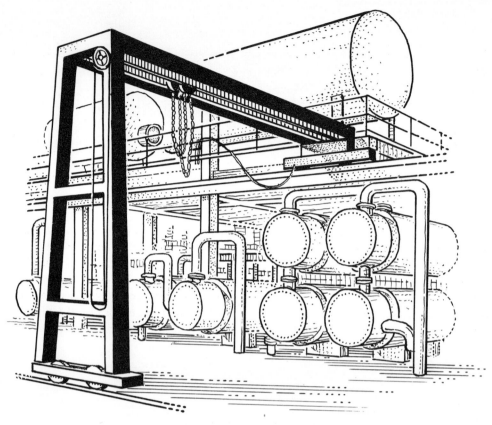

FIG. 7.8.   Gantry used with exchangers arranged in a row.

be shown on the plot plan.   Heaters are usually located near the unit edge, but the adjacent roadway should not be obstructed during tube-pulling.

Pumps, Compressors, and Other Mechanical Equipment. Mechanical equipment must be so placed that servicing is not complicated by problems of access to any part of the apparatus.   Pumps which are usually located outdoors are most advantageously placed in a row adjacent to road or accessway so that parts or entire units can be removed to the shop by conventional lifting equipment. A usually satisfactory free space around pumps is 3 ft. Space must be allowed for removing pistons and rods of reciprocating pumps.   Be sure to allow for piping in planning the spacing, for the 3 ft should be free of all interference, including piping.   Small pumps can be placed much closer where access to the whole unit is possible from the front end.   An operating aisle at the driver ends of pumps in a row should be a minimum of 5 to 6 ft.

Compressors are installed inside buildings, and these buildings must be planned to provide the space for servicing engine and compressor cylinders (Fig. 7.9).   A bridge crane is often provided for this purpose.   The free space at the compressor cylinder-end must allow for re-

moving the piston and for manipulating other parts removed from the unit.   The spacing of the several compressors will vary with size and style.   A common minimum is 8 to 10 ft.

Before deciding final access space around any mechanical apparatus, the designer should be certain someone familiar with its maintenance has studied the servicing and parts removal problem.   Centrifuges, filters, crystallizers, grinders, and many other types of mechanical equipment can be studied by the same criteria already outlined.   The main danger is allocating too little space.   It is rare that anyone will complain of too much.

Other Space Requirements.   In addition to the spaces needed between and around equipment, rail facilities, roadways, unit service or accessways, work areas, and pipeways are also needed.   Roadways and accessways through process units should not be less than 20 ft wide. Work areas for material and equipment storage during maintenance work on one or more units must be sized by maintenance people.   The work area is an important space that must not be forgotten.   Space for pipeways must also be planned for horizontal runs of pipe on racks or sleepers between and within units.

PIPING LAYOUT AND ARRANGEMENT

The piping designer can begin layout and arrangement of piping when he receives approved plot plans, flow diagrams, piping specifications, vessel drawings, vendor prints, and instrument location lists. A number of procedures are equally satisfactory. One common method is to prepare freehand sketches of all lines in plan by means of colored crayon directly on prints of the plot plan. Sometimes the plot plan is enlarged photographically or sketch-drawn to a large scale particularly for a project where piping is extensive. These sketches are also useful in taking-off materials for early placement of orders. When models are used, the designer studies the equipment on the model, using freehand sketches when needed. As he proceeds, he directs the model maker to install the piping in accordance with his sketch or oral instruction. The great advantage in this method is that the designer as well as his supervisors and operating company representatives can see the results unfold rather than wait for prints of a completed study.

When all inter-connecting piping between equipment is determined and shown schematically, the number of lines in various areas may be counted and the sizes of support racks for multiple horizontal runs checked with that estimated during the plot planning stage. After selecting the final width and height of the pipe racks or pipe ways, yard piping, consisting of parallel runs of piping, may be laid-out in more accurate detail. When models are used, the model builders may proceed with the racks and horizontal piping runs. Exact scaling in this type of drafting is usually wasteful if personnel performing piping layout are experienced.

After piping is arranged in plan (or completed on the model), piping runs from the racks to vessels, mechanical equipment, and heat exchange apparatus may be sketched (or may be installed by the model-builders).

Freehand sketching in the initial layout of piping often is an efficient and simple method of quickly determining the best arrangement. Freehand sketches can be redrawn by less experienced personnel, relieving senior designers for more important work. If models are used, all sketching should be reduced to the absolute minimum or the value of the model in reducing drafting costs will be lost.

There are no fixed rules for piping layout which may be applied arbitrarily to all plants, but some useful generalizations are possible and these will be helpful in laying out and arranging piping. These appear in the following sections categorized according to equipment or phase of piping design.

Nozzle Orientation

Nozzle elevations on vessels are invariably set by process engineers because position is determined during process design. Orienting nozzles within a given elevation, however, is the job of the piping designer so that ideal piping arrangements can be realized. Alternate nozzle orientations may be possible for manufactured

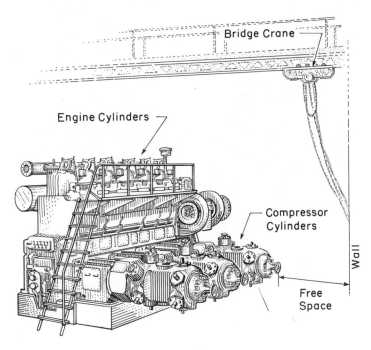

FIG. 7.9. Space needs at compressors.

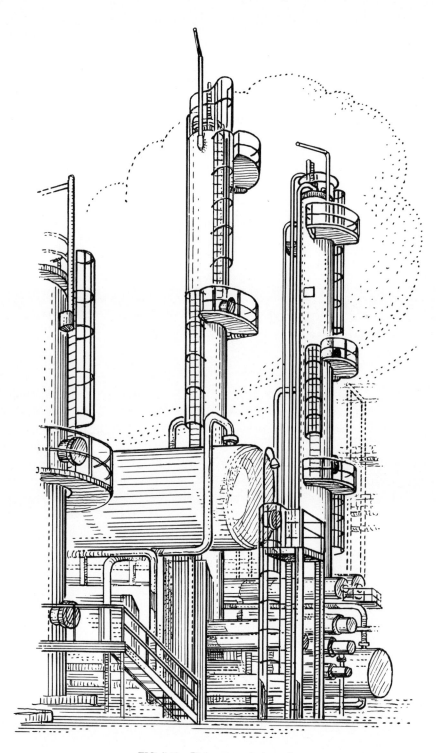

FIG. 7.10.   Piping at vertical vessels.

equipment, including heat exchangers, certain pumps, compressors, and other mechanical equipment. The piping designer should take advantage of these possibilities to improve the arrangement of connecting piping.

Vessels

VERTICAL VESSELS (Fig. 7.10)

Locate vertical runs of piping at the outside of platforms or elevated structures. It simplifies piping and minimizes chances for interferences. If piping is planned to pass through platforms or floors, extra work is incurred, and the needed openings often precipitate jurisdictional disputes between organized labor crafts participating in the construction. Piping from the upper portions of vertical vessels should be supported close to the vessel by attachments located on the vessel. This reduces the relative movement between the supports and the piping.

Run piping under rather than directly over platforms.

Orient all manholes and ladders in the same area if possible. This leaves uncluttered areas free of structural steel along the entire length of a vessel.

Layout lines from top of a vessel first. These are often large lines and the remaining spaces are clearly seen.

Provide clear space from the top of a vessel to ground for lowering parts and valves during maintenance by means of a davit.

Provide vent, drain, steam-out, and relief valve connections for most vessels. Check with project engineer if not shown on flow diagram to determine if they were forgotten.

HORIZONTAL VESSELS

Locate inlet and outlet at opposite ends, with inlet on top and outlet on bottom unless otherwise specified.

Locate liquid level indicator or controller near the center, remote from disturbances introduced by inlet or outlet nozzles.

Pressure connection should be located at top of drum and temperature connection at bottom near outlet or in outlet line.

Steam-out connection should be tangential and near bottom, opposite vent or manhole.

Locate manhole for easy access which may be in top, sides, or ends.

Heat Exchangers

Keep piping out of lifting area and spool for easy removal of "break-out" flanged piece. This avoids handling of long lengths of pipe during maintenance (Fig. 7.11).

Exchanger specifications usually give flow direction. If not given, pass cooling water through tubes entering at bottom channel connection and out the top. When so arranged the exchanger will retain water upon failure of water pressure. In general, the heated stream is passed up through the exchanger and the cooled stream down. Condensing vapors, including steam, enter at the top and discharge from the bottom (usually the shell side).

Exchangers in series or parallel arrangement can be conveniently stacked up to a height of about 12 ft (Fig. 7.8).

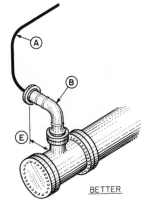
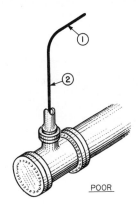

FIG. 7.11. Piping at heat exchangers.

Better

A. Piping out of lifting "removal" area.
B. "Break-out" flanged piece.
C. Stress reduced.
D. Piping can be supported from grade.
E. Adequate dimension to clear equipment.

Poor

1. Piping should not run directly over equipment.
2. No connections provided for removal of equipment.
3. Possible stress problem.
4. Piping support problem when equipment removed.

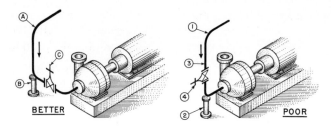

FIG. 7.12. Piping at pumps.

Better	*Poor*
A. No interference from overhead piping.	1. Piping should not be run directly over equipment. Fouls maintenance or repair removal.
B. Support actually relieves stress from pipe weight.	2. Support poor. Useless until equipment removed.
C. Valve in better operating position—also better for maintenance.	3. Congestion in direct front of equipment.
D. Stress reduced to equipment.	4. Valve wheel projects into walk ways from any location except if turned into equipment.
	5. Equipment will be stressed from bending.

Locate valves and instruments toward access aisles.

Check for access to bolts on pipe and valve flanges.

Avoid center line of exchanger for lines that must turn right or left. If nozzle is on center line, direct line to right or left at nozzle.

Pumps

Do not run piping directly over pumps. It makes maintenance of pump difficult (Fig. 7.12).

Avoid piping congestion in front of pumps (Fig. 7.12).

Locate pumps in-line and as nearly as possible opposite to the source of the fluid each is to pump. This reduces piping.

Anchor or support suction and discharge lines to prevent excessive stresses on pump casing. See section on "Piping Flexibility."

Suction lines should be as short as possible, with a minimum number of bends. This keeps pressure-drop down and helps to assure trouble-free pump operation. Use long-radius elbows for bends. Avoid sudden decreases in line size.

Use eccentric reducers with sloping side downward to avoid gas pockets (Fig. 7.13).

Slope suction line ¼ in. to the foot toward the pump when the pump is below the source and in the opposite direction when the pump is above the source.

Elbows entering double suction pumps should be

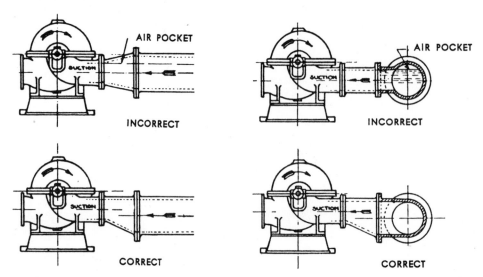

FIG. 7.13. Use of eccentric reducers. (Reproduced by permission from "Standards of the Hydraulic Institute," copyright 1955 by Hydraulic Institute, 122 East 42nd St., New York 17, N.Y.)

installed in the vertical to avoid unequal flow distribution.

Provide strainers in suction lines with cross-sectional areas three to four times that of line.

Avoid valves in suction lines unless needed for mani-folded pumps.

Provide approximately three diameters straight run of pipe between pump suction and last elbow. This avoids detrimental turbulence at inlet.

Place check valve between discharge nozzle and gate valve to prevent liquid backup when pump stops running.

Place relief valve between pump and gate valve of reciprocating pump. Steam-driven pumps with stalling pressures below allowable cylinder pressure do not re-quire relief valves.

Specify flat-face companion flange for pumps having flat-face flanges.

Auxiliary piping, such as gland oil, and cooling water piping should be arranged so that servicing and inspec-tion of the pump will not be impaired.

Compressors (Fig. 7.14)

Arrange engine-driven reciprocating compressors with crankshafts in single line. Opposed cylinder motor-driven compressors should be set so that cylinders run along both sides of building.

Avoid changes in piping direction and where nec-essary use long-radius elbows.

Avoid sharp bend at inlet of reciprocating compressors.

Use 10° bends on discharge line of centrifugal com-pressors when bends are required.

Do not use overhead piping at reciprocating ma-chines. Excessive vibration becomes a problem with overhead piping. Support compressor piping separately from other piping.

Install strainers in suction lines.

Anchor and support piping where necessary to prevent excessive strain on casing of centrifugals and cylinders of reciprocating compressors. Discharge line to surge drum should have unsupported elbow so pipe can bend with thermal stress, and relieve strain on compressor casing or cylinder.

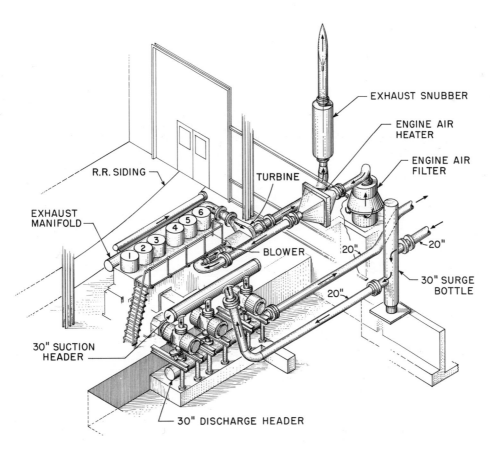

FIG. 7.14. Piping at compressors. (Adapted by permission of Clark Bros. Co., Inc., Division of Dresser Industries, Inc.)

Provide short spool pieces at compressors for easy removal during servicing. This is particularly important for centrifugal compressors having top halves that must be lifted for access to impellers.

Use check valves in discharge lines of centrifugals discharging into same header to protect from surging. Install as closely as possible to compressor.

Piping specifications should include instructions for cleaning all process piping at compressors.

Avoid loops or pockets in suction piping to prevent collection of condensate that can cause great damage to compressors.

Locate headers convenient to gas supply and parallel to row of compressors, with plenty of room for scrubbers when required. Locate headers first, because these set height of compressor house floor. Position from building may be dictated by pulsation damper size.

Locate stop valves at headers in easily accessible spots.

Locate safety valves near building with vent firmly anchored.

Place discharge surge chambers on sleepers under compressor cylinders.

Locate suction surge chambers on top of cylinders, using shortest possible connection.

Outlet jacket water should flow up to outlet header to avoid air traps.

Piping of accessories is important. Data are furnished on separate drawings based on manufacturers' recommendations.

Surge chambers should be placed so that they can easily be removed.

Stress and vibration analysis is important (see sections on this subject).

Locate scrubbers convenient to headers.

Tankage and Transfer Pump Manifolds

Design maximum flexibility in transfer pump manifold.

Locate pumps as near as possible to tankage.

Use single suction lines from tanks. Manifold at pumps when necessary.

Keep suction lines short.

Use separate lines for each product and isolate product lines with double-block valves with bleed in between blocks.

Provide bucket strainers at all transfer pumps.

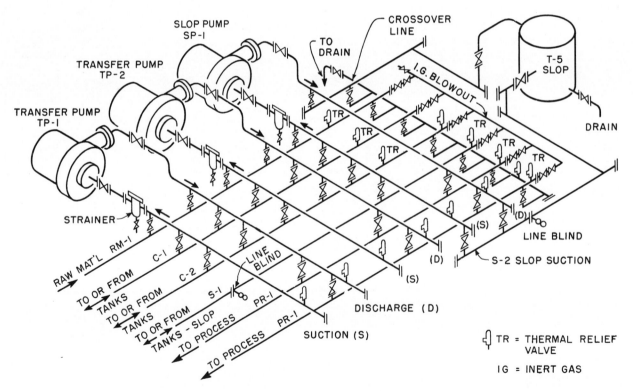

FIG. 7.15. Transfer pump manifold.

Provisions should be made for draining and/or purging all lines reasonably free of liquid. (See Fig. 7.15. Fluid may be purged via cross-over near Tank T-1 into line S-1, and via cross-over at manifold the liquid may be purged into T-5 slop tank. Any of the lines C-1, C-2, or S-1 may be purged from the tank area into T-5. The overhead fill line eliminates possibility of flow-back from T-5.)

Use line blinds as a precaution against minor leakage between lines that would constitute a hazard or serious impairment of product quality.

Provide more flanges than normally used, because tankage piping is subject to frequent changes.

Valving for manifolds composed of piping 3 in. and under may be simplified by using three-way cocks. For larger lines it is more economical to use ordinary valves or cocks.

Plug cocks are ideal for manifold service because they can be operated faster and are less likely to leak than other valves.

All portions of lines which may be blocked-off by valves when full of liquid should be provided with thermal relief valves (small ½ in. or ¾ in. spring-type relief valves located as shown in Fig. 7.15). The relieving pressure is set well above pumping pressures, but below fitting and pipe-failure pressures.

Piping in hazardous storage areas should be buried, or fireproofed if above ground. Above-ground piping that is fireproofed is generally preferred because buried pipe constitutes a maintenance problem. All valving should be steel, not cast iron.

Liquid Loading Facilities (See Fig. 7.16)

Number of connections is governed by fluids to be handled.

Tank cars and trucks may be unloaded from the bottom by gravity or by pumps. The safest unloading method is through the top, as shown in Fig. 7.16. In this manner accidental movement of the tank car will not result in the escape of all its contents. Bottom unloading of tank trucks is acceptable. A remote control valve for emergency shut-off should be provided at tank cars when bottom loading is employed.

Use mechanical swivel joints for flexibility of loading and unloading connections. Hose connections may be used instead, particularly for truck loading docks.

Place meters in piping near eye level.

Prefabricated assemblies for both railroad and truck loading docks may be purchased.

Valves

Locate key process valves for easy accessibility.

Control valves and relief valves should be located for easy repair or replacement (see Chapter 10 for details on control-valve piping).

Hand-operated valves requiring regular operation should be located so they can be operated without chain operators.

Block valves requiring only infrequent use should be accessible from platforms, if possible without additional expense. Otherwise, use extension stems or chain operators for valves above 7 ft 3 in. level.

Place block valves at header in all small branch lines.

Locate emergency valves in remote spot, such as back side of control room, so that operator can close valves

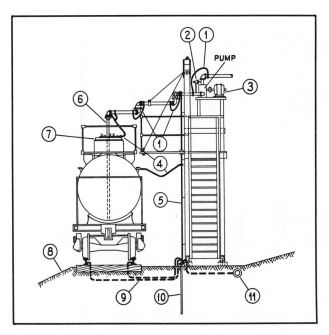

FIG. 7.16. Railroad tank car loading rack, showing piping and associated protective equipment. (Reproduced by permission from Factory Mutual Engineering Division, *Handbook of Industrial Loss Prevention*, McGraw-Hill Book Co., New York, 1959.)

1. Bonding wire attached with ground clamp.
2. Relief valve by-pass.
3. Explosion-proof motor.
4. Insulated flexible grounding cable. Not less than No. 4. Attached to tank car with ground clamp.
5. No. 4 stranded cable secured to platform column.
6. Non-ferrous tube.
7. Safety dome cover.
8. Ground slopes away from important facilities.
9. Bare copper conductor.
10. Ground rod driven to permanent moisture level.
11. Water main, if available.

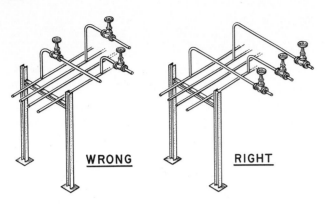

FIG. 7.17. Location of valves at pipe rack.

during fire or other emergency while being in relative safety as well as in a position to leave area should this be necessary.

Good quality valves are designed to sustain loading of attached piping, but low-pressure valves should not be used with heavy piping unless loads are analyzed, since stresses produced may exceed capacity of valve body. In all cases, avoid loads on valves which will cause distortion and prevent satisfactory operation.

Locate valves outside of pipe racks for ease of accessibility (see Fig. 7.17).

Do not locate valves in vertical runs where the take-off is from a header or vessel since this traps liquid and vessel or header cannot be completely drained during repair or inspection (see Fig. 7.18).

Piping (General)

CLEARANCES—MINIMUM

Above platforms and in buildings	7 ft
Above grade	9 ft
Over roadways	15 ft to 22 ft (depends on maintenance equipment)
Above grade for pipe on sleepers	12 in. to 18 in.
Vertical runs adjacent to equipment or structures—horizontal clearance	12 in.

LINE SIZES

(a) Do not use ⅜ in., 1¼ in., 2½ in., 3½ in., 4½ in., 5 in., and 7 in. (not commonly stocked).

(b) Drains, vents, and sample connections should be ¾ in. minimum.

(c) Overhead lines should not be less than 1 in., except for instrument leads. Smaller lines are difficult to support.

Spacing. Assume all lines are flanged and arranged with 1 in. between flanges. Run with flanges staggered, since this usually permits ample room for servicing when insulated and provides rapid means of estimating pipe rack widths (see Table 7.2, page 184).

Length of Lines. Keep as short as possible, especially alloy lines or lines requiring minimum pressure drop, such as pump suction and vacuum lines.

Run all lines above ground except water lines, sewer lines, or certain designated compressor lines.

Arrange lines running in same elevation so that bottom of lines are in the same plane. Insulated lines are supported on shoes to protect insulation which changes elevation slightly relative to bare pipe (see Fig. 7.19). Moderate temperature lines may be placed at same bottom-of-pipe elevation by cutting bottom portion of insulation as shown in Fig. 7.19.

Never change direction without changing elevation. Establish one elevation for East-West lines and another for North-South, usually 18 to 30 in. above, below, or both. This rule avoids interferences and confusion of lines (see Fig. 7.20).

Changes in direction of underground lines should be made with a change in elevation except for large mains.

Provide drains at low points in piping and equipment and at pockets in piping. Complete draining of all systems must be possible.

Provide vents at high points in piping where gas pockets may occur.

Provide ¾ in. utility-hose connections for water, steam and air. Locate these so that equipment can be reached by hose lengths not in excess of 50 ft.

Avoid pipe in trenches. When trenches are necessary, provide ample room around flanges and valves. About 6 in. is needed for effective maintenance. Trenches must be drained.

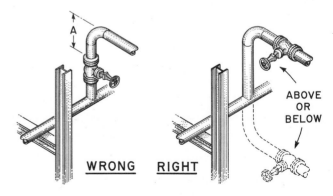

FIG. 7.18. Avoid locating valves in vertical runs. (Liquid will be trapped in the section A.)

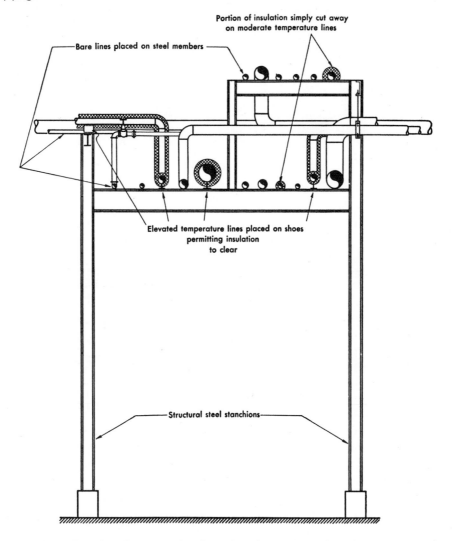

FIG. 7.19. Typical outdoor pipe rack. (Reproduced by permission of copyright holder, M. W. Kellogg, Co., from *Design of Piping Systems,* 2nd ed., John Wiley and Sons, New York, 1956.)

PIPE SUPPORT AND YARD PIPING

In the orderly arrangement of piping within a process unit and interconnecting piping between units, inevitably a number of lines are most conveniently grouped at the same elevation. These so-called banks or racks of pipe, often referred to as yard piping, must be supported. Such piping may be run at 12 to 18 in. above grade on concrete sleepers as in Fig. 7.21, below-grade (underground), and overhead on individual supports or common racks (Fig. 7.19 and 7.22). Installation costs increase in the order named. Thus, usually only water piping (in cold climates) and sewers and drains are installed underground in the modern process plant. Pipe on sleepers is most desirable in areas with little equipment, but within the usual process area overhead piping is preferred because ease of access to process equipment is assured by this arrangement. Figures

7.22, 7.23, 7.24, and 7.25 show conventional type yard-pipe racks and details.

As noted on Fig. 7.22, it is advantageous to use modular dimensions and standardized structural members and anchor bolts. Unless there is significant movement in the piping, and the size of piping requires intermediate support, the side stringers may usually be omitted. The 20-ft spacing between bents is generally satisfactory for most process plant piping. Allowable spacings may be calculated, using the beam relationships,[2] $S = 1.2(Wl^2/Z)$ where $S =$ maximum bending stress, psi; $Z =$ section modulus, in.[3]; $l =$ pipe span in ft; $W =$ total unit weight, lb/ft. The design of pipe supports must follow the rules of the American Institute of Steel Construction or American Concrete Institute.

Uninsulated piping is usually supported directly on a ¾-in. or 1-in. diameter bar welded to the steel member. Shoes are provided for insulated lines. Hot lines require

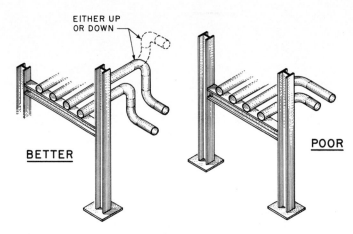

FIG. 7.20. Change elevation with change in direction.

guides and restrictions against vertical movement at some points. Figure 7.26 presents details of inexpensive procedures for providing pipe supports and anchors by field fabrication.* Elbow extensions to grade at pump suctions and other similar supports can also be field-fabricated, often from scrap structural steel or pipe. The piping designer must indicate where supports, guides, stops, anchors, and hangers are required. Most organizations will have standards for the usual cases. Heavy lines and lines which cause thermal expansion and vibration problems require special attention to the supporting, restraining, and bracing of piping. These problems and their solutions have been thoroughly discussed.[2]

Rules for Yard Piping Layout

Kern has given some common-sense suggestions for yard piping layout which are abstracted below in check list form (see Fig. 7.27).[1]

Heavy lines should be placed when possible near columns to reduce size of structural steel.

Place utility lines in the center because they serve equipment on both sides of bank. Process lines connecting to both sides of yard bank should be adjacent to the utility lines.

Process lines connecting to equipment to right of yard bank should be on right side, while those connecting to left side, on left.

If a large number of lines require two elevations, place utility lines in top bank and process lines in bottom. However, process lines which connect nozzles at higher elevation than top bank should be located in top bank.

Locate horizontal loops over yard as shown in Fig. 7.27.

These rules should prove to be helpful starting points,

* See Figs. 2.23, 2.24, and 2.25 for examples of manufactured hangers, supports, and expansion joints.

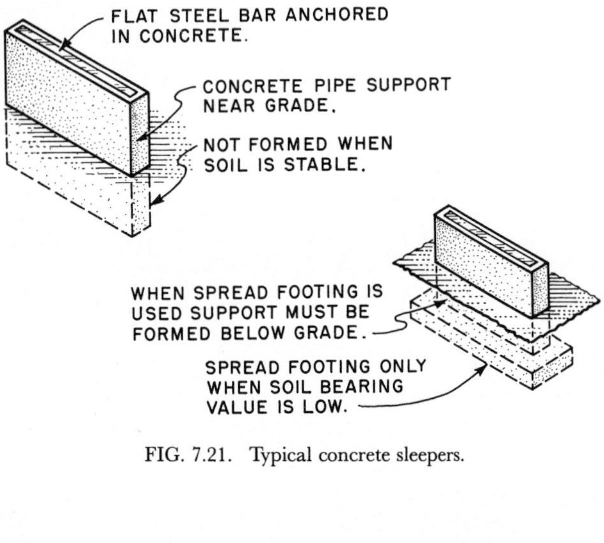

FIG. 7.21. Typical concrete sleepers.

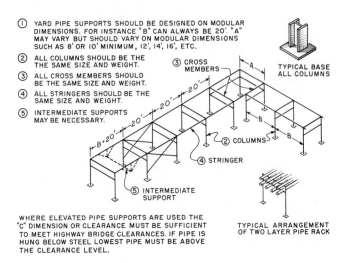

FIG. 7.22. Pipe rack arrangement.

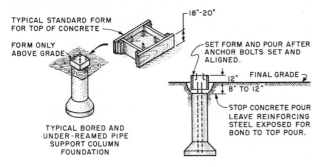

FIG. 7.23. Pipe rack foundations.

but the designer should always be governed not only by savings in piping and structural steel, but also by savings in construction and maintenance labor. Multiple-layer runs of piping, for example, may conserve steel and horizontal space, but single layer pipe racks generally save more in erection costs than the additional cost of structural steel.

PIPING STRESS ANALYSIS

The three conditions which cause significant stresses in piping are: (1) internal or external pressure; (2) external loads caused by the mass of the pipe, valves, fittings, and the fluid contained in the pipe and insulation, or dynamic loads such as wind or earthquake; (3) thermal expansion of the pipe. The selection of valves, fittings, and pipe to withstand pressure stresses has already been outlined in Chapter 2 and 3 in which the "ASA Code for Pressure Piping" has been discussed. The weight loads of the piping are for most cases controlled by employing adequate supports in accordance with standardized construction practices. The analysis of stresses

produced by thermal expansion is more intangible. Although thermal stresses can be anticipated; they cannot be precisely evaluated by simple methods. Fortunately, by following the rule of always using 90° bends or elbows when changing direction, most process piping systems have a good deal of inherent flexibility which renders harmless the thermal stresses produced during operation and shut-down. There are always, however, a certain number of lines within any process unit which require careful and competent stress analysis in order to make certain that piping reactions on equipment are within acceptable limits and that stresses will not develop sufficiently in the pipe to cause failure and joint leakage.

Accurate flexibility analysis is advisable in the following situations: (1) unusual piping arrangements for

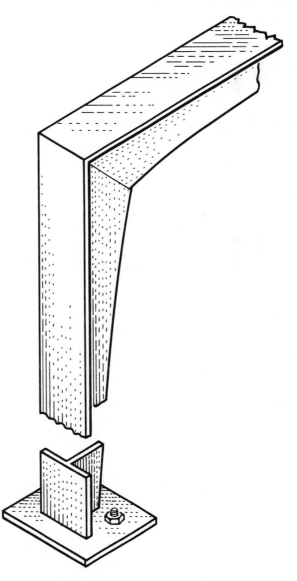

FIG. 7.25. Light prefabricated structural steel members may be used as pipe supports.

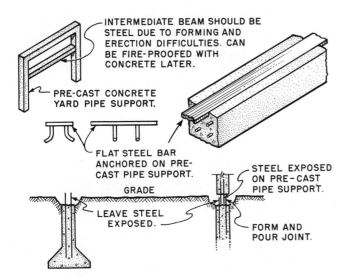

FIG. 7.24. Reinforced concrete pipe supports.

which approximate methods of analysis yield uncertain results; (2) hazardous lines at high temperatures or pressures; (3) stiff configurations requiring mandatory analysis in accordance with the piping code (see below).

What is the concern of the piping layout designer with stress analysis? First, accurate stress analysis is a job for experts. Much of the work now is done by using electronic computing machines, but the results require expert interpretation. It may be concluded, therefore, that the piping layout designer in general will refer complex problems of stress analysis to experts within his organization. But it would be intolerably costly to subject each line in the plant to complete analysis; thus a share of the burden in providing adequate flexibility for piping belongs to the piping designer because, by care in arranging piping, he can assure that the majority of the lines are safely within the stress limitations and will not require thorough analysis. To accomplish this task, he must have some understanding of the basic facts related to piping flexibility and be able to design flexible piping through proper layout.

Each time piping changes directions and is left free to move at the point of change, the system becomes more flexible. Figure 7.28 illustrates various offsets and loops which provide flexibility in piping. Sketches are arranged from left to right in increasing order of flexibility. The effect of configuration on a U-bend is also demonstrated. By making bends in more than one plane, the flexibility is further enhanced.

If the designer has a means of knowing when a given configuration is sufficiently flexible to be below the maximum allowable stress, he has a tool that will enable him to develop configurations which in most cases will be satisfactory and will not require thorough analysis. Fortunately, the "ASA Code for Pressure Piping" has given an empirical rule, which is widely used, for determining when a formal stress analysis is required for a two-anchor system of uniform pipe size. The rule states that formal analysis is required if the following approximate criteria are not satisfied:

$$\frac{DY}{U^2(R-1)^2} \le 0.03$$

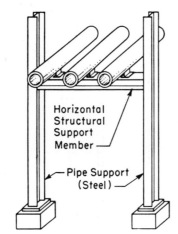

(a) Structural Steel Pipe Supports

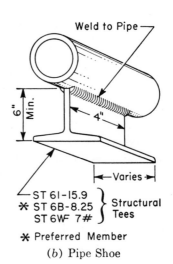

(b) Pipe Shoe

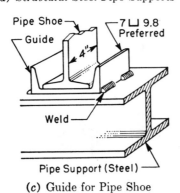

(c) Guide for Pipe Shoe

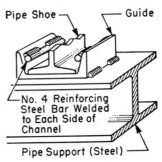

(d) Method for Restricting Vertical Movement

FIG. 7.26. Field-fabricated pipe supports and anchors. (Reproduced from H. F. Rase and M. H. Barrow, *Project Engineering of Process Plants,* John Wiley and Sons, New York, 1957.)

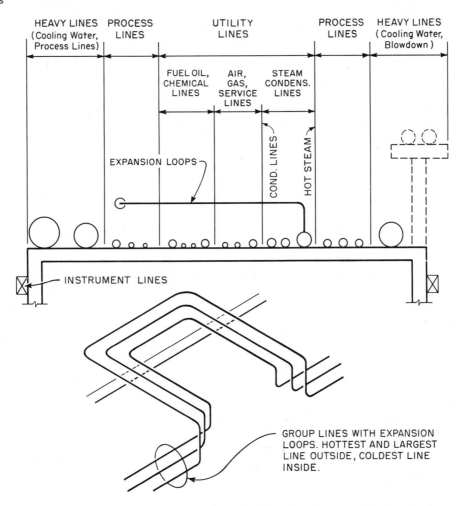

FIG. 7.27. Line arrangement on a one-level rack. [Reproduced by permission from Robert Kern, *Petroleum Refiner,* **39,** No. 12, p. 142 (1960). Copyright Gulf Publishing Co., Houston, Texas.]

where
D = nominal pipe size, inches.

Y = resultant of restrained thermal expansion and net linear terminal displacements, in inches = $[(eL_x)^2 + (eL_y)^2 + (eL_z)^2]^{1/2}$. Net linear terminal displacement should be added to eL term corresponding to direction in which it occurs (see Example 8.4).

U = anchor distance (length of straight line joining terminal or anchor points), feet.

R = ratio of developed pipe length to anchor distance, dimensionless.

L_x, L_y, L_z = projected lengths in designated directions, feet.

e = unit linear thermal expansion (see Fig. 7.29).

This equation has been plotted in handy graphical form (Fig. 7.30).[2] The chart has been simplified for carbon steels. Instead of calculating Y/U, it may be

noted from the geometry of configuration that $U = (L_x^2 + L_y^2 + L_z^2)^{1/2}$. Thus $Y = eU$ or $Y/U = e$. Now e is a function of temperature for a given material (Fig. 7.29). This function has been superimposed on the lower portion of the chart so that temperature may be used instead of Y/U. For those cases involving terminal displacements, Y/U must be calculated as in Example 7.4.

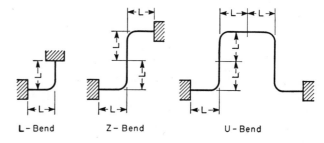

FIG. 7.28. Common expansion bends. (Reproduced from H. F. Rase and M. H. Barrow, *Project Engineering of Process Plants,* John Wiley and Sons, New York, 1957.)

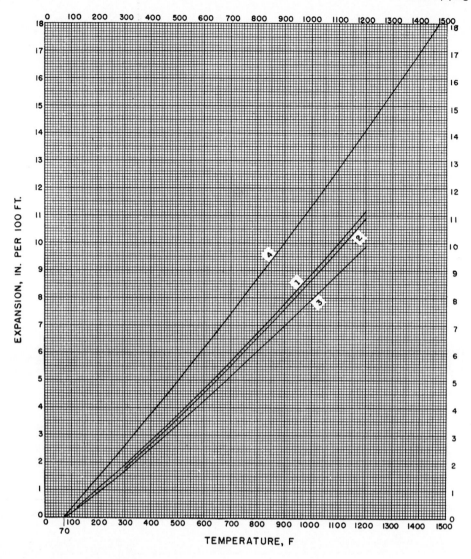

FIG. 7.29.   Thermal expansion for carbon and alloy steels.   (Reproduced by permission from M. W. Kellogg Co., *Design of Piping Systems,* 2nd ed., John Wiley and Sons, New York, 1956.)

1. Carbon, carbon-½% molybdenum, and ½% chromium-½% molybdenum steels.
2. 1% to 3% chromium-½% to 1% molybdenum steels.

3. 4% to 10% chromium-½% to 1½% molybdenum steels.
4. 18% chrominum-8% nickel steels (AISI types 302, 303, 304, 321, and 347.)

Example 7.1. *

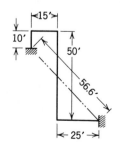

Material: ASTM A106, Gr A
Design temperature: $T = 900°$F
Unit expansion from 70°F: 0.078 in./ft (Fig. 7.29)

Type of service: Oil piping
Nominal pipe size: $D = 10$ in.
Developed length: $L = 100$ ft
Anchor distance: $U = 56.6$ ft
$U/D = 5.66$
$L/U = 1.77$
From Fig. 7.30, $R = 1.68$
$R < L/U$; hence, formal calculations are not mandatory.

*Examples 7.1 to 7.4 reproduced by permission: M. W. Kellogg Company, New York, *Design of Piping Systems,* John Wiley and Sons, New York, 1956.

Example 7.2.

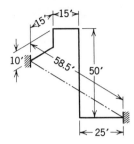

Material: ASTM A106, Gr A
Design temperature: $T = 900°$F
Unit expansion from 70°F: 0.078 in./ft
Type of service: Oil piping
Nominal pipe size: $D = 10$ in.
Developed length: $L = 115$ ft
Anchor distance: $U = 58.5$ ft
$U/D = 5.85$
$L/U = 1.97$
ASA B31.1 Code Criterion Fig. 7.30, $R = 1.67$
$R < L/U$; hence, formal calculations are not
  mandatory.

Example 7.3.

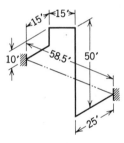

Material: ASTM A106, Gr A
Design temperature: $T = 900°$F
Unit expansion from 70°F: 0.078 in./ft
Type of service: Oil piping
Nominal pipe size: $D = 10$ in.
Results identical with those of Example 7.2.

Example 7.4. Displacement of Anchor Points.

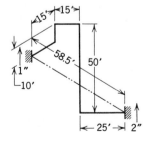

Material: ASTM A106, Gr A
Design temperature: $T = 650°$F
Unit expansion from 70°F: 0.052 in./ft
Type of service: Oil piping

Nominal pipe size: $D = 10$ in.
Developed length: $L = 115$ ft
Anchor distance: $U = 58.5$ ft
Expansion and terminal displacements:
  x-direction: 0.052×40 $= 2.08$ in.
  y-direction: $(0.052 \times 40) + 2 - 1 = 3.08$ in.
  z-direction: 0.052×15 $= 0.78$ in.
  $Y = (2.08^2 + 3.08^2 + 0.078^2)^{1/2}$ $= 3.8$ in.
$U/D = 5.85$
$Y/U = 0.0065$
$L/U = 1.97$

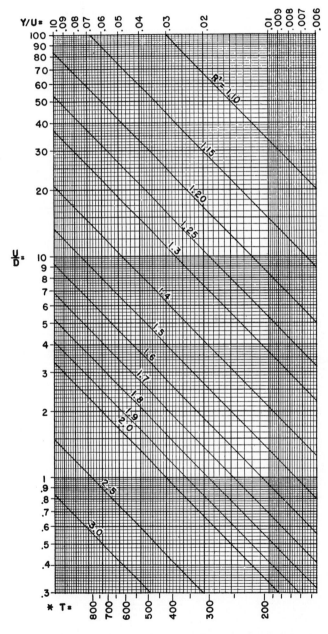

FIG. 7.30. ASA empirical rule for determining need of formal stress analysis. (Reproduced by permission from M. W. Kellogg Co., *Design of Piping Systems,* 2nd ed., John Wiley and Sons, New York, 1956.)

ASA B31.1 Code Criterion Fig. 7.30, $R = 1.61$
$R < L/U$; hence, formal calculations are not
mandatory.

The Code requirement illustrated in these four examples is a crude but helpful device. Many piping
designers, however, will want to learn more about stress
analysis so that they may improve their own abilities.
Fortunately, there is an excellent book on this subject
written in the lucid style common only to competent
practitioners. This book, *Design of Piping Systems,* by the
staff of M. W. Kellogg Company,[2] is strongly recommended for a reference book. Many designers will want
to master the simplified methods for flexibility analysis
which are described therein, for they will prove useful
aids in fashioning more economical piping systems. In
general, most designers will find it expedient to leave
the detailed analysis of complex problems to experts,
who maintain proficiency in the field by day-to-day
practice and constant study of the literature.

Expansion Joints

In the rare instances where conventional bends for
increasing flexibility and reducing stresses cannot be
used, and a straight or nearly straight run of pipe must
be firmly anchored at both ends, expansion joints are
used. Complete descriptions of types of expansion
joints, calculation methods for establishing movement
demands, and method for establishing purchasing requirements, is given in *Design of Piping Systems.*[2]

Vibration in Piping Systems

Any piping connected to equipment subject to pulsating flow, such as reciprocating compressors, may
produce excessive vibrations, that can be transmitted to
other equipment, structures, and foundations. The
forces are at times so great that they can cause fracture
of the piping. As described in *Design of Piping Systems:*[2]

A compressor of the reciprocating type is a source of periodic
pressure excitation at a frequency (cps.) equal to the rotational
speed (rps.) multiplied by the number of cylinders for single
action and twice the number of cylinders for double action for
any given stage. If this frequency approaches the acoustic frequency (natural frequency) of the connected piping system,
acoustic resonance in the form of large periodic pressure surges
will appear. Apart from possible adverse effects on machinery
and its operation, these pressure pulsations can be transmitted
directly to the foundations and buildings, and via bends acted
on by periodically variable forces, or through the connecting
pipe itself to other vessels, structures, or foundations.*

The natural frequencies of a configuration of piping can
be calculated by methods which have been well described.[2] Resonance of a piping system can often be
prevented during the design stage by proper placing of
both fixed and elastic supports. In addition, or when
these methods do not seem promising, pulsation damp-
eners may be planned. Surge tanks are routinely used
for reciprocating compressors to reduce pulsations and
for reducing hydraulic hammer in liquid systems. Vibration is a complex problem that is not wholly calcula-
ble and requires the attention of an expert who has the
benefit of years of experience. The piping designer will
find that the standard procedures used by his company
for compressor piping and other similar potential vibra-
tors will produce a workable configuration or one that
can be easily made so through detailed analysis.

REFERENCES

1. Kern, Robert, *Petroleum Refiner,* **39,** No. 12, p. 142 (1960).
2. M. W. Kellogg Co., *Design of Piping Systems,* 2nd ed., John Wiley and
   Sons, New York, 1956.

   * Reproduced by permission: M. W. Kellogg Company, New York,
   Design of Piping Systems, John Wiley and Sons, New York, 1956.

8

Pipe fabrication

After piping drawings have been prepared, piping materials must be procured and fabricated for erection. To fabricate piping means to assemble the pieces, such as elbows, tees, and flanges, into manageable sections which can be fitted together into the finished plant piping system. It is the purpose of this chapter to describe the role of the piping designer in this procedure and to give background in fabrication methods sufficient for understanding the problems that may arise.

SCREWED PIPING

All threaded (screwed) joint piping and small compression-joint tubing are fitted in the field. In these small sizes craftsman measure, cut, thread, and fit the piping runs to the equipment. Although this piping is dimensioned on drawings, it is practical for the pipe to be fitted in the field, since pipe 2 to 2½ in. or smaller is easily handled by a cut-and-fit technique. Variations or tolerances, within practical limits can be established by notes on the drawings, rather than by precise dimensions between each fitting. This is the usual procedure followed in plumbing practice, where exact dimensions are rarely shown on drawings. Because of the minor space needs of small piping, precise location is not important.

FIELD OR SHOP FABRICATION

The term fabrication is used in piping most commonly when referring to larger size pipe (3 in. and above) which is joined by welding and flanged joints. Because of the size and importance of such lines and the permanent nature of a welded joint, sections of this pipe must be assembled and welded with a degree of precision that requires careful layout of work, control of welding operations, the use of jigs and templates, and more precise tools than those necessary for threaded pipe. These conditions suggest work at ground level and in an organized shop-type atmosphere found in a permanent pipe-fabricating shop or, at times, in a well-planned temporary field shop. Portable tools and equipment are now manufactured that permit most shop methods to be duplicated in the field, except simultaneous heat-treating of large tonnages of fabricated pipe.

The factors which determine the extent of shop and field fabrication are varied and may be governed by the erector's previous experiences, piping size and type, and erector's agreements with organized labor in the area. If all piping is field-fabricated, the erector must maintain a staff of personnel experienced in the usual production-line procedures followed in fabricating shops. In addition, tools and equipment ordinarily only found in shops will be required. A more extensive field organization is also needed for material control, material handling, and the performance of certain technical operations. Additional shelter in the form of temporary sheds or buildings is necessary for part of the fabricating operations, for warehousing of certain materials, and for housing the added technical and clerical personnel.

Erectors who perform many large jobs each year and maintain a wide assortment of equipment and materials often find conditions that are favorable for complete field fabrication. There seems to be, however, no general rule which is rigidly followed. This is a competitive field and shop fabricators are constantly proving their value by providing competent workmanship at low cost. Thus the varied experiences of erectors with shop and field fabrication will often have a strong influence on their decision.

Piping size and type can have a significant influence on whether piping should be shop or field fabricated. Alloy piping often requires involved techniques for heat treating which are more difficult to perform in the field than in the shop. There are times, however, when the alloy piping is so extensive that it may be economical to provide the necessary apparatus in the field, since a certain amount of field fabrication must be performed and receive the same type of heat treatment as the rest of the pipe. Pipe in sizes 3 and 4 in. is easier to handle and often proves attractive for field fabricating.

Erectors must make agreements with local organized labor, and at times these agreements fix the amount of piping that must be field fabricated by establishing a break point on the size rather arbitrarily. These agreements are sought by the local unions in order to insure work for their craftsmen, because shop-fabricated pipe may be prepared in some other area of the country. Table 8.1 shows three alternatives generally used for fabricating process-plant piping. There are, of course, many other possible arrangements, but these are typical.

DRAWINGS FOR FABRICATION

The piping drawings described and illustrated in Chapter 6 identify all lines by a line number, and each section of pipe to be fabricated as a unit (spool) is designated by a related number, as shown in Figs. 6.3 and 6.4. The fabricator prepares shop drawings of each

spool called spool drawings. If the prime contractor is going to do the fabricating, his designers will prepare the spool drawings, or he will contract these drawings to another firm.

By examining Figs. 6.3 and 6.4 it will be noted that a complete piping drawing shows the location of all flanges, welds, and valves, and gives adequate dimensions to describe the piping system completely. Careful study of the piping drawings, however, reveals that this type of drawing would not be the most convenient device for a worker who is mass-producing sections of fabricated piping. His interest at the moment would be the particular section upon which he is working and he would want dimensions arranged in a manner that would be consistent with his mode of fitting and welding the pipe. It is thus only natural that the pipe fabricating industry has evolved a system most convenient for these needs.

Each section of piping that is to be fabricated and shipped as a unit is detailed on a spool drawing, such as shown in Figs. 8.1 or 8.2. These drawings are for identical units of pipe, the only difference being that one is a double-line drawing and the other a single-line. Double-line drawings are used only for those units of piping that are so complex that single-line drawings would not be adequate.*

The mark number shown on the drawing is also painted on each finished spool for the convenience of the erection crew. The marks are based on the line numbers appearing on the original piping drawing and indicated under the terms identification. The drawing illustrated is for a unit of piping that is part of line number P-205 appearing on drawing 36, and it is piece

* Most shops use single-line drawings exclusively.

TABLE 8.1
TYPICAL PIPE FABRICATION METHODS

Method	Pipe Size	Where Fabricated	Material Listing	Material Procured	Spool DWG'S Prepared*	Valves & All Other Appurtenances
I	2½″ & Below (screwed)	Jobsite	Designer	Designer	Not required	Designer
	3″ & 4″	Jobsite	Designer	Designer	Designer	Designer
	6″ & Above	Shop	Shop	Shop	Shop	Designer
II	2½″ & Below (screwed)	Jobsite	Designer	Designer	Not required	Designer
	3″ & Above	Shop	Shop	Shop	Shop	Designer
III	2½″ & Below (screwed)	Jobsite	Designer	Designer	Not required	Designer
	3″ & Above	Jobsite	Designer	Designer	Designer	Designer

* See Figs. 8.1 and 8.2.

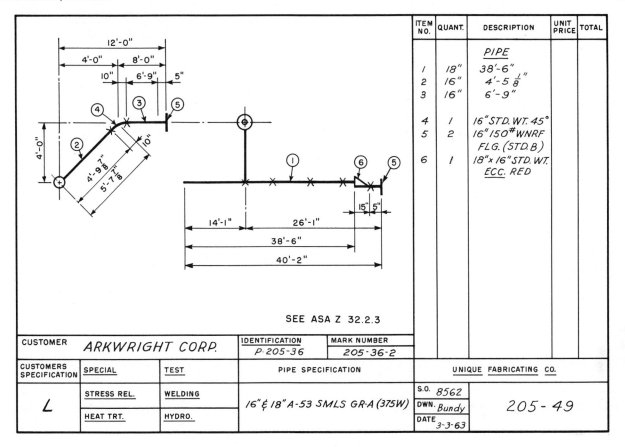

ITEM NO.	QUANT.	DESCRIPTION	UNIT PRICE	TOTAL
		PIPE		
1	18"	38'-6"		
2	16"	4'-5 $\frac{1}{8}$"		
3	16"	6'-9"		
4	1	16" STD. WT. 45°		
5	2	16" 150# WNRF FLG. (STD. B)		
6	1	18"x 16" STD. WT. ECC. RED		

SEE ASA Z 32.2.3

CUSTOMER *ARKWRIGHT CORP.*	IDENTIFICATION P·205-36	MARK NUMBER 205-36-2

CUSTOMERS SPECIFICATION	SPECIAL	TEST	PIPE SPECIFICATION	UNIQUE FABRICATING CO.
L	STRESS REL.	WELDING	*16"¢18" A-53 SMLS GR-A (375W)*	S.O. 8562
	HEAT TRT.	HYDRO.		DWN. *Bundy*
				DATE 3-3-63

205-49

FIG. 8.1. Single-line spool drawing.

number 2. These mark numbers or spool numbers will also appear on the piping drawing so that the erection crews can determine the sequence of pieces relative to equipment (Figs. 6.3 and 6.4).

It will also be noted in Figs. 8.1 and 8.2 that a bill of materials appears on each spool drawing. This enables the fabricator to place the orders for the correct amount of materials and also gives the welder a handy summary of the materials he must assemble for the particular spool. Thus the spool drawing is a complete description of a unit of piping, but it is a particular kind of description oriented to the people who must purchase and assemble the unit. The major dimensions appearing on the piping drawings are subdivided so that they are more convenient for fabricating purposes. Each item of pipe or fittings that becomes part of the unit is labeled and described in the bill of materials.

It is important to recognize the piping designer's influence on pipe fabrication. He not only lays out and arranges the piping, but he also decides the extent of the units of fabricated pipe by assigning spool numbers and designating the field welds. Clearly this procedure demands a thorough knowledge of pipe fabricating techniques, cost, and shipping problems. The entire

12-in. P-101-A line shown in Figs. 6.3 and 6.4 could possibly be fabricated as one spool. Railroad shipment would limit the size to about 30 ft by 9 ft. If ocean shipment is planned, then the size may be dictated by ocean freight charges which often depend on cubic measure, and pieces must be chosen so as to minimize the volume. If piping is large and alloy steel, it is fabricated as completely as possible in order to avoid excessive field welding and field heat-treating. Thus there are many factors that must be evaluated, and the piping designer should always consult with traffic people and the project engineer especially before "spooling" the major lines.

PIPE FABRICATION PROCEDURES

Pipe fabrication methods are governed to a considerable extent by the rules of the "American Standard Code for Pressure Piping." Section 6 of the Code, particularly, offers limits for design and rules for fabricating pipe and for qualifying welders. In addition to the various rules of the Code, the Pipe Fabrication Institute (PFI) has prepared a series of standards (see Chapter 3) for pipe fabrication which supplement the ASA Code.

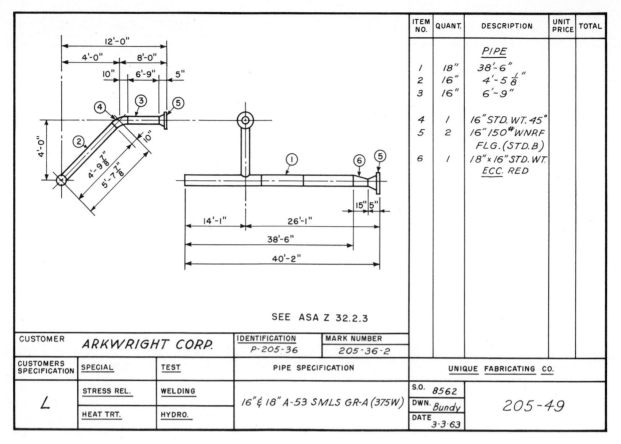

ITEM NO.	QUANT.	DESCRIPTION	UNIT PRICE	TOTAL
		PIPE		
1	18"	38'-6"		
2	16"	4'-5 $\frac{1}{8}$"		
3	16"	6'-9"		
4	1	16" STD. WT. 45°		
5	2	16" 150# WNRF FLG. (STD. B)		
6	1	18" × 16" STD. WT. ECC. RED		

SEE ASA Z 32.2.3

CUSTOMER			IDENTIFICATION	MARK NUMBER			
	ARKWRIGHT CORP.		P-205-36	205-36-2			

CUSTOMERS SPECIFICATION	SPECIAL	TEST	PIPE SPECIFICATION	UNIQUE FABRICATING CO.	
	STRESS REL.	WELDING	*16" & 18" A-53 SMLS GR-A (375W)*	S.O. *8562*	*205-49*
L				DWN. *Bundy*	
	HEAT TRT.	HYDRO.		DATE *3-3-63*	

FIG. 8.2. Double-line spool drawing.

Additions and revisions to these standards are being issued continuously to pipe fabricators who are members of the Association.

The general intents of these standards are:

1. To protect both purchaser and fabricator by providing consistent specifications and services.

2. To work toward standardizing fabricating methods.

3. To establish limits of responsibility and to fix tolerances.

When more completely established, or accepted, the rules will perhaps be comparable to the practices in shop fabrication of structural steel as defined by the American Institute of Steel Construction Handbook.

It is not the purpose of this section to repeat these rules but only to describe briefly the physical processes involved in pipe fabrication so that the piping designer can be aware of the efforts required to produce the finished product of his design.

Welding Processes

Pipe fabrication and welding are almost synonymous terms, for the welding of various fittings and pipe sections to produce a unit of fabricated pipe is the main job performed in a pipe shop. The companion to welding, flame cutting, is perhaps the second most important tool of the piping fabricator.

The ideal welding process is one that produces a joint with the same physical and chemical characteristics as the parent parts. In order to accomplish this goal, the metal parts must be heated to their melting points around the area where they are to be joined. By adding filler metal from a welding rod or by pressing the parts, the metal coalesces so that, upon cooling, a solid and strong joint is produced. The final characteristics of any metallic system are greatly affected by the metal composition, the rate of cooling, and the pre-heat or post-heat treatments. In early welding processes the sharp drop in temperature away from the weld area created quite a problem. The steels in use at that time had a higher carbon content than modern steel, and the sharp drop in temperature caused brittle and weak welds. In addition, welding rod compositions were not consistent and the air in contact with the molten metal caused the formation of slag-producing oxides. Today, welding processes have been greatly improved and most of the earlier problems have been solved, but success still depends upon the use of experienced craftsmen. This is so important that the "American Standard Code for Pressure Piping" and all other national codes require qualification tests for welders performing work on pipe that is to be designated as manufactured in accordance with a code. These tests require the welder to perform welds in specific positions with designated materials under a number of typical circumstances.

ELECTRIC-ARC WELDING PROCESS

Electric-arc welding is the most used welding procedure for pipe fabrication. Heat for fusing the metals is produced by an arc between two electrodes or between an electrode and the piece being welded.

There are three frequently used methods for arc-welding in fabricating metallic piping for process plants. These are called shielded metal-arc, submerged-arc, and gas tungsten-arc.

Shielded Metal-Arc Welding (Fig. 8.3). "Shielded metal-arc welding is an arc welding process wherein coalescence is produced by heating with an electric arc between a covered metal electrode and the work. Shielding is obtained from decomposition of the electrode covering. Pressure is not used and the filler metal is obtained from the electrode."* Manual shielded metal-arc welding is the most common welding process in pipe fabrication. The materials to be joined are tack-welded to hold them in position and then the main weld is made by depositing a series of beads or layers in the joint until the weld is completed. In producing a good weld, the exact procedure must be carefully selected, proper current and electrode determined, slag removed after each pass, and careful heat treatment performed when required. The arc is visible and the operator must wear protection against the radiation from the flame and from flying hot metal and slag spatter.

Submerged-Arc Welding. "Submerged-arc welding is an arc welding process wherein coalescence is produced by heating with an electric arc or arcs between a bare metal electrode, or electrodes, and the work. Pressure is not used and filler metal is obtained either from the electrode or from supplementary welding rod. The welding zone is covered or shielded by a blanket of granular, fusible material known as the melt, flux, or welding composition." The electrode is consumable bare wire which is continuously fed into position through defusible material. Usually this procedure is an automatic process used for circumferential pipe welds on heavy-wall piping and piping 8 in. or larger in diameter. The welding apparatus is fixed in position and the work is rolled at the desired speed. Both the electrode and flux are continuously fed. The flux is a fusible material which adheres to the molten weld metal and serves both as a flux and inert blanket. Submerged-arc welding is also used in manufacturing pipe from rolled plate. It produces the longitudinal welded seam in the rolled plate.

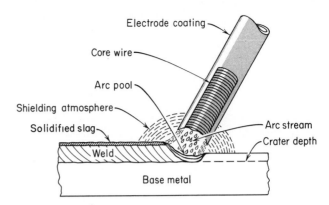

FIG. 8.3. Schematic representation of the shielded metal arc. (Reproduced by permission from *Welding Handbook*, Sec. 2, 4th ed., American Welding Society, New York 17, N.Y., 1958.)

Gas-Tungsten Arc Welding (Fig. 8.4). "Gas-tungsten arc welding is an arc welding process wherein coalescence is produced by heating with an electric arc between a single tungsten electrode and the work. Shielding is obtained from a gas or gas mixture. Pressure may or may not be used, and filler metal may or may not be used."† In this process the arc is struck between the work and cooled tungsten electrode. Filler metal is added manually or by machine into the hot zone. Helium, argon, or other gas, depending on the metal, is continuously discharged near the arc to blanket the molten weld area. On the first pass and until the inside of the pipe is sealed off by the weld, the pipe must be filled with the same blanketing gas so that the inside of the first pass will be protected. Gas-tungsten arc welding is most often used for difficult-to-weld metals such as aluminum, magnesium and high chromium-nickel steels.

Oxy-Acetylene Welding (Fig. 8.5). In oxy-acetylene welding the heat is supplied by the combustion of a mixture of oxygen and acetylene. The welding rod is introduced manually by the welder. The equipment for oxy-acetylene welding is very inexpensive and easily moveable. This technique is restricted to pipe sizes 2 in. and smaller and has almost entirely been displaced by metal-arc welding in the pipe fabrication field.‡ Oxy-acetylene cutting torches, however, continue to be a major means of cutting pipe. The cutting torch has a central oxygen jet surrounded by a number of small oxy-acetylene heating flames. The central oxygen flame is guided along the line where the cut is to be made. The

---

* Quotations in this section reproduced by permission: *Welding Handbook*, Sec. 2, 4th ed., American Welding Society, New York, 1958.

† Filler metal is always used for pipe welding. A gap of $\frac{1}{32}$ to $\frac{1}{16}$ in. is left between the two pieces to be welded. This gap is filled on the first pass.

‡ Because of the impurities introduced by the torch, oxy-acetylene welding is permitted only in rare cases for pipe fabrication.

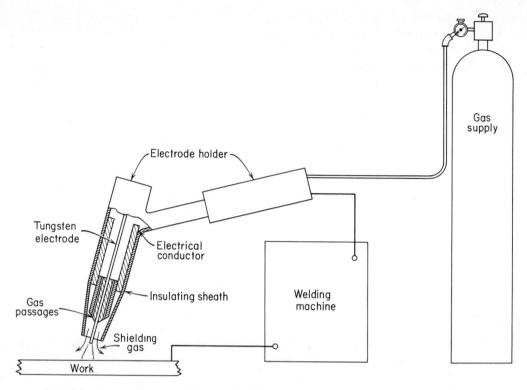

FIG. 8.4. Gas-tungsten arc welding process. (Reproduced by permission from *Welding Handbook,* Sec. 2, 4th ed., American Welding Society, New York 17, N.Y., 1958.)

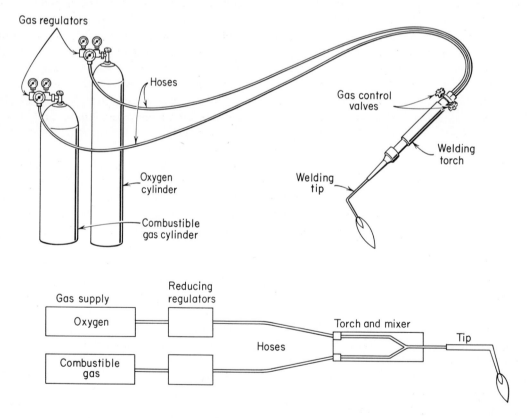

FIG. 8.5. Oxy-acetylene welding. (Reproduced by permission from *Welding Handbook,* Sec. 2, 4th ed., American Welding Society, New York 17, N.Y., 1958.)

high temperature coupled with an excess oxygen atmosphere rapidly melts and oxidizes the metal, producing a clean cut. These torches are used both manually and in automatic cutting machines.

WELDING PLASTICS

The increased use of plastic piping for corrosive services at moderate pressures and temperatures in process plants makes it essential to know welding processes related to plastics. Plastic pipe may be joined by using a specially prepared cement dissolved in a solvent or it may be joined by one of several welding techniques employing heat. Heated-tool welding uses some method for rapidly heating the ends of the pipe to be joined, such as a radiant hot plate. Upon heating, the plastic softens and the pipes can be joined by applying pressure. Hot gas-welding, similar in technique to oxy-acetylene welding for metals, is also used. A heated stream of air or gas supplies the heat to melt the pieces being joined and the filler rod of plastic. Since the use of plastics in process piping is a rapidly changing field, exact specifications and rules for welding plastics should be carefully checked with the manufacturer of the material.

Heat Treatment and Stress Relieving

Heat treatment and stress relieving of piping is necessary to relieve stresses caused by temperature differentials from the hot welding zone into the cooler zones, to maintain or restore the original crystalline structure, and to maintain certain desired properties such as corrosion resistance. The heat treatments used in welding piping are pre-heating, interpass heating, and post-heating or stress-relieving. Pre-heating reduces the rate at which the weld metal cools and aids in producing a more ductile weld which can shrink in the final cooling without cracking. Common pre-heating temperature is 400°F but in many cases 200°F has proved adequate.

Interpass heating maintains the work at an elevated temperature during welding, usually around 400°F. Interpass heating is generally a continuation of the pre-heating technique.

For small piping pre-heating may be carried out by the use of an oxy-acetylene torch, but with larger piping it is necessary to wrap the pipe with electric heating units, employing inductive, conductive, or radiant heating. Any sharp drop in temperature during welding is undesirable and by pre-heating the work and maintaining this heat, the hot zone of the weld is extended and the temperature drop is more gradual.

Many metals, especially thick specimens, require after welding a high post-heating treatment called stress-relieving. This procedure relieves the stresses produced by welding by softening the heat-affected zone next to the weld. But the correct temperature and mode of stress-relieving for each alloy must be carefully followed because some desirable properties such as fatigue strength can be reduced by excessive temperatures. Table 8.2 shows common practices with a number of well-known metals. Piping designers need to be familiar with heat-treating processes to be used on the pipe they are designing. Although stress-relieving can be done in the field by portable induction or conduction heaters, the most satisfactory method is in large furnaces located in a shop where careful control is possible. The designer should recognize that each field weld he specifies may have to be heat-treated in a special manner, requiring extra equipment and technicians. On the other hand, it is not desirable to avoid all field welds, for substituting flange joints not only increases the cost of the piping but also the maintenance problems. Each joint in a piping system, when not required for disassemblying equipment, is just another potential point for leakage.

Welding Symbols

The American Welding Society has established standard symbols for welding details. Some of the more common ones that are of interest to piping designers are included in Table 8.3.

Layout and Fabrication

Those interested in a detailed description of layout and fabrication details should read Chapter 10 of the *Standard Manual on Pipe Welding*[1] and then visit a pipe-fabricating shop to see the techniques being applied. The major steps given below serve to indicate the scope of each operation as described in the Manual.[1]

I. Preparation of Material
    Pipe is cut to proper lengths and beveled. Other materials are assembled.

II. Layout and Assembly
    Involves the fitting of the parts for the subassembly or unit of piping prior to welding. The largest piece is placed on table and other pieces are arranged relative to it. The pieces are held together by holding devices, jigs, and tack welds. Tack welds are made in sufficient number and size to prevent distortion during welding. Highly skilled mechanics are needed for layout and assembly, and all work must be checked prior to final welding. Adequate allowances must be made for shrinkage of the assembly during welding.

TABLE 8.2
THERMAL TREATMENTS FOR WELDMENTS*

Material	Soaking Temperature, °F.	Time Hr./In. Thk.
Carbon steel	1100–1250	1
C–½% Mo steel	1200–1325	2
1% Cr–½% Mo steel	1250–1350	2
2% Cr–½% Mo steel	1300–1375	2
2¼% Cr–1% Mo steel	1325–1400	2
5% Cr–½% Mo (Type 502) steel	1350–1400	2
9% Cr–1% Mo steel	1375–1425	3
12% Cr (Type 405) steel	1300–1350	2
12% Cr (Type 410) steel	1350–1400	2
15% Cr (Type 430) steel	1400–1450	4
Austenitic Mn steels	1900–2000, air cool	1
Austenitic Cr–Ni steels	1900–2000, air cool	1
Low alloy Cr–Ni–Mo steels	1100–1250	1
2 to 5% Ni steels	1100–1150	1
9% Ni steel	1000–1100	2
Monel	1100–1250	1
Inconel	1100–1250	1
Nickel	1100–1250	1
Cast iron	1050–1250	1

* Reproduced by permission: *Welding Handbook,* 3rd ed., American Welding Society, New York, 1950.

III. Welding

One of the procedures already described is employed.

IV. Post-Heat Treatment

V. Aligning

Welding and subsequent heat treatments often cause misalignment in piping assemblies. This misalignment can be corrected by:

(a) Controlled heating and cooling of areas next to the weld which contracts the pipe, deflecting it to that side heated and cooled. This requires great skill and can be applied only to metals not injured by the heating and cooling cycle.

(b) Machining flange faces or butt ends of pipe so that ends are square with normal axis and conform to desired over-all dimensions.

(c) Cold pulling or applying force gradually on assemblies not badly misaligned. This can be used when the stresses introduced are not detrimental.

VI. Final Dimensional Check

This is done before removing piece from work table.

VII. Cleaning and Marking

(a) Weld projections on inside of pipe at joints are chipped. These projections are avoided by backup rings.

(b) Inside of pipe is cleaned with mechanical cleaners.

(c) Special cleaning methods may be dictated by the type of process fluid which will flow in the pipe.

(d) Each assembly is marked with proper spool number.

VIII. Inspection and Testing

(a) Assembly is checked for compliance with purchaser's specifications and the ASA Code.

(b) Tests include:

(1) Radiographic inspection of welds.

(2) Examining portions of welds.

(3) Magnetic particle inspection and fluorescent penetrant inspection of welds (both of these techniques reveal discontinuities in welds).

(4) Hydrostatic test of assembly—this is done in accordance with Code regulations.

TABLE 8.3

STANDARD WELDING SYMBOLS

(Reproduced by permission from: *Welding Handbook,* 4th ed., Sec. 2, American Welding Society, New York, 1958.)

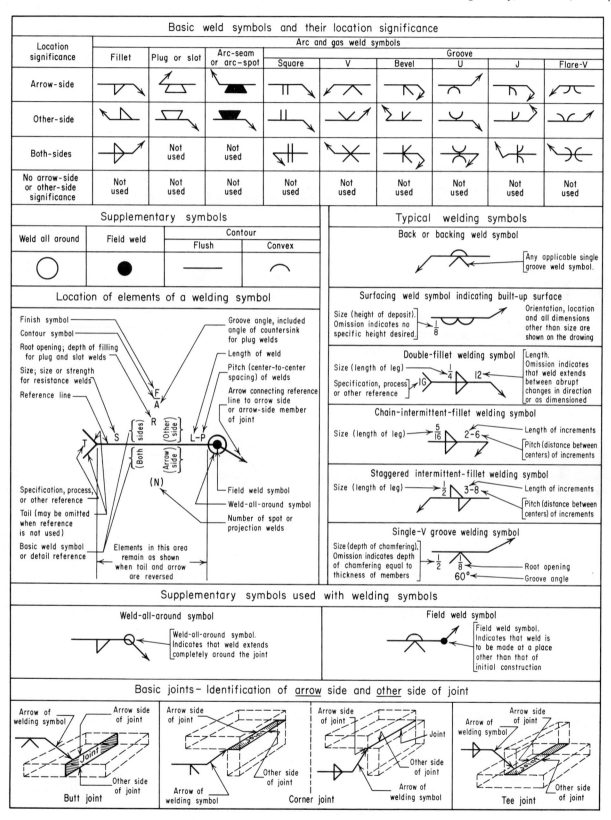

TABLE 8.3 (*Concluded*)
STANDARD WELDING SYMBOLS

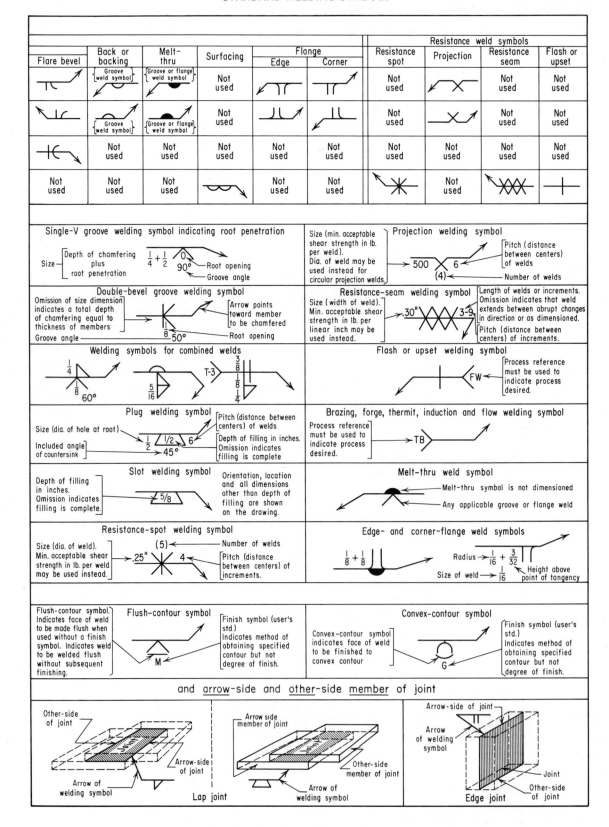

ERECTION OF PIPING

These series of fabricating operations produce with precision and speed finished pipe assemblies which are shipped to the job site to be erected by the prime contractor. If the piping designer planned his original layout and arrangement with care, the erection of the assemblies will proceed rapidly. The pieces are hoisted into place and carefully aligned, using clamps. Then field welds are made or flanges are bolted and tightened. Great care is exercised in preventing dirt or other foreign materials from entering the pipe.

Much of this work is done above ground and the workers need adequate scaffolding to work freely. They will also work more rapidly and competently if the piping designer arranged the piping with adequate clearance for rapid assembly.

REFERENCES

1. Heating, Piping and Air Conditioning Contractors National Association, *Standard Manual on Pipe Welding*, 2nd ed., New York, 1951.
2. *Welding Handbook,* Sec. 2, 4th ed., American Welding Society, New York, 1958.

9

Utility and underground piping

Process piping is required for so many types of services that only general rules may be given for its design. But utility piping such as water, steam, sewage, and blowdown is designed for known fluids, and there are a number of useful and specific observations that can be given as aids in design.

WATER

Water piping will be discussed in accordance with the type of water service.

Process Water

Process water is used in the process as a reactant, solvent, or direct-contact cooling agent. Often highly purified water is required to prevent product contamination. In these cases, treated and demineralized water or uncontaminated steam condensate is used. When water comparable to distilled water is not needed, raw water direct from wells or surface courses may be employed.

Usual system pressure: Depends on process pressures.
Common piping materials:
   Pipe: 2″ and below—galvanized steel or wrought iron.
            3″ and above—steel.
   Fittings: 2″ and below—threaded malleable iron or steel.
            3″ and above—welded steel.
   Valves: Brass, bronze, cast iron with non-ferrous trim.

Service Water

Service water is used for washing down areas, washing out equipment and for testing equipment in some processes. It is also employed in irrigation and maintenance. Raw water direct from a natural source may be used for this purpose, or part of the service may be partially treated fresh water from the sanitary water treating system.

Usual system pressure: 30 to 40 psig.
Common piping materials: Same as for process water (also cast-iron pipe and fittings).

Sanitary Water

This service supplies plumbing fixtures and plant drinking water. Fresh water is treated sufficiently to be potable. If the source is from deep wells, the water need only be chlorinated. If a separate service is provided for drinking, the sanitary system water need not be treated, except for removal of matter which may clog small piping. In order to prevent contamination, sanitary water systems should never be connected into any process system or water system which, in turn, is connected to part of the process. When this rule is not followed, there is always the chance that flammable gases and poisonous gases or liquids may be carried in the water system and liberated in a nonhazardous area or at a plumbing fixture.

Usual system pressure: 20 to 30 psig.
Common piping materials: Same as for service water, except that copper tubing with solder joint fittings are used

inside buildings installed in accordance with plumbing codes.

Cooling Water

Water-cooled condensers and coolers are the largest consumers of water in a process plant. In some localities plants can be located adjacent to large seemingly inexhaustible sources of water which enable the water to be used once and then discarded. These circumstances are becoming ever more rare. Even swiftly flowing rivers soon reach their capacity for cooling as they become lined with plants on both sides.

Cooling towers have become the common means of supplying a reliable source of cooling water. Treated water from which deposit-causing minerals have been removed or their solubility increased is circulated over a tower through which air is blown. A small portion of the water is evaporated causing rapid cooling of the remainder which is reused for cooling service. Makeup water must be added but the amount is small in a well-designed system. The water is supplied to the users at 80 to 90°F and returned to the cooling tower at a maximum of 120°F.

Usual system pressure: 50 to 75 psig. The system pressure must be adequate to overcome pressure drop produced in the cool-water and warm-water piping, including static heads at equipment and cooling tower, as well as pressure losses through controls, valving, and equipment. The warm water enters the cooling tower at 35 to 40 ft above grade.

*Common piping materials:** Cast-iron pipe and fittings, cement-lined steel pipe in large sizes above 24 in., or reinforced concrete pipe. Concrete pipe must not be used in areas that could be penetrated by acid, for acid disintegrates the concrete.

Jacket Cooling Water

Demineralized water or condensate must be used for internal-combustion engine jacket cooling. Small separate water systems are used for this purpose. Treated jacket water is cooled either in water-cooled or air-cooled exchangers.

Fire-Protection Water

Fire water headers must be planned independently from other water systems because they should not be used for other purposes. A cooling tower basin can be used as the reservoir for emergency water, but the standard elevated tank is the most common storage for emergency water. Piping details at elevated tanks are shown in Figs. 9.1, 9.2, and 9.3. Plans should be made for emergency connection to other large reservoirs, such as raw water storage ponds, rivers, or lakes. A typical fire-pump installation is shown in Fig. 9.4. Dual-drive pumps or two pumps (one gasoline-driven) are used to insure operability in case of power or steam failure.

Line sizes at fire pumps must be calculated with care as outlined in Fig. 9.5 and Table 9.1. Arrangement of piping and selection of types of valves and fittings should be in accordance with the rules of the National Bureau of Fire Underwriters (New York) National Fire Protection Association (Boston) or Factory Mutual Engineering Division (Norwood, Mass.). Valving should follow American Water Works Association requirements for fire-protection systems. Hose connections on hydrants and external connections to the system should be se-

* Enamel-lined cast-iron pipe has been successfully used for brackish water service. The enamel lining prevents attack of the iron by marine organisms.

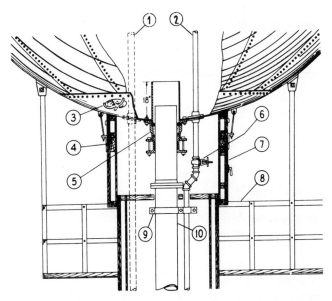

FIG. 9.1. Details of pipe connections to bottom of steel gravity tank with pipe riser.

1. Inside type brass overflow pipe, if used.
2. Hot water circulating pipe.
3. Handhole for removing sludge.
4. Wooden frostproof casing when required.
5. Acceptable expansion joint.
6. Approved OS&Y gate valve.
7. Door in frostproof casing.
8. Walkway.
9. Brace for hot water circulating pipe.
10. Pipe riser.

(Reproduced by permission from Factory Mutual Engineering Division, *Handbook of Industrial Loss Prevention*, McGraw-Hill Book Co., New York, 1959.)

218

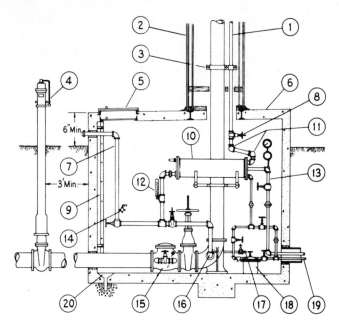

FIG. 9.2. Valve pit and pipe connections at base of tank on independent tower. Tank has a pipe riser and steam-heated gravity circulating heating system.

1. Hot water circulating pipe.
2. Frostproof casing.
3. Pipe clamps. Locate at about 25 ft. intervals. Loose fit around hot water circulating pipe.
4. Indicator post gate valve. May be replaced with OS&Y gate valve pit on yard side of check valve if space is not available.
5. Hatch cover.
6. Valve pit.
7. Drain pipe.
8. OS&Y gate valve.
9. Ladder.
10. Tank heater with relief valve set at 120 lb.
11. Four-elbow swing joint.
12. Thermometer.
13. Steam supply pipe.
14. Drain cock. ½ in.
15. Approved check valve with by-pass.
16. Cold water circulating pipe.
17. Pipe to mercury gage.
18. Steam trap.
19. Condensate return.
20. Valve pit drain.

(Reproduced by permission from Factory Mutual Engineering Division, *Handbook of Industrial Loss Prevention*, McGraw-Hill Book Co., New York, 1959.)

lected to match the threading used by the nearest municipal system so that equipment will be interchangeable in the event of a major disaster.

Usual system pressure: 100 to 150 psig.

Common piping materials: Cast-iron pipe with outside bitumastic coating. Bell and spigot or mechanical joints are used. Clamps should be provided on all packed joints where a blowout might occur because of

water hammer. Monolithic concrete blocks should be installed at bends to prevent blowouts.

Sprinkler Systems

Sprinkler systems are usually subcontracted to the sprinkler manufacturer and must often be approved by an insurance company. Piping and fittings are galvanized throughout.

Foam Systems

Foam systems are effective in smothering certain types of hydrocarbon fires. One economical method of providing such protection in hazardous areas is to install water connections to which can be attached portable containers of foam-producing agents. The plant fire water system can then be used to supply the necessary water for producing the foam.

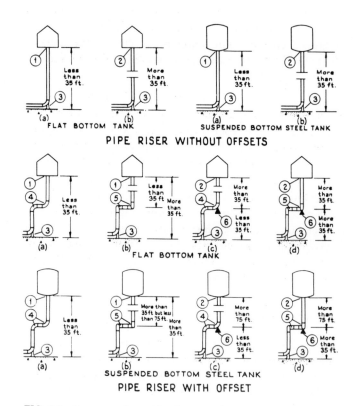

FIG. 9.3. Support and provision for expansion for pipe risers.

1. Rigid connection.
2. Expansion joint.
3. Base elbow.
4. Two-elbow swing joint.
5. Four-elbow swing joint.
6. Support.

(Reproduced by permission from Factory Mutual Engineering Division, *Handbook of Industrial Loss Prevention*, McGraw-Hill Book Co., New York, 1959.)

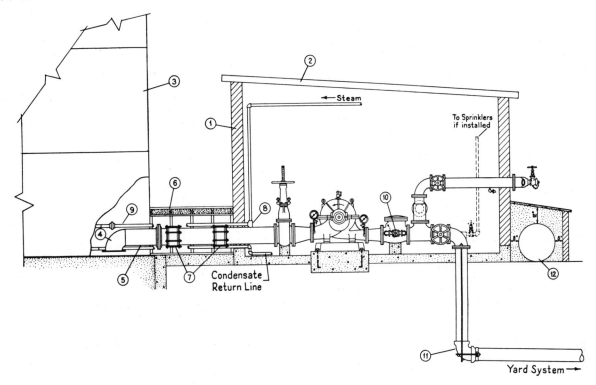

FIG. 9.4. Horizontal shaft fire pump installation—suction from above-ground tank.

1. Minimum size of pump room 12 ft. by 16 ft. by 10 ft. high for one pump. Larger rooms are required for multiple pump installations.
   Walls—Noncombustible construction—brick, tile, concrete, or corrugated asbestos on steel frame. (Latter suitable only where climate makes heating unnecessary.)
   Floor—Concrete.
2. Noncombustible roof—otherwise install sprinklers.
3. Factory Mutual approved aboveground suction tank. Foundations and heating details of suction tanks should be in accordance with Factory Mutual specifications for water tanks for fire protection.
4. Entrance elbow and vortex plate. Furnished as part of Factory Mutual approved tank.
5. Suction pipe. Size—See Fig. 9.5.

6. Frostproof casing. See 3 above.
7. Flexible couplings.
8. Steam trap.
9. Heating coil in tank. See 3 above.
10. Suction tank filling by-pass. May be used when primary supply from public mains. Provide separate fill pipe to suction tank when primary supply from gravity tank or other limited source.
11. Discharge pipe. Size—See Table 9.1.
    Minimum earth cover—See Fig. 9.7.
12. When gasoline engine is driver, enclose fuel tank outside pump room in sand-filled masonry or concrete enclosure.

(Reproduced by permission from Factory Mutual Engineering Division, *Handbook of Industrial Loss Prevention,* McGraw-Hill Book Co., New York, 1959.)

Usual system pressure: Same as for fire water.

Common piping materials: Black or galvanized pipe with welded or screwed steel fittings for connections into apparatus. Use forged steel fittings within the fire zone to avoid failure during a fire.

Notes on Layout and Arrangement of Water Systems

HEADERS

Process units are most conveniently serviced by means of headers, as shown in Fig. 9.6, page 222. Dead spaces in these headers should be provided with valves for periodic flushing.

CAPACITY

Plans should be made for future expansion, including space for expanding the cooling tower and pumping system. A design capacity of approximately 20 to 25 per cent above maximum operating capacity is commonly used.

SIPHONS

Avoid true siphons wherever possible. Velocities above 2 ft/sec reduce operating difficulties caused by changes in elevations.

CHANGES IN DIRECTION

Use long-radius bends for changes in direction.

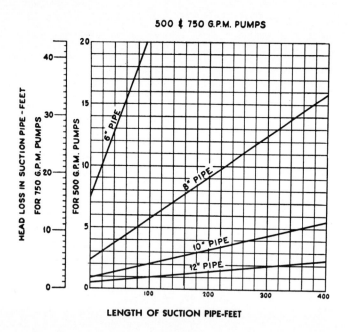

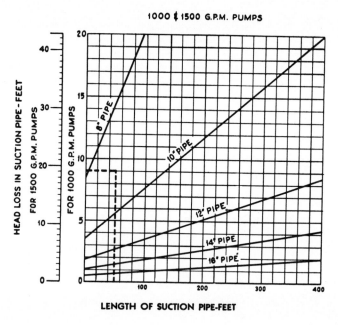

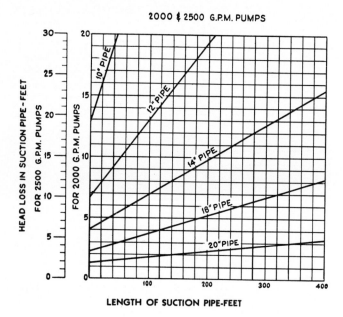

Example:

Assume 1000 gpm. pump taking suction under lift with pumping water level at 150% of rated pump capacity 6 ft. below pump center line. Suction pipe is to be 50 ft. long. What is the minimum size suction pipe required so that the maximum allowable suction lift of 15 ft. will not be exceeded?

Maximum allowable head loss in suction pipe = 15 ft. — 6 ft. = 9 ft.

Enter chart for 1000 gpm. pump at 9 ft. head loss in suction pipe. Move along ordinate to 50 ft. This point falls between the curves for 8 and 10 in. pipe.

Therefore 10 in. pipe is the minimum size suction pipe required so that the maximum allowable suction lift of 15 ft. will not be exceeded.

FIG. 9.5. Charts for determining size of pump suction pipe. These charts:

1. Are based on flow of 150% of rated capacity of pump and head loss calculated on the basis of C = 70 from Williams & Hazen Hydraulic Tables.
2. Have a head loss at 0 length of suction pipe which is made up as follows:

> 28% due to loss in foot valve,
> 56% due to loss in three elbows, and
> 16% due to velocity head.

This head loss at 0 length may be changed in accordance with these percentages when a foot valve is not required or there is a difference in the number of elbows.

(Reproduced by permission from Factory Mutual Engineering Division, *Handbook of Industrial Loss Prevention,* McGraw-Hill Book Co., New York, 1959.)

TABLE 9.1
SIZING LINES AT FIRE PUMPS

(Reproduced by permission from Factory Mutual Engineering Division, *Handbook of Industrial Loss Prevention*, McGraw-Hill Book Co., New York, 1959.)

Size of Pump (gpm.)	No. Hose Valves (5)	Size of Relief Valves (inches)	Size of Waste Pipe (1) (inches)	Size of Test Valve (inches)	Size of Discharge Pipe (2) (inches)	Size of Priming Pipe (3) (4) (inches)	Priming Water Required = A + B. See (4)		
							A. For Pump and Fittings	B. For Suction Pipe	
							Gallons	Diameter of Pipe (inches)	Gallons per Foot
500	2	3	5	1¼	6	2½	40	6	4.4
750	3	3½	6	1½	8	3	63	8 / 10	7.4 / 12.2
1000	4	4	8	2	8	3½	75	12	17.6
1500	6	5	8	2½	10	4	115	14	24.0
2000	6	6	10	2½	10	4	140	16	31.4
2500	8	6	10	2½	12	4	175	20	49.0

(1) If more than one elbow installed, use next larger size pipe.

(2) In special cases of long runs of pipe, the next larger size may be required.

(3) When suction pipe is not over 25 ft long. For longer suction pipes, larger priming connection may be required.

(4) When continuous automatic priming is provided, the size of the priming pipe and gallons of priming water required should be in accordance with Art. 2506 b, Factory Mutual Bulletin 25.23.

(5) When 2 hose valves are required, use 4-in. pipe between the detachable hose header and the connection to the discharge pipe; when 3 or 4 are required use 6-in. pipe; when 6 or 8 are required use 8-in. pipe. When this pipe is over 15 ft long increase one pipe size.

WATER HAMMER

Water hammer caused by closing valves is overcome by shutting off valves slowly, especially during the last 20 per cent of travel. Large valves should be provided with by-passes to aid in closing, and the by-pass should be closed last. Power failure at the water pumps can also cause water hammer as the water reverses and returns to the low pressure area created at the pump by the power failure. This problem is avoided by installing air or water surge drums, relief valves, or an open overflow. The plant shown in Fig. 9.6 employs 20-ft by 80-ft towers filled with water at all times. If the pressure in the line drops below the static head of the tanks, check valves located between the tanks and the water-line open and release stored water in the line.

ANCHORS

Dead ends and changes in direction of bell and spigot pipe must be properly anchored to protect against blow-outs (Fig. 9.7).

DEPTH OF COVER

Underground pipe must be laid below the frost line. Sufficient depth must also be provided to prevent dam-age from impact shock created by moving vehicles (see Tables 3.7 through 3.9). Shannon proposed the following formula for calculating depth of the frost line:[1]

$$d = 1.65 F^{0.468}$$

where: d = depth of frost line, in inches.

F = freezing index.

The freezing index can be approximated by using mean monthly temperatures.

$$F = (32n - \Sigma T_m)30.2$$

where: n = number of months during which temperature is less than 32°F.

ΣT_m = sum of mean temperatures during each of these months.

More accurate predictions are described in the original article. Graphical estimating data for areas of the United States can be obtained from Fig. 9.8. Usually 2½ to 3 ft of cover is adequate in warm climates for protection against mechanical damage and excessively low and high temperatures.

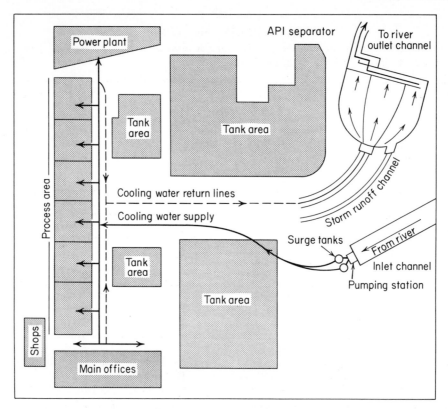

FIG. 9.6. Once-through cooling water system, Tidewater Oil Company, Delaware Refinery. [Reproduced by permission from W. C. Uhl, *Petroleum Processing*, p. 134 (July 1957.)]

STEAM

Steam Distribution

Process plants have at least three steam systems: high pressure, exhaust, and condensate. The design of each system and the operating pressure ranges are functions of plant consumption, power plant demands, type of process equipment served, location of users, and fuel economics. A typical process plant system is given in Table 9.2.

In this typical plant, superheated steam is generated at 600 psig. The major portion of this steam is used to drive turbine-driven electric power generators. The turbines operate noncondensing at 150-psig exhaust with some superheat remaining. The total exhaust from the generators is used in the low-pressure system, along with makeup steam from the 600-psig header. Makeup is necessary in nearly any such system, because only rarely can electric power generation be exactly balanced to meet plant steam needs.

The exhaust from the low-pressure system users such as turbine drives for pumps is directed into the 25-psig exhaust system. This exhaust steam is used for many low-temperature heating duties and its pressure is maintained constant by makeup from the 150-psig system.

The various pressure levels are maintained by automatic controls. Each of the systems must be provided with relief valves so that the system pressure will not exceed the maximum design pressure if the control instruments fail.

The exhaust system or a low-pressure steam system may be utilized for condensate return, provided sufficient slope in piping can be obtained. Manufactured condensate-accumulating and pumping units, complete with controls on an integral base, can be obtained. * Strategic location of several of these units in a plant can eliminate the problems of designing piping for gravity flow. The piping for the return system can usually be simplified when the fluid is pumped.

Piping Materials for Steam

Section I of the "ASA Code for Pressure Piping" is widely used as a guide in selecting piping and valving for steam systems in process plants. Although the code permits cast-iron valves, fittings, and pipe for low-pressure and low-temperature steam, use of cast iron is questionable in process areas. Piping system fabricating costs are often increased by using cast iron which offsets

---

* Often two condensate systems are used, a high and a low pressure.

BELL AND SPIGOT CAST-IRON PIPE

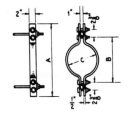

Pipe clamp

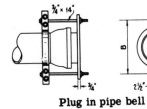

Plug in pipe bell

Anchor strap (1/2" x 2" stock).

Pipe Size	A	B	C
4"	15"	7 1/4"	5"
6	17	9 1/2	7 1/8
8	19 1/2	11 3/4	9 1/2
10	21 1/2	14	11 1/2
12	24	16 1/2	13 1/2

Pipe Size	A	B
4"	12 1/4"	9 3/4"
6	14 1/2	12
8	16 1/2	14 1/4
10	19	16 1/2
12	21 1/4	19

Pipe Size	A	B	C	D
4"	9 3/4"	12 1/4"	1 3/4"	2 1/2"
6	12	14 1/2	2 7/8	3 1/2
8	14 1/4	16 3/4	4	4 3/4
10	16 1/2	19	4 1/2	5 3/4
12	19	21 1/4	6	6 3/4

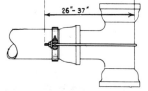

Tee - Spigot to pipe bell

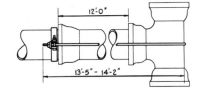

Tee - Bell to pipe spigot

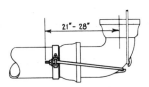

1/4 Bend - Spigot to pipe bell

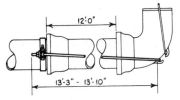

1/4 Bend - Bell to spigot of pipe

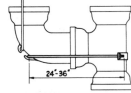

1/4 Bend - Spigot to bell of tee

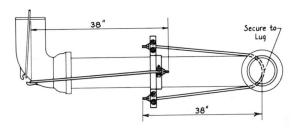

1/4 Bend - Bell to 1/4 bend long spigot

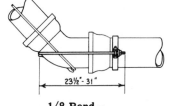

1/8 Bend - Spigot to pipe bell

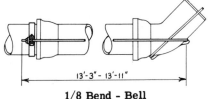

1/8 Bend - Bell to pipe spigot

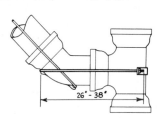

1/8 Bend - Spigot to bell of tee.

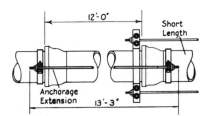

Anchorage extension - Beyond bell of a short length of pipe connected to a bend or tee

Plug - 1/4 bend

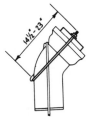

Plug - 1/8 bend

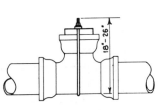

Plug - Side outlet of tee

FIG. 9.7. Anchoring bell and spigot pipe. (Reproduced by permission from Factory Mutual Engineering Division, *Handbook of Industrial Loss Prevention*, McGraw-Hill Book Co., New York, 1959.)

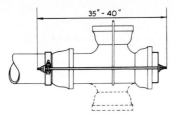

Plug - Run outlet of
spigot tee or cross

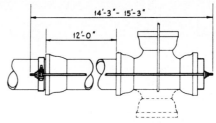

Plug - Run outlet of
bell tee or cross

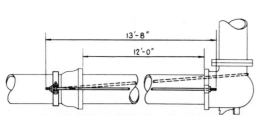

Hydrant - Bell to pipe spigot

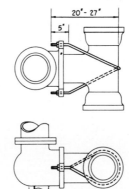

Hydrant - Bell
to spigot of tee

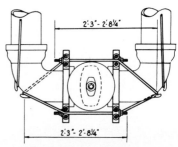

Indicator post valve -
Bell to 1/4 bend spigots

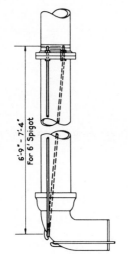

Flange and spigot -
spigot to 1/4 bend bell

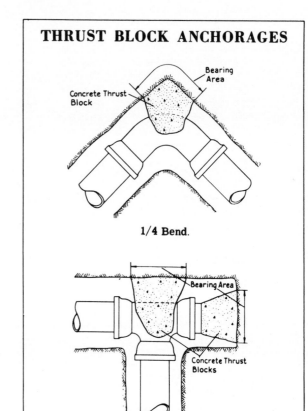

THRUST BLOCK ANCHORAGES

1/4 Bend.

Tee and plug.

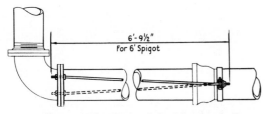

Flange and spigot - Spigot to pipe bell

FIG. 9.7. (*Continued*)

STANDARDIZED MECHANICAL JOINT CAST-IRON PIPE

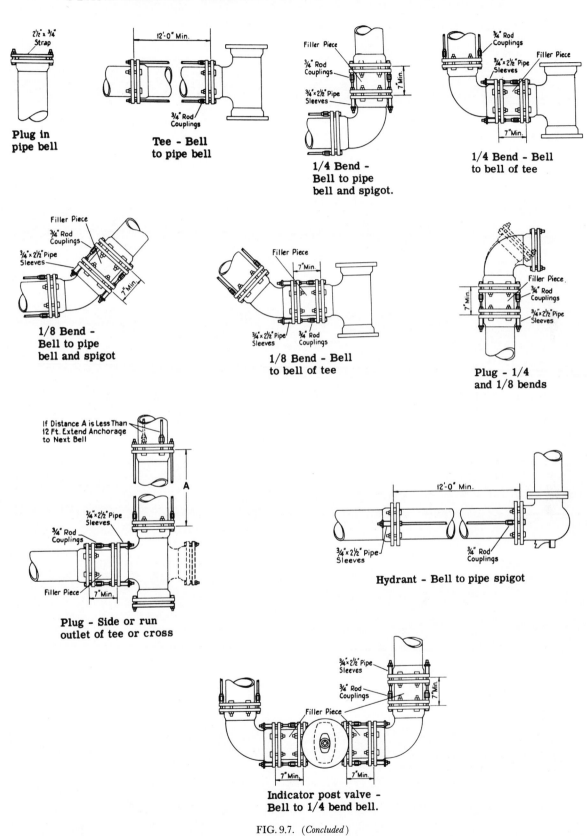

Plug in pipe bell

Tee - Bell to pipe bell

1/4 Bend - Bell to pipe bell and spigot.

1/4 Bend - Bell to bell of tee

1/8 Bend - Bell to pipe bell and spigot

1/8 Bend - Bell to bell of tee

Plug - 1/4 and 1/8 bends

Plug - Side or run outlet of tee or cross

Hydrant - Bell to pipe spigot

Indicator post valve - Bell to 1/4 bend bell.

FIG. 9.7. *(Concluded)*

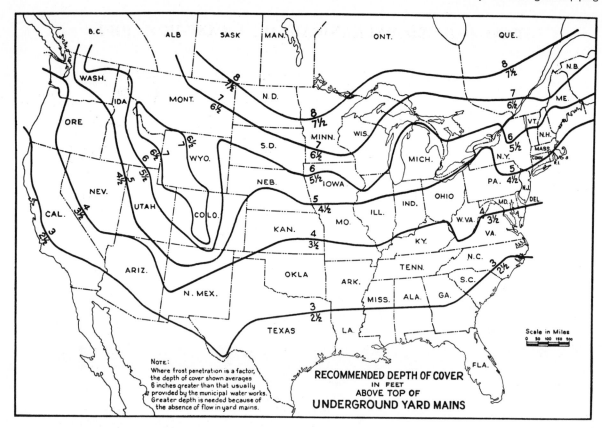

FIG. 9.8.   Recommended depth of cover above underground yard mains.   (Reproduced by permission from Factory Mutual Engineering Division, *Handbook of Industrial Loss Prevention,* McGraw-Hill Book Co., New York, 1959.)

the savings gained from lower material costs.   Further, cast iron will fail earlier than steel as a result of a fire. The subsequent failure of the steam system can cause severe hazards to personnel already endangered by the fire.   In the steam-generating area which is non-hazardous, cast-iron valves and accessories may be used.

Water Hammer in Steam Systems

As steam loses heat in a pipe, condensate forms.   If this condensate is not removed or is trapped in low points of the system, the high-velocity steam will gather the condensate as a slug and slam it against the nearest obstruction.   The obstruction, which may be a regulating valve, a bend in the line, or a steam trap, receives the full impact of the slug.   The result is a loud noise similar to that produced by hammering on pipe.   The force producing the noise often causes serious damage and even failure of portions of the piping system.

Draining Steam Systems

Condensate must be removed periodically from steam distributing systems to prevent water hammer and to assure a dry steam supply.   In addition, condensate must be removed from all process units heated by steam so that the most effective heating will be realized. Steam traps are installed at all natural drainage points in steam mains, such as the bottom of loops, just before risers, and ahead of reducing and shut-off valves.   Long steam lines having no natural drainage points are provided with drip pockets at intervals of 300 to 800 ft.   A

TABLE 9.2
TYPICAL PROCESS PLANT STEAM SYSTEM

System	Steam	Exhaust	Service
High Pressure	600 psig, 700°F	150 psig	Turbo-generators, Electric Power
Low Pressure	150 psig	25 psig and condensing	Plant Services, Power, Process Heating
Exhaust	25 psig	Condensate	Plant Services, Building Heating
Condensate	Liquid, 5–10 psig		Return to Steam Generation Plant and Waste

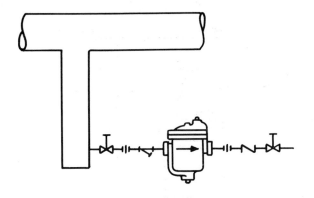

FIG. 9.9. Drip pocket in steam main. (Reproduced by permission of Armstrong Machine Works, Three Rivers, Mich.)

steam trap is then installed in this drip pocket (see Fig. 9.9). Long runs of steam pipe require expansion bends, and it is most convenient to provide a vertical bend toward grade (Fig. 9.10) which also serves as a convenient drip pocket. Main steam distribution headers and all saturated steam and exhaust steam piping should be run with a gradient to facilitate condensate drainage toward the drip pockets.

Steam traps are required at the discharge of all process heating units such as kettles, reboilers, and heat exchangers. When saturated steam is supplied to steam-driven mechanical equipment such as reciprocating engines, water separators should be provided before the throttling valves. This insures that dry steam reaches the engine. A steam trap is located below the separator to remove periodically the condensate that accumulates.

Typical steam trap hook-ups and rules for designing installations are given in Fig. 9.11.

Planning a Steam Distributing System

A plot plan of the process plant serves as a convenient means of preliminary layout and planning of a steam distributing system. A list of all steam users should be prepared, clearly showing the amount and type of steam to be consumed by each user. Maximum anticipated consumption should also be noted. By locating these users on the plot plan, the total quantities of steam in each pressure class are then calculated. These data, together with the plot plan, aid the designer in locating high-pressure exhaust and condensate lines. In addition, he can determine if shortages of lower pressure steam may exist and provide the necessary pressure-reducing stations for introducing high-pressure steam to the low-pressure system (Fig. 9.12). Steam lines can then be sized in accordance with the methods of Chapter 4 and detailed layout and arrangement begun.

Careful analysis of the stresses and generous allow-ance for expansion of steam lines are essential in steam piping design. Steam turbines require particular attention because they are high-speed, precision machines. Pipe reactions at the turbine can cause misalignment, resulting in poor operation or actual mechanical damage. Expansion joints in steam lines at turbines should be provided with tie-rods to limit the elongation of the joint and absorb the actual thrust created by internal pressure, thus preventing transmission to the turbine flange. These joints should be located so that the movement of the pipe places the expansion in shear instead of tension or compression. Useful rules for steam turbine piping have been developed by the National Electrical Manufacturers Association.[2]

VALVING AT DRIVERS

Gate valves should be located in the horizontal run between header and the driver consuming the steam. On live steam lines locate a gate valve in the horizontal run, but at the header, and locate a globe valve at the driver. For exhaust lines provide only a gate valve at the driver.

CONDENSATE BLOW-OFF

A ½-in. minimum condensate blow-off line should be provided for draining steam drivers prior to use. The line should be conducted to the nearest drain, but should not be submerged below the water level in the drain.

PLAN VIEW
(a) HORIZONTAL BEND
Requires support at end of loop.

ELEVATION
(b) VERTICAL LOOP ABOVE MAIN
Application of insulation is difficult.
Support is a problem under high
wind loads.

ELEVATION
(c) VERTICAL BEND BELOW MAIN
Requires minimum space.
Does not require additional support.
Provides convenient drip leg.

FIG. 9.10. Expansion bends in steam mains.

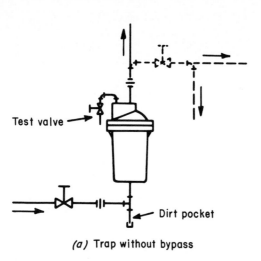

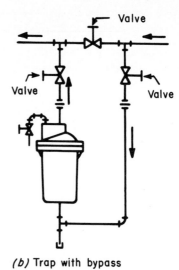

(a) Trap without bypass   *(b)* Trap with bypass

FIG. 9.11.   Typical steam-trap hookups.   (Reproduced by permission of Armstrong Machine Works, Three Rivers, Mich.)

1. Bypasses should be provided for traps located at process equipment requiring continuity of service when trap is removed for repairs or replacement.   Also provide bypasses on traps discharging condensate into condensate return systems.
2. Locate bypass valve above the trap to avoid loss of prime caused by a leak in bypass valve.
3. Use a strainer ahead of the trap when steam is excessively dirty.
4. Provide gravity drainage to trap.
5. Keep discharge line short, especially in cold climates.
6. Multiple trap discharges into a single header, which is or may be under pressure, should be provided with stop valves and check valves.
7. Provide 6-in. minimum dirt pocket below the trap as a means of collecting scale and pipe cuttings.
8. Install test valve in trap outlet or discharge line for checking operation of traps having discharges which are not visible.
9. Provide shutoff valves before all traps.   This makes inspection of traps possible.
10. Trap discharge piping should be designed for maximum discharge pressure.
11. Discharges to the atmosphere should be so arranged that hot discharges do not strike personnel.
12. Provide air eliminators on traps handling flows that contain large volumes of air.

FUELS

Liquid Fuel

Liquid fuels for the process plant are usually conventional fuel oils or hydrocarbon-base waste products. Circulating systems are used to distribute oil from a storage tank to the users.   Complete pumping systems can be purchased as a unit for attaching to distribution and return lines.   Strainers must be provided to prevent foreign matter from entering the burners.   When heating is required to reduce viscosity, the lines are steam-traced or steam-jacketed, and the necessary steam condensate traps are provided.   The usual materials are steel pipe and accessories.   Cast-iron piping with mechanical joints may be used for underground fuel piping, but steel pipe coated and wrapped to prevent soil corrosion is more common in process plants.

Gaseous Fuel

Gaseous fuels require the same piping materials as any hydrocarbon gas.   If the system is underground, cast iron with mechanical joints (bell and spigot in some applications) or coated steel pipe may be used. Above ground systems, however, are the most common for gaseous fuels.   The pipe can be run along with other process piping on racks, utilizing carbon-steel pipe and valves.   Hazards from leaks are minimized by the above ground location.

Fluidized Solid Fuels

The process plant rarely uses solid fuels of any kind for process heating.   Solid fuels are used at times for auxiliary services such as steam generation.   When solid fuels are fluidized the fuel is ground in ball mills and blown into the combustion zone pneumatically. Piping, valves, and flanges are usually carbon steel. Light-weight pipe is sometimes used for air transmission and for the fluidized fuel.

Waste-Product Fuels

Waste-product fuels are combustible substances which are difficult to reclaim or materials which must be burned for disposal.   Such materials may require pretreating and heating before they can be handled with common ferrous piping materials. The piping material selected must depend on the composition of the waste product.

Piping at Burners

Piping designers must arrange and select piping connections to burners. Oil burners require atomizing gas, and in process plants steam is usually used for this purpose. The piping designer must locate the oil-storage tank at a safe distance from the furnace (50 ft for usual tank). Valves controlling the fuel supply to burners should be arranged for remote control. Steam-smothering and water-quenching emergency valves should be located near the fuel valves.

PLANT AIR

Air systems may be provided for power tools and for use in blowing fluids, operating various cleaning services, and emptying piping. Air pressure for each system varies with the service. Usual pressures are from 50 to 100 psi. Simple reciprocating electric or steam-driven compressors or rotary compressors are used. Pressure is maintained by on-off controls or clearance-pocket controls (reciprocating machine) and suction valve unloaders. Atmospheric air is usually taken in through a combined filter and silencing element. The discharge pressure will govern the number of stages. When more than one stage is required, inter-cooling is necessary. After-cooling of the air condenses moisture which is

removed along with lubricating oil traces by a drip separator. Water piping is necessary for cylinder cooling and inter- and after-coolers. If the machine is driven by an internal combustion engine, water piping will be required for engine jacket and lubricating oil system cooling.

An accumulator or receiver is provided either near the machine or somewhere in the system. Several accumulator-receivers may be provided to smooth out pressure fluctuations. All such receivers are provided with pressure-relieving devices and drain valves to withdraw moisture and lubricating oil. For some services, air-dryer units may be required. Piping for plant air systems is usually galvanized steel or nongalvanized steel with cast iron, malleable iron, or steel fittings.

INSTRUMENT AIR

Instrument air may be independent from the plant-air system, but similar equipment must be provided. Separate independent small compressors for each process area generally are more economical than a complete plant instrument-air system. It is common practice to provide an air dryer for instrument-air systems.

WASTE DISPOSAL

Process plant wastes may be solids, liquids, or gases. A large proportion is usually waste water contaminated

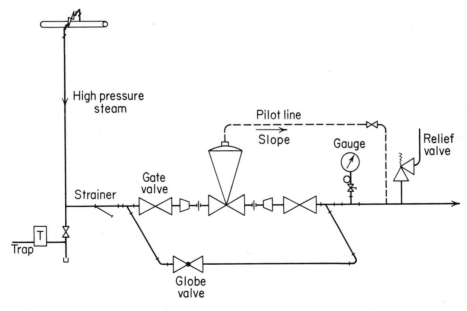

FIG. 9.12. Pressure-reducing station. (Reproduced by permission from Charles T. Littleton, *Industrial Piping,* McGraw-Hill Book Co., New York, 1951.)

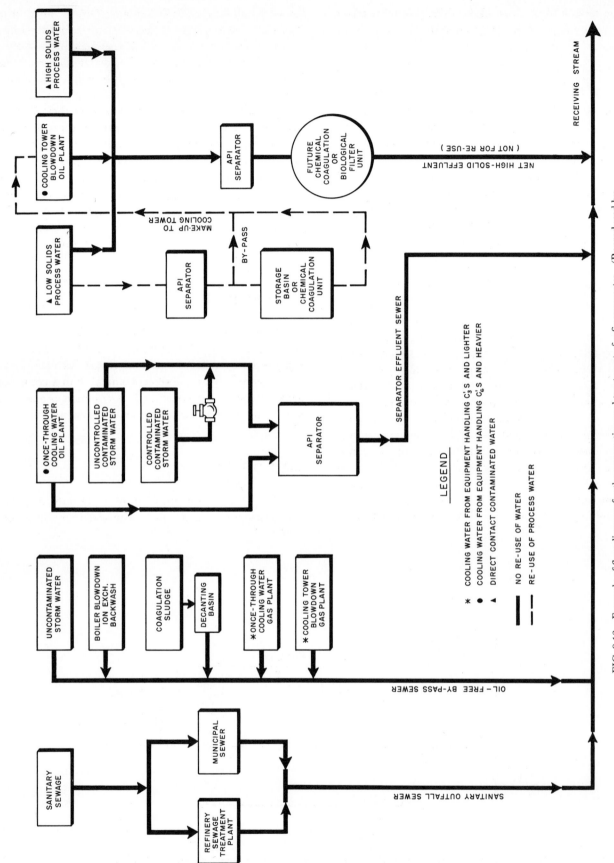

FIG. 9.13. Example of flow diagram for the segregation and treatment of refinery wastes. (Reproduced by permission from "API Manual Disposal of Refinery Wastes," Vol. 1, *Waste Water Containing Oil*, 6th ed., 1959.)

LEGEND

* COOLING WATER FROM EQUIPMENT HANDLING C_3'S AND LIGHTER
● COOLING WATER FROM EQUIPMENT HANDLING C_6'S AND HEAVIER
▲ DIRECT CONTACT CONTAMINATED WATER

――― NO RE-USE OF WATER
----- RE-USE OF PROCESS WATER

with varying amounts of products or by-products. Disposal of these waste waters constitutes a major problem in process plant design and operation. Before discharging into surface waters, contaminants must be removed and the water rendered harmless to living organisms in and around any lake, stream, or bay. Some useful recommendations are given in the following sections which will aid in realizing this goal through proper sewer design and layout.

Segregation of Waste Waters

There are various kinds of waste waters in a process plant and it is usually advantageous to separate the major types by providing sewer systems for each. The advantages of such an arrangement can be seen in Fig. 9.13 which is a schematic diagram for a large oil refinery. Process water contaminated by direct contact with oils in the process is handled in a "dirty-water" sewer. Oil-bearing process streams contain emulsions that require special handling which is not necessary for many of the other waste waters, and it is thus advantageous to segregate the oily streams from the clean streams. The oil is removed by means of a separator (Fig. 9.14) and the clear water discharged to the receiving stream. When flows are large, it is often desirable to segregate process water that has a high percentage of solids from that containing only small amounts of solids. The latter can be readily freed of oil and solids and used as cooling-tower makeup water.

Surface drainage from paved areas contains oily materials which must be removed in a separator similar to that used for process waters. In large plants surface drainage is handled in a separate system. In smaller plants the surface drains are discharged into the same system that carries the process waste water.

Every plant will discharge certain oil-free water wastes. These include uncontaminated surface drain-

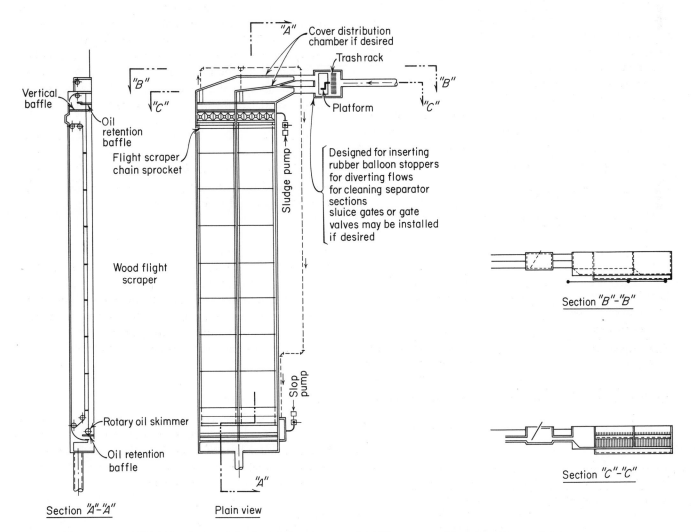

FIG. 9.14. An example of general arrangement for API separator. (Reproduced by permission from "API Manual Disposal of Refinery Wastes," Vol. 1, *Waste Water Containing Oil,* 6th ed., 1959.)

age, boiler blowdown, water-treating plant wastes, and certain inorganic process wastes. These waters can be collected in a single system which may bypass the oil separating stage.

Sanitary sewage can be discharged into a municipal sewer system, or the effluent from toilets can be conducted to a septic tank, the clear discharge of which is combined with waste water from lavatories and discharged into the receiving stream. Discharge into the dirty-water sewer is also practiced but not always successfully because the emulsion-producing characteristics of the detergents present in most sanitary sewage cause difficulties in the oil-water separators.

Large refineries and chemical plants produce many other troublesome wastes which are often segregated for special treatment. Waste caustic and acid streams can be collected and mixed under controlled conditions prior to discharging into the main sewer. Plants that produce large amounts of dilute acid wastes employ acid-resistant sewer pipe for conducting the waste to an equalizing basin. From there it passes to a neutralizing system and thence to a settling basin before being discharged into public waters or to other units for further treatment when necessary.

Clearly, the complexities of waste disposal indicate the need for expert appraisal and planning. The piping designer can aid in this task by understanding the need to adhere closely to instructions for segregating various types of sewage, even though seemingly better piping layouts are possible by combining many discharges.

Notes on Sewer Design

The following notes will aid in the layout and planning of sewer systems.

PIPING MATERIALS

Material	Service
Cast Iron, Bell and Spigot Pipe and Fittings	Hot fluids, heavy soil loads
Steel Pipe and Fittings	(a) For pressure piping of hot, volatile fluids to nearest junction box (b) For process fluids above 350°F
Vitrified Tile	(a) General cold wastes in areas of moderate soil loads (b) Acid wastes (acid-proof tile also used) (c) Large sewers (above 24 in.)
Lead Pipe	Acid wastes
Reinforced Concrete	(a) Non-acid wastes (b) Large sewers (above 24 in.) (c) Drains for cold wastes
Soil Pipe	Building drains and plumbing (ASA A40.1)

Manholes and catch basins are made of reinforced concrete. Those in acid-waste service must be lined with acid-resisting brick.

SIZING SEWER SYSTEM

Details of sizing sewer lines have been discussed in Chapter 4. All such data are based on municipal practice with water sewage. It is assumed that the water flow will carry any lighter materials. A gradient adequate for 2.5 to 3.0 ft/sec is the criterion for assuring that solids will be swept out of the system. Of course, the gradient will be influenced by the terrain and the final discharge point. Oversizing is not only uneconomical but also dangerous. Excessive amounts of hazardous gases may accumulate in oversized systems.

In sizing a sewer system to carry surface drainage, local weather bureau statistics should be examined to determine maximum rainfall intensities. Surface water run-off depends on the plant site contours and gradients, the type of soil, extent of paved and roofed areas, and the type of terrain surrounding the plant. Since process areas are usually paved or heavily compacted, the percentage of run-off (percentage of rainfall not absorbed by the soil) is usually considered to be 100 per cent. The run-off in paved areas enclosed by firewalls may be governed by fire hose capacity rather than rainfall and both should be checked.

Figure 9.15 presents rainfall intensity data which may be employed for estimating purposes when accurate local data are not immediately obtainable. The run-off is calculated from the intensity. Since 1 in./acre is approximately equal to 1 cu ft/sec, then

$$Q = CIA$$

where Q = run-off in cubic feet per second.

C = fraction of rainfall not absorbed by soil (1.0 for paved areas).

I = intensity of rainfall in inches per hour. Data on 15-minute storms based on two-year frequencies are widely used.

A = drainage area in acres. (43,560 sq ft in 1 acre).

CHANGE IN DIRECTION

To prevent clogging and to facilitate cleaning, sewer lines are run without change in direction between manholes on major branch lines (laterals). Sub-laterals are run straight wherever possible, but they are often joined to the lateral by a Y-connection.

MANHOLES

Manholes permit access to sewers for gauging and cleaning sludge from the system. They also provide observation points for isolating unusual leaks of oils or chemicals from the process. The following rules will assure that these goals are realized.

Install manholes:

1. In straight runs of sewer mains: every 250 ft maximum.
2. At point of connection of branches (laterals) to main sewers.

3. At dead ends and any other spots to facilitate cleaning.
4. At points of change in grade.

Provide:

1. Solid covers with vents—these vents shall be conducted to safe location when flammable materials are vented (Fig. 9.16).

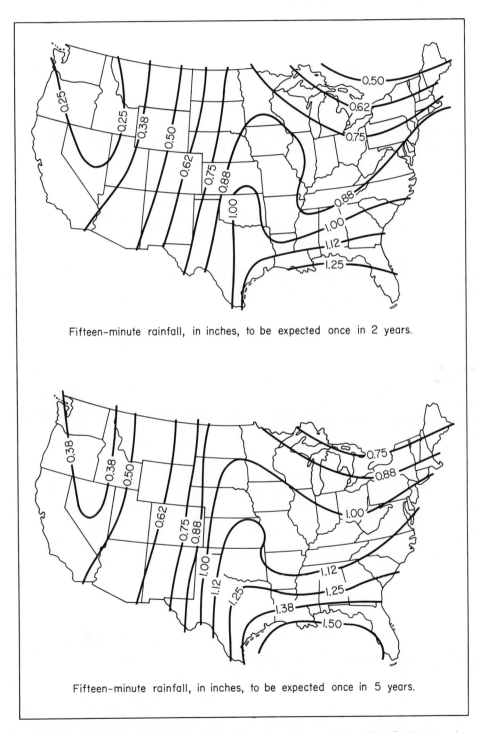

Fifteen-minute rainfall, in inches, to be expected once in 2 years.

Fifteen-minute rainfall, in inches, to be expected once in 5 years.

FIG. 9.15. Rainfall intensity-frequency data. (Reproduced by permission of Clay Products Association, "Clay Pipe Manual," Chicago, Ill., 1951.)

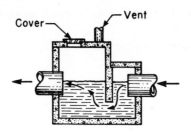

FIG. 9.16. Sewer manhole with seal. (Reproduced from H. F. Rase and M. H. Barrow, *Project Engineering of Process Plants,* John Wiley and Sons, New York, 1957.)

2. Water seal manholes—install in cooling water and waste water service to prevent spreading of fires and explosions. Water-sealed intakes are preferred since oil layers are not trapped by the seal. The seal can be designed as in Fig. 9.16, or the inlet pipe can be a right angle bend and be immersed 6 to 8 in. below the water level.

CATCH BASINS

Catch basins (Fig. 9.17) are installed in both paved and unpaved areas for the collection of surface drainage and process water containing solids. The sediment will collect in the basin and can be periodically removed. Slotted cast-iron or steel-grating covers are used. The outlets of catch basins are sealed as shown in Fig. 9.17 or by employing a wall-type cleanout seal or outlet baffle with water inlet beneath the water surface. Each of these arrangements requires a 180° change in water direction which assures effective deposit of solids in the basin.

BLOWDOWN

Disposal of materials discharged by automatic relieving devices or by manually operated blowdown or dropout valves is an important design problem which the piping designer must assist in solving. Cold, nonhazardous materials are piped to knockout drums for recovery. Hot and hazardous material, a major portion of which is usually vapor, must be disposed of by burning. Relief valves on top of high towers can often be vented safely to the atmosphere. All other hazardous materials are manifolded into discharge lines that terminate at burning pits or flare stacks. Burning pits which produce smoke require large free areas remote from operating units and residential communities, and thus flare systems are the more common means of disposal.

A typical waste-gas flare system for a refinery is shown in Fig. 9.18. The main line, which collects gases from all units, passes through a 10-ft by 30-ft underground drip or entrainment separator, a water seal tank, and thence to the flare where the gases are burned. Large emergency flows, high enough to break the 12-in. seal, are directed into the auxiliary flare for burning. This flare also handles gases from the compressors during failures. Two lines, running separately from operating units, are made to bypass the entrainment and water seal section. One line contains corrosive gases (HCl) which would dissolve in the water seal, causing rapid corrosion of the tank. The other is an emergency dropout line from hydrogenation reactors, the flow of which when discharged would be beyond the capacity of a reasonably sized seal system. It is possible to conduct all discharges to the auxiliary flare for short periods when repairs must be made to the principal flare.

The illustrated system emphasizes the necessary components of good design—overload capacity, segregation of corrosive gases and excessive dropout flows, and provision for operating during repair.

Notes on Piping for Blowdown Systems

(1) Local Venting of Relief Valves (Fig. 9.19).
  (a) Discharge riser should terminate a minimum of 10 ft above any platform within 20- to 40-ft radius.
  (b) A protecting cover for preventing water accumulation in the discharge line should be installed. A ⅜-in. weephole in the ell should be provided to assure drainage of any water.
  (c) Support discharge piping to prevent excessive stresses on relief valve.
  (d) Install steam-snuffing lines in discharge lines when desired for extinguishing fires. Line should not be oversized so that steam flow would hamper relief valve discharge.

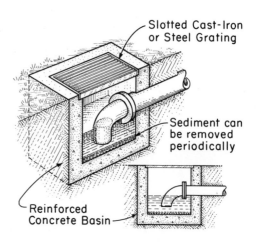

FIG. 9.17. Catch basin.

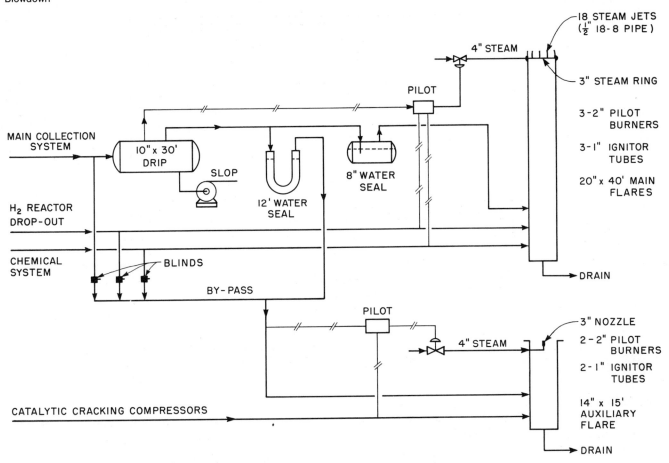

FIG. 9.18. Waste-gas flare system. [Adapted by permission from D. L. Cleveland, "Proceedings of API," *Refining*, **32M,** III, 146 (1952).]

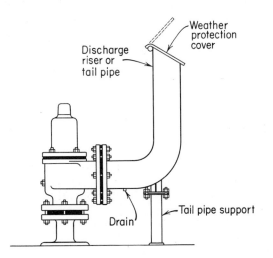

FIG. 9.19. Local venting at relief valves. [Reproduced by permission from Nels E. Sylvander and Donald L. Katz, "The Design and Construction of Pressure Relieving Systems," *Engr. Research Bulletin No. 31,* University of Michigan (1948).]

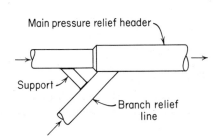

FIG. 9.20. Entrance of branch lines to main header. [Reproduced by permission from Nels E. Sylvander and Donald L. Katz, "The Design and Construction of Pressure Relieving Systems," *Engr. Research Bulletin No. 31,* University of Michigan (1948).]

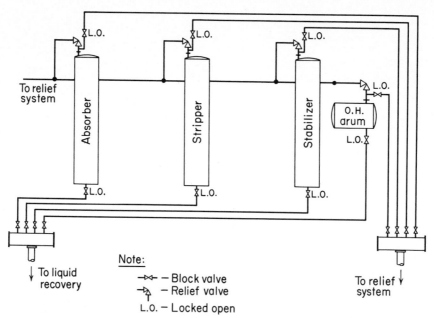

FIG. 9.21. A manual vapor-blowdown and liquid-dropout system. [Reproduced by permission from Nels E. Sylvander and Donald L. Katz, "The Design and Construction of Pressure Relieving Systems," *Engr. Research Bulletin No. 31,* University of Michigan (1948).]

(2) Blowdown

(*a*) Slope lines gradually toward knockout drums to assure removal of condensed fluids.

(*b*) Study expansion and anchoring problems with care. The sudden discharge of hot fluids into cold lines can cause unusually severe stresses which the system must withstand. Support must be provided at points of impact of the high velocity gas stream.

(*c*) Corrosive fluids should be handled in a separate system. Such fluids can be neutralized in a special quench tower before disposing.

(*d*) Toxic gases are often isolated in a separate system to avoid diffusion throughout the plant in a common system.

(*e*) Fluids that become highly viscous or solidify at air temperature should be handled in separate steam-traced headers.

(*f*) Use 45° angle inlets for branch lines into headers (see Fig. 9.20).

(*g*) Size piping so that pressure drop will not exceed 10 per cent of the set pressure of the relief valve. The discharge line from a relief valve should never be smaller than the outlet size of the relief valve. Relief piping requires careful study by process engineers who are experts in safe practices. The piping designer should not attempt to size such systems.

(3) Dropout Piping (Fig. 9.21)

Dropout valves are provided on equipment for use in rapid depressuring of equipment during an emergency, such as a fire. If this is done before the pressure rises to the point at which relief valves would discharge, the possibility for an explosion is greatly lessened.

(*a*) Locate fast-acting dropout valves at safe distance from unit and provide hydraulic or pneumatic devices for remote operation.

(*b*) Provide separate headers when large quantities are involved.

REFERENCES

1. Shannon, W. L., *J. New England Water Works Assoc.,* **59,** 356 (1945).
2. "Standards for Mechanical-Drive Steam Turbines," National Electrical Manufacturers Association, Pub. No. SM 20, 155 East 44th Street, New York, N.Y.
3. Uhl, W. C., *Pet. Processing,* p. 134 (July, 1957).

10
Instrument piping

One or more elements of most instruments are located in a section of piping, and piping designers must be able to plan piping connections which will assure proper performance of the instrument. Data and recommendations for this purpose are presented in the following sections. No attempt is made to discuss instrument operation except in cases where such discussion will aid in piping design decisions.

INSTRUMENT ELEMENTS

The elements of an instrument installation are shown diagramatically in Fig. 10.1. They are a primary element which senses the condition of interest, the controller which translates the measured variable into a control signal, and the control element which responds to the signal and changes the control agent in a manner to produce the desired condition in the controlled medium. An example of a controlled system might be the temperature control of a process fluid being heated by hot oil in an exchanger. The primary measuring device would be a thermocouple installed in the piping of the fluid leaving the exchanger. The controller would be a temperature-indicating or recording controller which would transmit a pneumatic signal to an automatic control valve (the control element). If the temperature of the process fluid leaving the exchanger decreased, the control valve would open wider and permit more hot oil to flow into the exchanger to further heat the process fluid.

However, not all instrument installations involve controlling a condition. Some are used only for indicating, in which case the control element is not present and an indicating instrument takes the place of the controller in the diagram.

In either control or indicating systems, piping is associated with the installation. This piping may be divided into three categories: (*a*) primary piping for conveying a characteristic of the process fluid to recording or indicating instrument, (*b*) control valve piping, (*c*) pneumatic piping for conveying signal to the control element (control valve) which provides the corrective action. Pneumatic piping is also used for transmitting primary signals for long distances to remotely mounted recording or indicating instruments.

PRIMARY PIPING

All primary piping should have the same pressure and temperature rating as the process piping to which it is connected. Globe valves or needle valves are used for block valves in sizes up to ¾ in., because they are less expensive in this range than gate valves.

Orifice Meter Piping

The orifice meter is the simplest, cheapest, and most popular device for measuring flow rates in process plants. An orifice plate (Fig. 10.2) is inserted in the line and restricts the flow area, causing a pressure change which is related to the flow. The venturi tube, the flow nozzle, and the pitot tube are similar in principle, and are used when specified by the instrument engineer for special situations as indicated in Table 10.1.

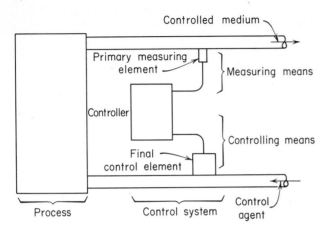

FIG. 10.1. Elements of a controlled system. (Reproduced by permission from D. P. Eckman, *Principles of Industrial Process Control,* John Wiley and Sons, New York, 1945.)

ORIFICE TAPS

1. For lines 2-in. to 12-in. nominal diameter, use manufactured orifice flanges provided with flange taps (½ in. for flanges up to 600-lb ASA and ¾ in. for 900-lb ASA and above). Figure 10.3 illustrates an orifice flange. If the instrument engineer specifies an orifice installation for a line less than 2 in., increase size to 2 in. for meter run. Slip-on type orifice flanges are preferred because they are simpler to install with a smooth inner surface. Welding neck flanges, when required, should be installed with the aid of jigs and clamps to assure perfect alignment.

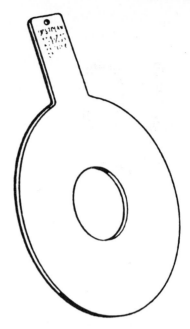

FIG. 10.2. Concentric orifice plate. (Reproduced by permission from L. K. Spink, *Principles and Practice of Flow Meter Engineering,* 8th ed., The Foxboro Company, Foxboro, Mass., 1958.)

2. For lines over 12 in., use vena-contracta taps with conventional flanges as shown in Fig. 10.4. The upstream tap should be located such that $M = D$, and the downstream tap should be located in accordance with Fig. 10.5. Full-flow pipe taps at 2½ diameters upstream and 8 pipe diameters downstream are also used.

TABLE 10.1

SPECIAL SITUATIONS REQUIRING USE OF METERS OTHER THAN CONCENTRIC ORIFICES*

Situations	Flow Nozzle	Venturi Tube	Area Meter (Rotameter)	Eccentric Orifice	Magnetic Meter
Small installations 1½″ line size or less			X		
Highly viscous fluids	X		X		X
High pumping costs		X			X
Fluid contains large amounts of solids	X (in vertical line with flow downward)	X			X
Fluid contains small amounts of solids or fluid is vapor with condensate	X (in vertical line with flow downward)			X (opening at bottom)	
Liquid with vapor or air				X (opening at top)	X
High velocity steam	X				

* X indicates suggested meter.

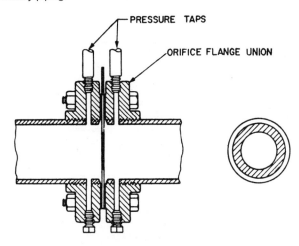

FIG. 10.3. Orifice assembly with flange taps. (Reproduced by permission from D. M. Considine, *Process Instruments and Controls Handbook*, McGraw-Hill Book Co., 1957.)

3. Vena-contracta, or other types of taps made in the pipe, are made as shown in Fig. 10.6, using ½-in. fittings. The tap hole should be drilled as follows:[1]

Nominal Diameter of pipe, in.	Maximum Diameter of tap, in.
2	¼
3	⅜
4 and above	½

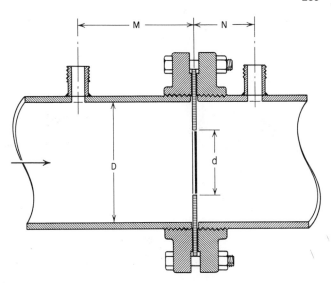

FIG. 10.4. Vena contracta taps. (Reproduced by permission from L. K. Spink, *Principles and Practice of Flow Meter Engineering*, 8th ed., The Foxboro Company, Foxboro, Mass., 1958.)

To insure meter accuracy, the tap must be flush with pipe and all burrs removed. The edges should be rounded slightly.

REQUIRED LENGTH OF STRAIGHT PIPE FOR ORIFICES

The accuracy of an orifice meter depends to a large degree on locating the orifice far from flow disturbances.

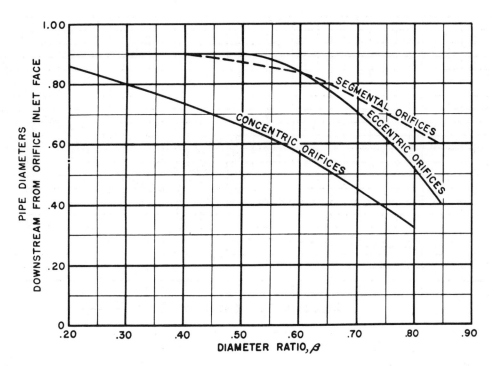

FIG. 10.5. Location of downstream vena contracta taps $\left(\beta = \dfrac{\text{Orifice diameter}}{\text{Inside pipe diameter}} \right)$. (Reproduced by permission from L. K. Spink, *Principles and Practices of Flow Meter Engineerng*, 8th ed., The Foxboro Company, Foxboro, Mass., 1958.)

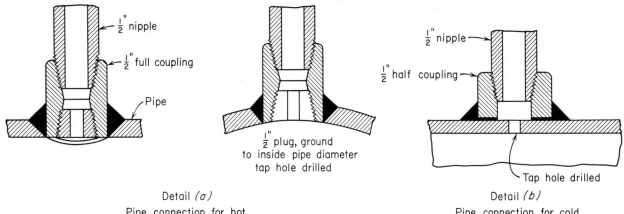

Detail *(a)*
Pipe connection for hot,
high pressure service

Detail *(b)*
Pipe connection for cold,
low pressure service

FIG. 10.6. Orifice taps in pipe. (Reproduced by permission from R. F. Stearns, R. R. Johnson, R. M. Jackson, and C. A. Larson, *Flow Measurement with Orifice Meters,* D. Van Nostrand Co., New York, 1951. Copyright, Esso Research and Engineering Co.)

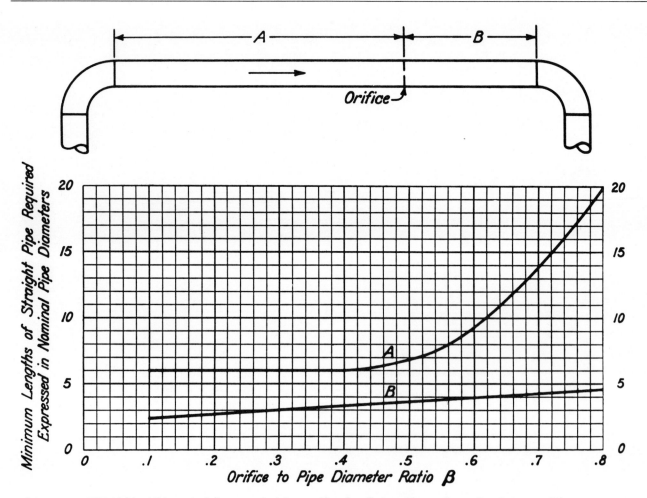

FIG. 10.7*A*. AGA standards for approach piping at orifices: Installation with one ell preceding the meter tube.

Note 1. When "Pipe Taps" are used, *A* should be increased by 2 pipe diameters and *B* by 8 pipe diameters.

Note 2. When the diameter of the orifice may require changing to meet different conditions, the lengths of straight pipe should be those required for the maximum orifice to pipe diameter ratio that may be used.

(Reproduced by permission from "Orifice Metering of Natural Gas," Gas Measurement Committee Report No. 3, American Gas Association, New York, 1955.)

Ideally, each orifice should be preceded by 50 to 80 pipe diameters of straight pipe and followed by at least 10 pipe diameters of straight pipe. These goals are not always attainable in process piping. The American Gas Association has through careful experimentation developed minimum criteria which are presented in Fig. 10.7. Lengths of straight pipe required after various disturbances are given. Although wide-open gate valves and cocks are not included, values can be approximated by using the curves for changes in pipe size. Straightening vanes should be used only when no other arrangement is economically possible.

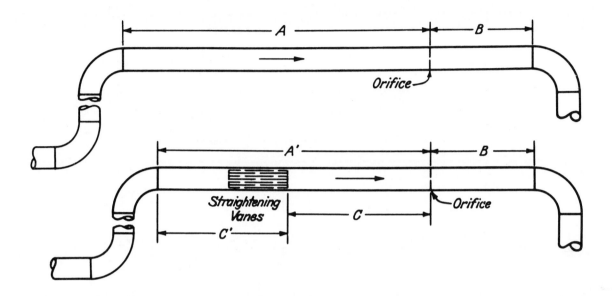

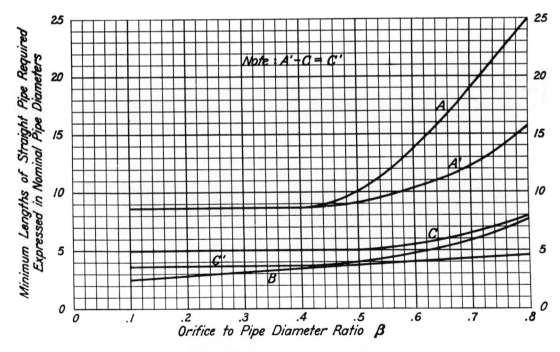

FIG. 10.7B. AGA standards for approach piping at orifices: Installation with two ells or bends preceding the meter tube (bends in same plane). These should be those required for the maximum orifice to pipe diameter ratio that may be used.

Note 1. When "Pipe Taps" are used, A, A', and C should be increased by 2 pipe diameters, and B by 8 pipe diameters.

Note 2. When the diameter of the orifice may require changing to meet different conditions, the lengths of straight pipe should

(Reproduced by permission from "Orifice Metering of Natural Gas," Gas Measurement Committee Report No. 3, American Gas Association, New York, 1955.)

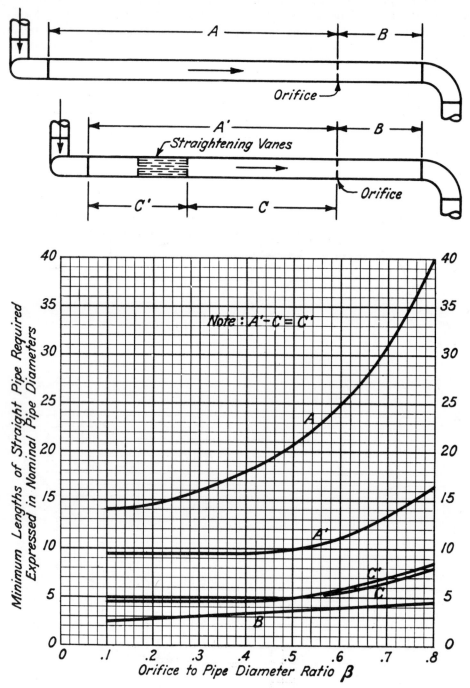

FIG. 10.7C.   AGA standards for approach piping at orifices: Installation with two ells or bends
preceding the meter tube (bends not in same plane).

Note 1. When "Pipe Taps" are used, *A, A′,* and *C* should be increased
by 2 pipe diameters and *B* by 8 pipe diameters.

Note 2. When the 2 ells shown in the above sketches are closely preceded by a third which is not in the same plane as the middle

or second ell, the piping requirements shown by *A* should be
doubled.

(Reproduced by permission from "Orifice Metering of Natural Gas,"
Gas Measurement Committee Report No. 3, American Gas Association,
New York, 1955.)

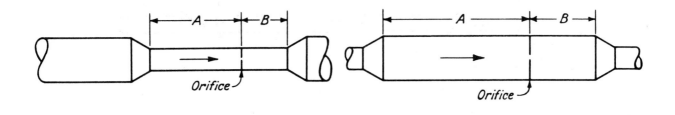

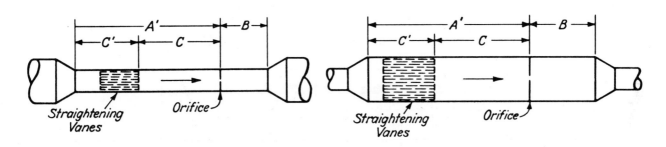

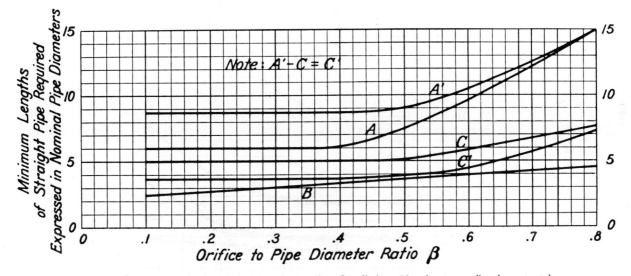

FIG. 10.7D. AGA standards for approach piping at orifices: Installation with reducer preceding the meter tube.

Note 1. When "Pipe Taps" are used, lengths A, A', and C should be increased by 2 pipe diameters, and B by 8 pipe diameters.

Note 2. Straightening vanes will not reduce lengths of straight pipe A. Straightening vanes are not required because of the reducers; they are required because of other fittings which precede the reducer. Length A is to be increased by an amount equal to the length of the straightening vanes whenever they are used (see bottom sketches).

Note 3. When the diameter of the orifice may require changing to meet different conditions, the lengths of straight pipe should be those required for the maximum orifice to pipe diameter ratio that may be used.

(Reproduced by permission from "Orifice Metering of Natural Gas," Gas Measurement Committee Report No. 3, American Gas Association, New York, 1955.)

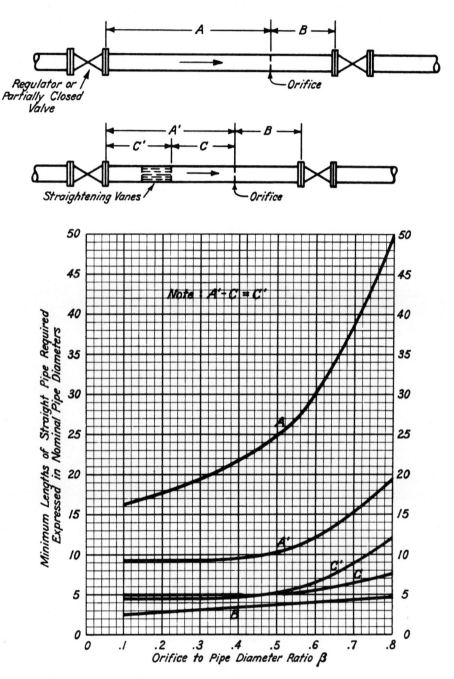

FIG. 10.7E. AGA standards for approach piping at orifices: Installation with valve or regulator preceding the meter tube.

The installation sketch above shows a valve which restricts the flow of the gas, such as a regulator or a partially closed gate valve, globe valve, or plug valve. However, if the gate valve, globe valve, or plug valve is wide open, it may be considered as not creating any serious disturbance and shall be located according to the requirements of the fitting immediately preceding it, but in no case to be located closer to an orifice than permitted by curves A, A', or B in the appropriate figure of installation sketches, Figs. 10.7 B, C, and D. Where there are no other fittings, installation sketch Fig. 10.7B shall apply in locating the above mentioned wide open valves.

(Reproduced by permission from "Orifice Metering of Natural Gas," Gas Measurement Committee Report No. 3, American Gas Association, New York, 1955.)

CLEARANCES NEEDED FOR ORIFICE METER RUNS

Figure 10.8 shows recommended clearances needed for easy installation and maintenance of connecting piping to meter.

PIPING LAYOUTS AT METER BODIES

In modern plants the meter body is located as close as possible to the orifice. The signal is transmitted from the locally mounted meter to a recording or indicating instrument in the control room. Two major types of sensing devices are employed—mercury-actuated and force-balance transmitters (Fig. 10.9). In the mercury manometer the differential pressure causes a movement in both legs. Movement in the large leg is a function of the total differential pressure and is transmitted by means of the float. The force-balance instrument operates by developing a force in opposition to the differential pressure. As P_2 in Fig. 10.9 increases, the balance beam covers more of the nozzle through which air is being released. This movement increases the air pressure in the balancing diaphragm and causes it to force the beam away from the nozzle. An equilibrium of force is thus established, which is represented by the air pressure attained by the system. This air pressure is

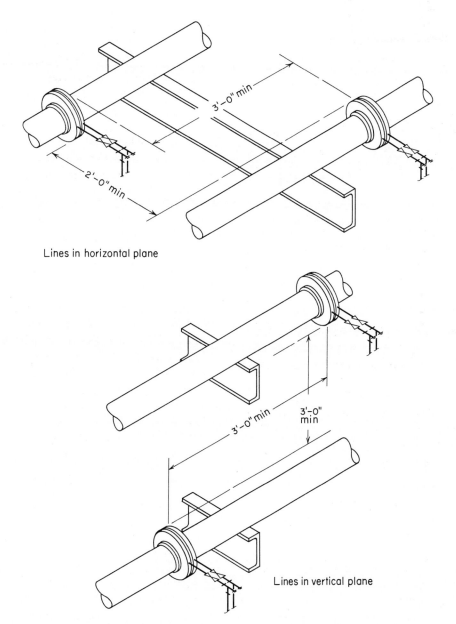

Lines in horizontal plane

Lines in vertical plane

FIG. 10.8. Recommended clearances for orifice meter runs. (Reproduced by permission from R. F. Stearns, R. R. Johnson, R. M. Jackson, and C. A. Larson, *Flow Measurement with Orifice Meters,* D. Van Nostrand Co., New York, 1951. Copyright, Esso Research and Engineering Co.)

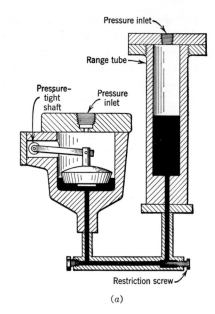

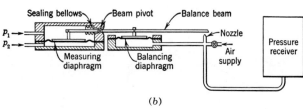

FIG. 10.9. Differential pressure-sensing devices. (a) Mercury-actuated. (b) Force-balance. (Reproduced by permission from D. P. Eckman, *Industrial Instrumentation,* John Wiley and Sons, New York, 1950.)

then readily transmitted to recording instruments and controllers. General rules applying to both types:

1. Use ½ in. minimum piping between orifice and meter.

2. Piping should have same specifications as process line.

3. Slope lines 1 in./ft to assure self-draining.

4. In laying-out piping, avoid arrangements that place excessive stresses on the meter body.

5. Protect piping from freezing, where necessary, by steam tracing or electric heating.

Mercury-Actuated Meter Piping

1. Location. Locate as close as possible to orifice (not over 50 ft). Select an easily accessible place free from vibration and excessively high temperatures. These requirements are usually met by mounting the meter with its center line about 4 ft above grade or platform, using a stand made of pipe or by mounting directly on yard piping supports.

2. Liquid meters. Arrange piping as shown in Figs. 10.10 and 10.11. By locating below the process line, gases or vapors can escape from the meter piping, and

metering error caused by differences in fluid densities will be avoided.

(a) Two sets of *shut-off valves* are provided, one at the process line and the other near the meter.

(b) The *bypass* valve is used for zeroing the instrument.

(c) *Liquid seals* are required to protect the meter from corrosive fluids. The mercury manometer on mercury-actuated meters produces a relatively large volumetric displacement during operation. If sealing fluids are used, seal pots of sufficient diameter are installed so that the change in level is negligible at the interface between process and seal fluids. The pots are installed close to the orifice, as shown in Fig. 10.11. Bypass lines are used for equalizing fluid levels during filling. Hot viscous process liquids must be steam-traced from the orifice through the top half of the seal pot.

(d) *Purge piping* can be specified in place of sealing systems. The purge fluid (liquid, gas, or air) is selected so that it will not affect the process. A small but continuous stream of purge fluid is discharged through the meter lines and into the main line, thereby preventing contact of the process fluid with the meter. On the system shown in Fig. 10.10 purge piping would be installed in the vertical portion of each lead line.

3. Steam meters. In measuring steam flow, the lead lines collect condensate and it is necessary to maintain a constant level of condensate in each lead line. This is done by means of condensate pots (syphon bottles), installed as shown in Fig. 10.12. These condensate pots should be located as close as possible to the orifice taps. Where freezing temperatures are common, lead lines and pots must be heated or a sealing liquid employed. An inert gas purge system may be used as a further alternative.

4. Gas and vapor meters. Gas meter piping is illustrated in Fig. 10.13. In vertical lines carrying wet gases, the flow should be downward if an orifice is to be installed. Corrosive gases which must not touch the meter body may be handled by using seals, or preferably, an inert gas purge similar to those described for liquids. Orifice and meter piping for gas metered for sale or purchase under contract should be installed in accordance with American Gas Association recommendations.[2]

Force-Balance Transmitter. Because of its small size the force-balance transmitter can be mounted at the pressure taps, as shown in Figs. 10.14 and 10.15. Volume displacement is so small that seal pots or condensing

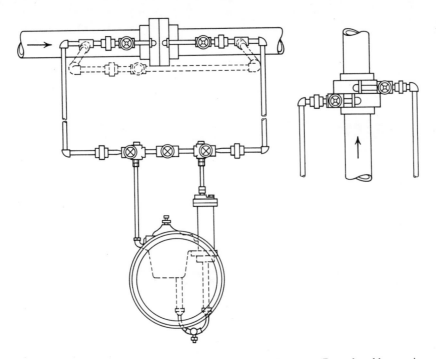

FIG. 10.10. Liquid flow-meter piping for mercury-actuated meter. (Reproduced by permission from L. K. Spink, *Principles and Practice of Flow Meter Engineering*, 8th ed., The Foxboro Company, Foxboro, Mass., 1958.)

chambers are not required. Stainless steel construction permits direct contact with many corrosive fluids. When extremely corrosive fluids or fluids with entrained solids must be measured, purges are installed. The amount of purge needed, however, is small compared to the mercury-actuated meter.

Two shut-off valves and a bypass for zeroing the instrument are required, as shown in the illustrations. The proximity to the process line is only restricted by the maximum operating temperature of the instrument, which is around 250°F for most models. The transmitter should be located below the taps for liquid or steam flow and above the taps for gas flow. A liquid can be sealed by filling the connecting lines with sealing fluid. If a gas-flow system must be sealed, the meter must be located below the taps as for liquid flow.

Other Differential Pressure Flow Meters

Flow nozzles, venturi tubes, and pitot tubes, though used less frequently than orifices, must be installed with equal care. Recommendations by the American Society of Mechanical Engineers have been reproduced and reviewed by Spink.[4]

Laboratory studies indicate that the venturi meter is less sensitive to disturbances caused by fittings and valves located up or downstream in the piping.[3] The recovery

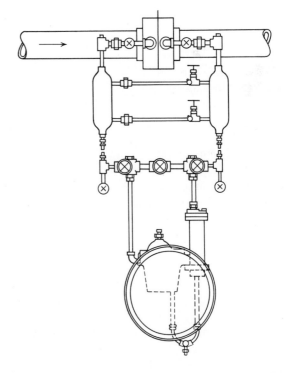

FIG. 10.11. Liquid-seal piping for mercury-actuated meter. (Reproduced by permission from L. K. Spink, *Principles and Practice of Flow Meter Engineering*, 8th ed., The Foxboro Company, Foxboro, Mass., 1958.)

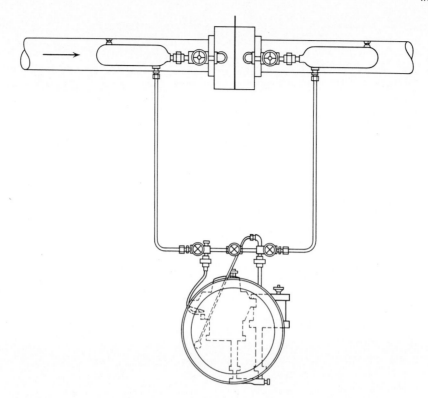

FIG. 10.12.   Steam-meter piping for mercury-actuated meter.   (Reproduced by permission from L. K. Spink, *Principles and Practice of Flow Meter Engineering,* 8th ed., The Foxboro Company, Foxboro, Mass., 1958.)

cone of the venturi usually automatically provides sufficient length of pipe on the downstream side prior to fittings or valves.   The upstream criteria should be set by the instrument engineer, based on studies of the several published tests.

Area Meters

The most common area meter, the rotameter, is shown in Fig. 10.16.   The pressure drop across the meter is maintained constant by allowing the area of the restriction to vary.   The rotameter consists of a tapered tube and float.   As the flow rate increases, the float moves upward, creating more area for flow.   The position of the float, which is directly related to the flow rate, may be read directly when the tapered tube is constructed of glass, or its position may be sensed electrically when metal tubes are used.

Certified drawings of rotameters are needed for planning the piping since there are many styles and sizes. An ideal rotameter installation is shown in Fig. 10.16. The bypass permits operating when the rotameter is removed from service for cleaning or repair.   The check valve reduces liquid water-hammer problems. Although rotameters are built for rugged service, piping

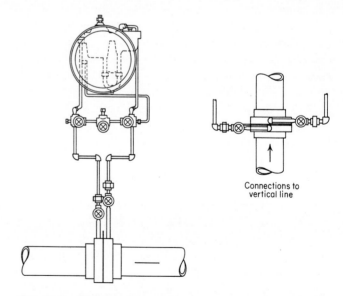

Connections to
vertical line

FIG. 10.13.   Gas-meter piping for mercury-actuated meter.   (Reproduced by permission from L. K. Spink, *Principles and Practice of Flow Meter Engineering,* 8th ed., The Foxboro Company, Foxboro, Mass., 1958.)

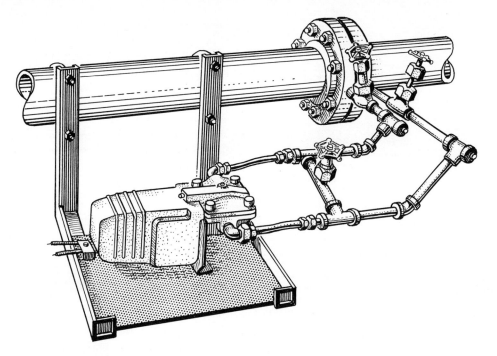

FIG. 10.14. Force-balance transmitter mounted for steam or liquid flow measurement.

should be arranged to avoid unnecessary stresses on the rotameter. Cleanout plugs should be freely accessible and ample room provided for removing the rotameter from the system. Where possible, the more easily removed flanged-meter is preferred.

Pressure-Instrument Piping

Pressure taps in piping for use in connecting to pressure gauges or remote pressure indicators and controllers should be fabricated as shown in Fig. 10.6. The usual connection is ½-in. nominal pipe size and should be located in a straight run of pipe for accurate measuring of the static pressure. A shut-off valve is provided at the tap, and the piping designer must indicate the connection and valve on the drawing. Needle valves are preferred for locally mounted pressure gauges because they protect against sudden pressure surges and dampen pulsations. Services which produce excessive pulsations are provided with manufactured pulsation dampers installed between the gauge and valve.

Corrosive fluids are isolated from pressure gauges by providing a flexible diaphragm between the gauge and the valve. Measured fluid is on one side and a transmission fluid on the other side in contact with the measuring element. When steam pressure is being measured, a syphon consisting of several turns of flexible tubing is installed between the valve and the gauge. The syphon fills with condensate and protects the gauge from high temperatures.

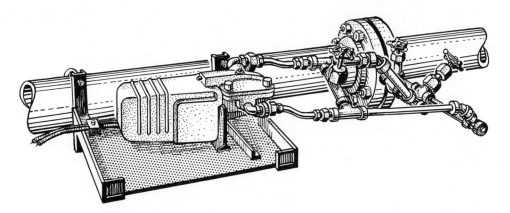

FIG. 10.15. Force-balance transmitter mounted for gas flow measurement.

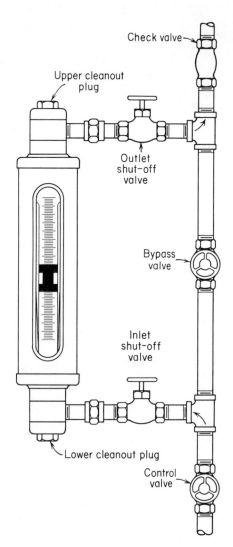

FIG. 10.16. Typical rotameter installation. (Reproduced by permission of Fischer & Porter Company, Warminster, Pa.)

Pressure gauges should be located where they will not be subjected to high temperatures which weaken soldered connections in the gauge and also cause measuring errors. Gauges installed at vibrating equipment should be protected from excessive vibrations by mounting the gauge on a wall or rigid panel and connecting it to the equipment with flexible tubing.

Recorders or controllers located in control rooms are actuated by pneumatic signals produced by a transmitter at the process line. The process fluid transmits the pressure to the transmitter directly or through a sealing liquid or diaphragm and sealing liquid combination. Electrically operated transducers for measuring pressure have diaphragms between the process fluid and the gauge armature and can be locally mounted if temperatures are not excessive.

Liquid-Level Instruments

When differential-pressure instruments are used for level measurement, connections similar to those described for fluid-flow measurement are used. The instrument engineer will specify the exact position of connections on the tank or vessel.

Float-type level instruments are mounted directly on the vessel, using two shut-off valves and a drain valve (½ in.) as shown in Fig. 10.17. Liquid level gauges require similar valving.

Temperature Instruments

The primary element for measuring temperature must be located in the piping or equipment. These elements are enclosed in a thermowell (Fig. 10.18) and no valving is required. These wells are produced either threaded or flanged. A 1-in. nominal pipe size connection is most common. Typical recommended thermowell installations are shown in Fig. 10.19. The piping designer must check the size and length of the well and be certain that it will fit into the piping or equipment.

PIPING FOR PNEUMATICALLY ACTUATED CONTROL VALVES

The most common flow-control device is the pneumatically actuated diaphragm control valve (Fig. 2.16). It is opened or closed by a stem attached to a plate or diaphragm which forms the movable side of a chamber. The force required for movement is produced by air pressure in the chamber, and in the opposite direction by a spring. Movement of the diaphragm, and thus the stem, causes the valve plug to open or close the valve orifice. Single-port valves are used for small capacities and for services requiring tight shut-off. Double-port valves are specified for large flow and services in which the normal position of the plug is partially open. The air pressure at the diaphragm is transmitted from the controlling instrument which, in turn, receives its impulse from the primary element, as in Fig. 10.17.

When the valve size exceeds 3- or 4-in. nominal pipe size and when operating pressures are high, it is usual to provide a separate air-power source at the valve to change diaphragm pressure in proportion to the pneumatic signal from the controlling instrument. This device is called a valve positioner and is a type of relay which converts the instrument signal to a pneumatic pressure sufficient in force to move the valve stem to the

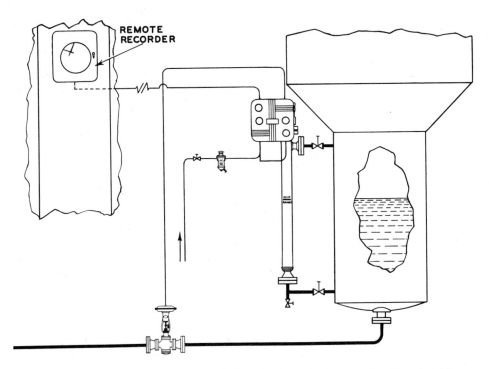

FIG. 10.17. Liquid-level instrument installation, showing valving at vessel. (Reproduced by permission of Mason-Nielan Division of Worthington Corporation, Norwood, Mass.)

desired position indicated by the instrument signal. When a positioner is used, separate source of air from the instrument-air system must be run to the control valve. A small reducing valve, filter, and drip-well is usually provided.

Pneumatic and hydraulic systems may be combined, in which case signals from the instrument are delivered pneumatically with power supplied by a hydraulic device.

Diaphragm-operated control valves must be installed in the upright position and provided with block and by-pass valves for removing when servicing. The Instrument Society of America has prepared recommended control-valve manifold designs which were developed

for maximum flexibility. Dimensions are given for 150-lb and 300-lb ASA flanges in Table 10.2, starting on page 253. Other pressure ratings should be arranged similarly, but the dimensions will, of course, not be the same. When minimum cost is the governing factor, the instrument engineer will specify modified arrangements.

PNEUMATIC PIPING

Connections between secondary-type instruments (transmitters) and the remote instrument, and between the instrument and some device such as a control valve,

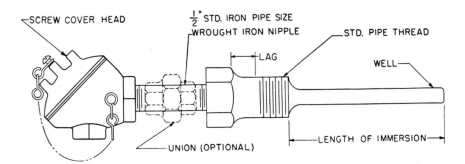

FIG. 10.18. A thermocouple assembly, showing the thermowell. (Reproduced by permission of Minneapolis-Honeywell Regulator Company.)

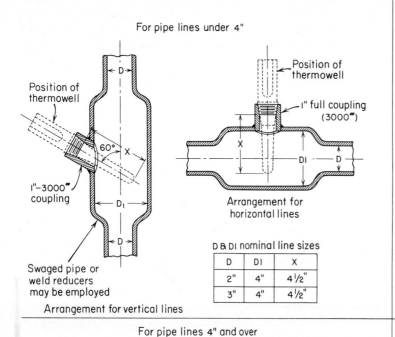

For pipe lines under 4"

D & DI nominal line sizes

D	DI	X
2"	4"	4½"
3"	4"	4½"

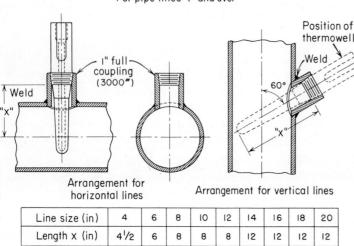

For pipe lines 4" and over

Line size (in)	4	6	8	10	12	14	16	18	20
Length x (in)	4½	6	8	8	8	12	12	12	12

For ells 3" and over (weld-ell illustrated)

Note:

Cast ells should be provided with a 2" diam. cast boss drilled and tapped for 1" diam. pipe. Where the boss in not cast on the ell, it should be built up by welding if the material is suitable for welding.

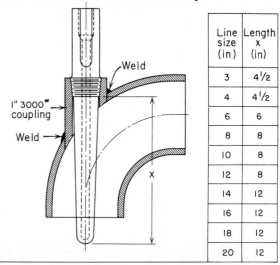

Line size (in)	Length x (in)
3	4½
4	4½
6	6
8	8
10	8
12	8
14	12
16	12
18	12
20	12

FIG. 10.19. Recommended thermowell installations. (Reproduced by permission from R. F. Stearns, R. R. Johnson, R. M. Jackson, and C. A. Larson, *Flow Measurement with Orifice Meters*, D. Van Nostrand Co., New York, 1951. Copyright, Esso Research and Engineering Co.)

1. Outer ends of thermowells should face upward in all installations.
2. Couplings should be machined or ground flush with inside contour of pipe and should be free from icicles and burrs.
3. Dimension "X" refers to length of well, from end of thread to tip of well.
4. For 3-in. and 4-in. lines, the elbow installation is preferable to the others shown and should be used except where an unreasonable cost is involved.
5. When special alloys are involved, welding should be in accordance with the applicable welding code and best recommended practice for the materials concerned.

require piping when the signal or power medium is a fluid. The usual fluid is air, and it is piped in plastic or copper tubing. Multiple-tube plastic assemblies are fabricated and assembled at the factory. The bundle is enclosed in an outer plastic or metallic jacket. The use of these factory-assembled bundles reduces erection labor to a fraction of that required for the separate individual runs of copper tubing which must be arranged on racks with almost continuous support. Multiple-tube assemblies do not require elaborate support; instead, galvanized metal troughs are used. Connections are made by compression-type fittings. Metallic tubing, polyethylene tubing, and piping are used to connect with the various instrument elements. Pneumatic tubing is commonly ¼-in. outside diameter. In planning

bundles, the instrument piping designer should provide spare tubes (about 20 to 25% in excess of immediate needs) when specifying the number of tubes in the bundle.

INSTRUMENT-AIR SYSTEMS

Nearly all modern instrument installations provide a source of filtered, dried air for instrument use. Air dryers utilize regenerative, chemical type desiccants. For large installations, automatic regeneration by steam heat or electric heat may be used. Refrigeration-type dryers also are becoming more common.

The modern process plant depends on control instruments almost completely for product purity or product control. The use of dry, filtered air removes the possibility of instrument failure and subsequent process upsets caused by foreign matter or mosisture in the pneumatic system.

Instrument-air system pressures are usually in the range of 40 to 50 psig. Control and signal pressures in pneumatic systems rarely need to exceed 30 psig. General instrument-air system pressure is regulated by compressor controls and/or pressure reducing instruments in the system. Each instrument-air user is provided with a small capacity independent spring-type pressure reducer, a filter, and drip-well. This necessary equipment is inexpensive and is further insurance against foreign matter entering the instrument.

Materials of construction for air systems are discussed in Chapter 9.

PIPING AT RELIEF VALVES

A relief valve is designed for the single purpose of protecting equipment by relieving pressure, and is the only valve which is designed for a continuously closed position (Fig. 2.17). These valves open automatically when the force on the seat exceeds that of the spring.

Discharge piping from relief valves has been discussed in Chapter 9. The relief valve itself should be located whenever possible in an accessible place so that special platforms for servicing are not required.

As a convenience during maintenance, a plug valve in a locked-open position is often installed before the relief valve. Under severe operating conditions that may demand frequent servicing, dual relief valves may be specified. Plug valves on these installations are interlocked so that both valves cannot be closed at the same time. Valves may be installed in relief-valve discharge lines which tie into a manifold. This is done when the instrument engineer is concerned about flow of discharge from one valve into one of the other lines being serviced. These valves must also be locked open.

Piping designers should be keenly aware of the critical role of relief valves in plant safety. The piping instructions given by the instrument engineer must be strictly followed so that the relieving system will function properly when necessary.

TABLE 10.2
STANDARD CONTROL-VALVE MANIFOLD DESIGNS
(Reproduced by permission: Instrument Society of America, Pittsburgh, Pa.)

MISCELLANEOUS DETAILS

1. Nine component pieces are shown, the use of which will allow the assembly of any of the six standard manifolds shown. Certain of the pieces are usable in more than one of the manifold designs and obviously none of the manifold designs requires the use of all the component pieces. All the component pieces can be fabricated by using conventional weld-type fittings and pipe. The inlet and outlet connections of the manifolds are such as to allow butt welding in a piping system.

2. The six manifold designs are designated Type I thru Type VI. The sequence in which these designs appear has no bearing on the preference or acceptability of the designs, this being determined by the designers' or users' particular application.

3. The block valves used in the designs may be either full manifold size gate valves or plug valves, since the face-to-face dimensions on these are the same. This will allow using the type of valve most suitable for a particular application.

4. By-pass valves shown are full manifold size globe valves up through 4-in. size and full manifold size gate valves in larger sizes.

5. ¾-in. couplings are shown to allow installing bleeders either up-stream or down-stream from the control valve. The particular application and the judgment of the user should determine whether both, one, or neither of the couplings are used for installing bleeders.

6. The clearances in the manifold designs are such as to allow for the removal of the control valve plug from a manifold size valve. Control valves with cooling fins were not considered, but valve positioners can be accommodated.

7. It is assumed that the piping systems in which these manifolds will be used have been properly sized and designed so that the variation between main line size and the control valve size is not excessive. Where the main line size is more than two sizes larger than the control valve size, it is considered economical to swage ahead of the manifold so that the manifold pipe size and the control valve fall within the above relationship. Swaging may also be considered at the manifold inlet to provide a manifold only one size larger than the control valve if dictated by economics or where the need for flexibility for installing a larger control valve is not likely. Because of this, references have been made throughout this report to manifold sizes rather than to main line sizes.

8. ¹⁄₁₆-in. is provided at each flange for gasket.

9. The encircled numbers indicate piece numbers, and dimension lines represent center lines of pipes comprising each piece.

TABLE 10.2 (*Continued*)
TYPE I CONTROL VALVE MANIFOLD

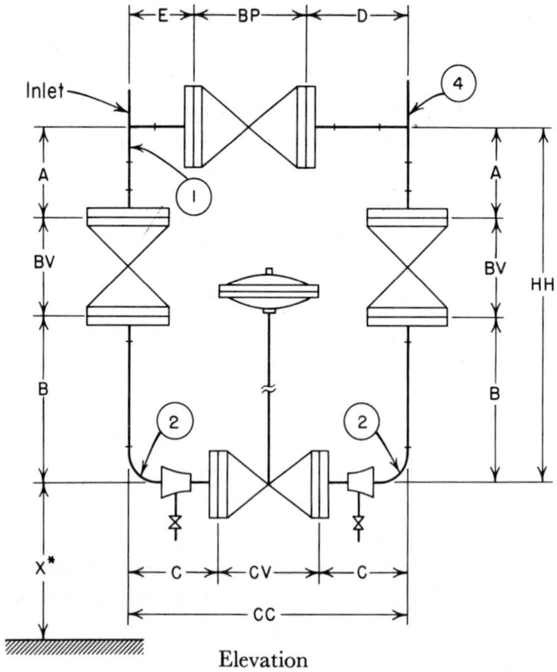

Elevation

See Detail 7 regarding line size.

Component Table
1—Piece No. 1
2—Piece No. 2
1—Piece No. 4

TABLE OF DIMENSIONS
(All dimensions given in inches.)

* If plug removal necessary.

Rating	Manifold Size	Control Valve Size	CV	BV	BP	CC	HH	A	B	C	D	E	X
With 300-lb ASA Flanges	2	1½	9¼	8½	10½	26¾	39	8¾	21⅝	8 11/16	10⅞	5¼	22⅝
	3	2	10½	11⅛	12½	29⅛	41½	9½	20¾	9¼	10	6½	26⅝
	3	1½	9¼	11⅛	12½	29⅛	41½	9½	20¾	9⅞	10	6½	26⅝
	4	3	12½	12	14	34⅞	42⅛	10⅞	19⅛	11⅛	13¼	7½	29½
	4	2	10½	12	14	34⅞	42⅛	10⅞	19⅛	12⅛	13¼	7½	29½
	6	4	14½	15⅞	15⅞	44⅜	53½	17⅛	20⅜	14⅞	18⅞	9½	39
	6	3	12½	15⅞	15⅞	44⅜	53½	17⅛	20⅜	15⅞	18⅞	9½	39
	8	6	18⅝	16½	16½	54½	56⅝	22	18	17⅞	26½	11⅜	45½
	8	4	14½	16½	16½	54½	56⅝	22	18	19 15/16	26½	11⅜	45½
With 150-lb ASA Flanges	2	1½	9¼	7	8	26¾	39	10¼	21⅝	8 11/16	13⅝	5	22⅝
	3	2	10½	8	9½	29⅛	41½	12⅝	20¾	9¼	13⅜	6⅛	26⅝
	3	1½	9¼	8	9½	29⅛	41½	12⅝	20¾	9⅞	13⅜	6⅛	26⅝
	4	3	12½	9	11½	34⅞	42⅛	13⅞	19⅛	11⅛	16⅛	7⅛	29½
	4	2	10½	9	11½	34⅞	42⅛	13⅞	19⅛	12⅛	16⅛	7⅛	29½
	6	4	14½	10½	10½	44⅜	53½	22½	20⅜	14⅞	24⅝	9⅛	39
	6	3	12½	10½	10½	44⅜	53½	22½	20⅜	15⅞	24⅝	9⅛	39
	8	6	18⅝	11½	11½	54½	56⅝	27	18	17⅞	31⅞	11	45½
	8	4	14½	11½	11½	54½	56⅝	27	18	19 15/16	31⅞	11	45½

TABLE 10.2 (*Continued*)
TYPE II CONTROL VALVE MANIFOLD

See Detail 7 regarding line size.

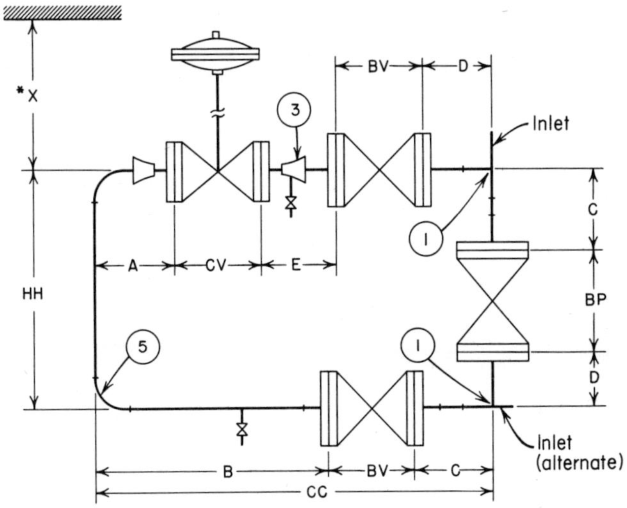

Elevation

Component Table
2—Piece No. 1
1—Piece No. 3
1—Piece No. 5

TABLE OF DIMENSIONS
(All dimensions given in inches.)

* If plug removal necessary.

Rating	Manifold Size	Control Valve Size	CV	BV	BP	CC	HH	A	B	C	D	E	X
With 300-lb ASA Flanges	2	1½	9¼	8½	10½	40¼	24⅝	8¹¹⁄₁₆	22⅞	8¾	5¼	8⁷⁄₁₆	36⅝
	3	2	10½	11⅛	12½	46⁵⁄₁₆	28⅝	9¼	25⁹⁄₁₆	9½	6½	8¹¹⁄₁₆	38½
	3	1½	9¼	11⅛	12½	46⁵⁄₁₆	28⅝	9⅞	25⁹⁄₁₆	9½	6½	9⁵⁄₁₆	38½
	4	3	12½	12	14	55⅞	32½	11⅛	32⅞	10⅞	7½	9⅛	39⅛
	4	2	10½	12	14	55⅞	32½	12⅛	32⅞	10⅞	7½	10⅛	39⅛
	6	4	14½	15⅞	15⅞	65½	42⅝	14⅞	32⅜	17⅛	9½	11½	50
	6	3	12½	15⅞	15⅞	65½	42⅝	15⅞	32⅜	17⅛	9½	12½	50
	8	6	18⅝	16½	16½	76⁵⁄₁₆	50	17⅞	37¹¹⁄₁₆	22	11⅜	11¹¹⁄₁₆	51¾
	8	4	14½	16½	16½	76⁵⁄₁₆	50	19¹⁵⁄₁₆	37¹¹⁄₁₆	22	11⅜	13¾	51¾
With 150-lb ASA Flanges	2	1½	9¼	7	8	38⅜	23⅜	8¹¹⁄₁₆	21	10¼	5	8³⁄₁₆	36⅝
	3	2	10½	8	9½	42⁷⁄₁₆	28⅜	9¼	21¹¹⁄₁₆	12⅝	6⅛	8⁵⁄₁₆	38½
	3	1½	9¼	8	9½	42⁷⁄₁₆	28⅜	9⅞	21¹¹⁄₁₆	12⅝	6⅛	8¹⁵⁄₁₆	38½
	4	3	12½	9	11½	48¾	32⅝	11⅛	25¾	13⅞	7⅛	8¾	39⅛
	4	2	10½	9	11½	48¾	32⅝	12⅛	25¾	13⅞	7⅛	9¾	39⅛
	6	4	14½	10½	10½	60¾	42⅛	14⅞	27⅝	22½	9⅛	11⅛	50
	6	3	12½	10½	10½	60¾	42⅛	15⅞	27⅝	22½	9⅛	12⅛	50
	8	6	18⅝	11½	11½	70⁹⁄₁₆	49⅝	17⅞	31¹⁵⁄₁₆	27	11	11⁵⁄₁₆	51¾
	8	4	14½	11½	11½	70⁹⁄₁₆	49⅝	19¹⁵⁄₁₆	31¹⁵⁄₁₆	27	11	13⅜	51¾

TABLE 10.2 (*Continued*)
TYPE III CONTROL VALVE MANIFOLD

See Detail 7 regarding line size.

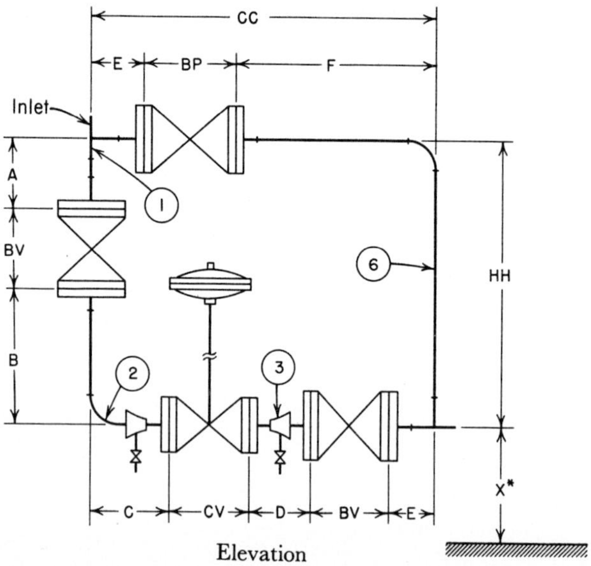

Elevation

Component Table

1—Piece No. 1
1—Piece No. 2
1—Piece No. 3
1—Piece No. 6

* If plug removal necessary.

TABLE OF DIMENSIONS
(All dimensions given in inches.)

Rating	Manifold Size	Control Valve Size	CV	BV	BP	CC	HH	A	B	C	D	E	F	X
With 300-lb ASA Flanges	2	1½	9¼	8½	10½	40⅜	39	8¾	21⅝	8 11/16	8 7/16	5¼	24½	22⅝
	3	2	10½	11⅛	12½	46 5/16	41½	9½	20¾	9¼	8 11/16	6½	27 3/16	26⅝
	3	1½	9¼	11⅛	12½	46 5/16	41½	9½	20¾	9⅞	9 5/16	6½	27 3/16	26⅝
	4	3	12½	12	14	52½	42⅛	10⅞	19⅛	11⅛	9⅛	7½	30⅞	29½
	4	2	10½	12	14	52½	42⅛	10⅞	19⅛	12⅛	10⅛	7½	30⅞	29½
	6	4	14½	15⅞	15⅞	66½	53½	17⅛	20⅜	14⅞	11½	9½	41	39
	6	3	12½	15⅞	15⅞	66½	53½	17⅛	20⅜	15⅞	12½	9½	41	39
	8	6	18⅝	16½	16½	76 5/16	56⅝	22	18	17⅞	11 11/16	11⅜	48 5/16	45½
	8	4	14½	16½	16½	76 5/16	56⅝	22	18	19 15/16	13¾	11⅜	48 5/16	45½
With 150-lb ASA Flanges	2	1½	9¼	7	8	38⅜	39	10¼	21⅝	8 11/16	8 3/16	5	25¼	22⅝
	3	2	10½	8	9½	42 7/16	41½	12⅝	20¾	9¼	8 5/16	6⅛	26 11/16	26⅝
	3	1½	9¼	8	9½	42 7/16	41½	12⅝	20¾	9⅞	8 15/16	6⅛	26 11/16	26⅝
	4	3	12½	9	11½	48¾	42⅛	13⅞	19⅛	11⅛	8¾	7⅛	30	29½
	4	2	10½	9	11½	48¾	42⅛	13⅞	19⅛	12⅛	9¾	7⅛	30	29½
	6	4	14½	10½	10½	60⅜	53½	22½	20⅜	14⅞	11⅛	9⅛	40⅝	39
	6	3	12½	10½	10½	60⅜	53½	22½	20⅜	15⅞	12⅛	9⅛	40⅝	39
	8	6	18⅝	11½	11½	70 9/16	56⅝	27	18	17⅞	11 5/16	11	47 15/16	45½
	8	4	14½	11½	11½	70 9/16	56⅝	27	18	19 15/16	13⅜	11	47 15/16	45½

TABLE 10.2 (*Continued*)
TYPE IV CONTROL VALVE MANIFOLD

See Detail 7 regarding line size.

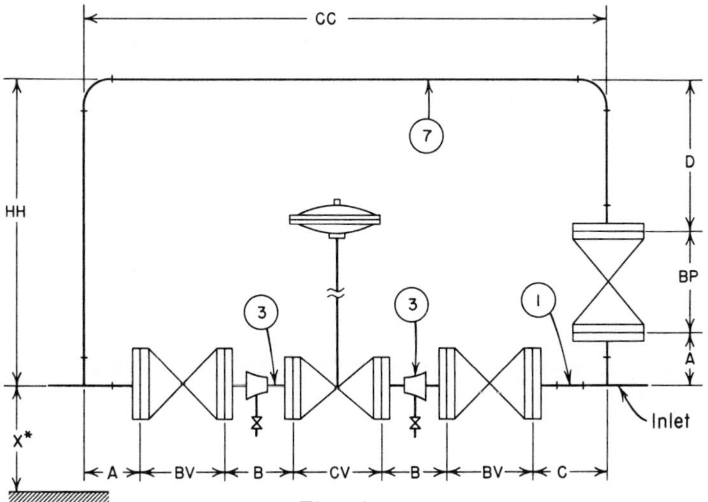

Elevation

Component Table
1—Piece No. 1
2—Piece No. 3
1—Piece No. 7

TABLE OF DIMENSIONS
(All dimensions given in inches.)

* If plug removal necessary.

Rating	Manifold Size	Control Valve Size	CV	BV	BP	CC	HH	A	B	C	D	X
With 300-lb ASA Flanges	2	1½	9¼	8½	10½	57½	39	5¼	8 7/16	8¾	23⅛	22⅝
	3	2	10½	11⅛	12½	66½	41½	6½	8 11/16	9½	22⅜	26⅝
	3	1½	9¼	11⅛	12½	66½	41½	6½	9 5/16	9½	22⅜	26⅝
	4	3	12½	12	14	73½	42⅛	7½	9⅛	10⅞	20½	29½
	4	2	10½	12	14	73½	42⅛	7½	10⅛	10⅞	20½	29½
	6	4	14½	15⅞	15⅞	96¼	53½	9½	11½	17⅛	28	39
	6	3	12½	15⅞	15⅞	96¼	53½	9½	12½	17⅛	38	39
	8	6	18⅝	16½	16½	108¾	56⅝	11⅜	11 11/16	22	28⅝	45½
	8	4	14½	16½	16½	108¾	56⅝	11⅜	13¾	22	28⅝	45½
With 150-lb ASA Flanges	2	1½	9¼	7	8	55¼	39	5	8 3/16	10¼	25⅞	22⅝
	3	2	10½	8	9½	62¼	41½	6⅛	8 5/16	12⅝	25¾	26⅝
	3	1½	9¼	8	9½	62¼	41½	6⅛	8 15/16	12⅝	25¾	26⅝
	4	3	12½	9	11½	69⅜	42⅛	7⅛	8¾	13⅞	23⅜	29½
	4	2	10½	9	11½	69⅜	42⅛	7⅛	9¾	13⅞	23⅜	29½
	6	4	14½	10½	10½	89¾	53½	9⅛	11⅛	22½	33¾	39
	6	3	12½	10½	10½	89¾	53½	9⅛	12⅛	22½	33¾	39
	8	6	18⅝	11½	11½	102⅝	56⅝	11	11 5/16	27	34	45½
	8	4	14½	11½	11½	102⅝	56⅝	11	13⅜	27	34	45½

TABLE 10.2 (*Continued*)
TYPE V CONTROL VALVE MANIFOLD

See Detail 7 regarding line size.

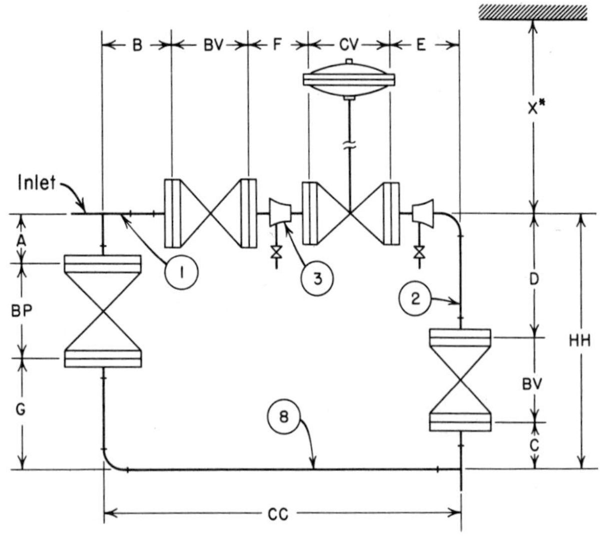

Elevation

Component Table
1—Piece No. 1
1—Piece No. 2
1—Piece No. 3
1—Piece No. 8

* If plug removal necessary.

TABLE OF DIMENSIONS
(All dimensions given in inches.)

Rating	Manifold Size	Control Valve Size	CV	BV	BP	CC	HH	A	B	C	D	E	F	G	X
With 300-lb ASA Flanges	2	1 1/2	9 1/4	8 1/2	10 1/2	43 7/8	35 1/2	5 1/4	8 3/4	5 1/4	21 5/8	8 11/16	8 7/16	19 5/8	36 5/8
	3	2	10 1/2	11 1/8	12 1/2	49 5/16	38 1/2	6 1/2	9 1/2	6 1/2	20 3/4	9 1/4	8 11/16	19 3/8	38 1/2
	3	1 1/2	9 1/4	11 1/8	12 1/2	49 5/16	38 1/2	6 1/2	9 1/2	6 1/2	20 3/4	9 7/8	9 5/16	19 3/8	38 1/2
	4	3	12 1/2	12	14	55 7/8	38 3/4	7 1/2	10 7/8	7 1/2	19 1/8	11 1/8	9 1/8	17	39 1/8
	4	2	10 1/2	12	14	55 7/8	38 3/4	7 1/2	10 7/8	7 1/2	19 1/8	12 1/8	10 1/8	17	39 1/8
	6	4	14 1/2	15 7/8	15 7/8	74 1/8	45 7/8	9 1/2	17 1/8	9 1/2	20 3/8	14 7/8	11 1/2	20 3/8	50
	6	3	12 1/2	15 7/8	15 7/8	74 1/8	45 7/8	9 1/2	17 1/8	9 1/2	20 3/8	15 7/8	12 1/2	20 3/8	50
	8	6	18 5/8	16 1/2	16 1/2	86 15/16	50	11 3/8	22	15 3/8	18	17 7/8	11 11/16	22	51 3/4
	8	4	14 1/2	16 1/2	16 1/2	86 15/16	50	11 3/8	22	15 3/8	18	19 15/16	13 3/4	22	51 3/4
With 150-lb ASA Flanges	2	1 1/2	9 1/4	7	8	43 5/8	33 3/4	5	10 1/4	5	21 5/8	8 11/16	8 3/16	20 5/8	36 5/8
	3	2	10 1/2	8	9 1/2	48 15/16	35	6 1/8	12 5/8	6 1/8	20 3/4	9 1/4	8 5/16	19 1/4	38 1/2
	3	1 1/2	9 1/4	8	9 1/2	48 15/16	35	6 1/8	12 5/8	6 1/8	20 3/4	9 3/8	8 15/16	19 1/4	38 1/2
	4	3	12 1/2	9	11 1/2	55 1/2	35 3/8	7 1/8	13 7/8	7 1/8	19 1/8	11 1/8	8 3/4	16 5/8	39 1/8
	4	2	10 1/2	9	11 1/2	55 1/2	35 3/8	7 1/8	13 7/8	7 1/8	19 1/8	12 1/8	9 3/4	16 5/8	39 1/8
	6	4	14 1/2	10 1/2	10 1/2	73 3/4	45 7/8	9 1/8	22 1/2	14 7/8	20 3/8	14 7/8	11 1/8	26 1/8	50
	6	3	12 1/2	10 1/2	10 1/2	73 3/4	45 7/8	9 1/8	22 1/8	14 7/8	20 3/8	15 7/8	12 1/8	26 1/8	50
	8	6	18 5/8	11 1/2	11 1/2	86 9/16	50	11	27	20 3/8	18	17 7/8	11 5/16	27 3/8	51 3/4
	8	4	14 1/2	11 1/2	11 1/2	86 9/16	50	11	27	20 3/8	18	19 15/16	13 3/8	27 3/8	51 3/4

TABLE 10.2 (*Concluded*)
TYPE VI CONTROL VALVE MANIFOLD

X = Clearance above centerline pipe
Y = Clearance below centerline pipe

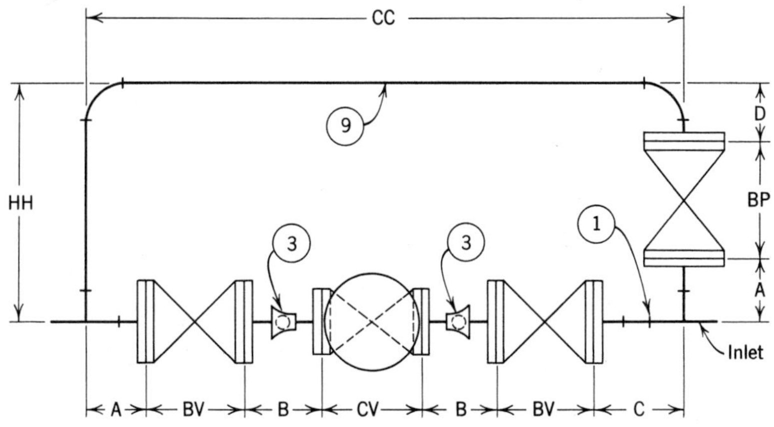

Plan View

Component Table
1—Piece No. 1
2—Piece No. 3
1—Piece No. 9

TABLE OF DIMENSIONS

See Detail 7 regarding line size.

(All dimensions given in inches.)

* If plug removal necessary.

Rating	Manifold Size	Control Valve Size	CV	BV	BP	CC	HH	A	B	C	D	X	Y
With 300-lb ASA Flanges	2	1½	9¼	8½	10½	57½	20⅝	5¼	8 7/16	8¾	4¾	22⅝	36⅝
	3	2	10½	11⅛	12½	66½	25¼	6½	8 11/16	9½	6⅛	26⅝	38½
	3	1½	9¼	11⅛	12½	66½	25¼	6½	9 5/16	9½	6⅛	26⅝	38½
	4	3	12½	12	14	73½	29	7½	9⅛	10⅞	7⅜	29½	39⅛
	4	2	10½	12	14	73½	29	7½	10⅛	10⅞	7⅜	29½	39⅛
	6	4	14½	15⅞	15⅞	96¼	35⅜	9½	11½	17⅛	9⅞	39	50
	6	3	12½	15⅞	15⅞	96¼	35⅜	9½	12½	17⅛	9⅞	39	50
	8	6	18⅝	16½	16½	108¾	40⅜	11⅜	11 11/16	22	12⅜	45½	51¾
	8	4	14½	16½	16½	108¾	40⅜	11⅜	13¾	22	12⅜	45½	51¾
With 150-lb ASA Flanges	2	1½	9¼	7	8	55¼	17⅝	5	8 3/16	10¼	4½	22⅝	36⅝
	3	2	10½	8	9½	62¼	21½	6⅛	8 5/16	12⅝	5¾	26⅝	38½
	3	1½	9¼	8	9½	62¼	21½	6⅛	8 15/16	12⅝	5¾	26⅝	38½
	4	3	12½	9	11½	69⅜	25¾	7⅛	8¾	13⅞	7	29½	39⅛
	4	2	10½	9	11½	69⅜	25¾	7⅛	9¾	13⅞	7	29½	39⅛
	6	4	14½	10½	10½	89¾	29¼	9⅛	11⅛	22½	9½	39	50
	6	3	12½	10½	10½	89¾	29¼	9⅛	12⅛	22½	9½	39	50
	8	6	18⅝	11½	11½	102⅝	34⅝	11	11 5/16	27	12	45½	51¾
	8	4	14½	11½	11½	102⅝	34⅝	11	13⅜	27	12	45½	51¾

REFERENCES

1. *Fluid Meters—Their Theory and Application,* 4th ed., American Society of Mechanical Engineers, New York, 1937.

2. "Orifice Metering of Natural Gas," Gas Measurement Committee Report No. 3, American Gas Association, New York (1955).

3. Pardoe, W. S., *Trans. ASME,* **65,** 337 (1943).

4. Spink, L. K., *Principles and Practice of Flow Meter Engineering,* 8th ed., The Foxboro Company, Foxboro, Mass., 1958.

Appendix

INDEX OF TABLES AND FIGURES IN APPENDIX

Physical properties of fluids

DENSITY

Density of Gases

Gas densities vary with both pressure and temperature.

$$\rho = \frac{zpM}{RT}$$

where ρ = density in lb/cu ft

z = compressibility, Fig. A.9 (page 278)

p = pressure, psia = psig + 14.7

M = molecular weight (Table A.1, page 266)

T = temperature, $^\circ R = {}^\circ F + 460$

R = gas constant, 10.71 (psi)(ft^3)/(lb-mole)($^\circ R$)

or
$$\rho = 0.09337\,\frac{zpM}{T}$$

Example A.1: Calculation of Density for a Pure Gas

Determine the density of propane gas at 355 psig and 238°F.

Compressibility factor from Fig. A.9 (page 278):

$T_c = 206^\circ F$ or $206 + 460 = 666^\circ R$ (Table A.1, page 266)

$P_c = 617.4$ psia (Table A.1)

$$T_r = \frac{238 + 460}{666} = \frac{698}{666} = 1.05$$

$$P_r = \frac{355 + 14.7}{617.4} = \frac{370}{617} = 0.8$$

$z = 0.8$ from Fig. A.9

Molecular weight, $M = 44$ (Table A.1)

$$\rho = \frac{(0.09337)(0.8)(370)(44)}{698}$$

$$= 1.74\,\text{lb/cu ft}$$

Example A.2: Calculation of Density for a Mixture

Determine the density, at 205°F and 1655 psig, of a mixture containing 80.2% methane (CH_4), 10.3% ethane (C_2H_6), 6.5% propane (C_3H_8), 2.0% *iso*-butane (i-C_4H_{10}), 1.0% *n*-butane (n-C_4H_{10}). Average critical properties and molecular weights must be calculated as demonstrated in the following table. Percentages are on mole or volume basis.

Component	Mole %	Molecular wt[1]	Fractional mol wt	Critical pressure psia[1]	Fractional Critical pressure	Critical temperature, $^\circ R$[4]	Fractional critical temperature
CH_4	80.2	16	12.8[2]	673	539.0[3]	344	275.9[4]
C_2H_6	10.3	30	3.1	708	72.9	550	56.7
C_3H_8	6.5	44	2.9	617	40.2	666	43.3
$i = C_4H_{10}$	2.0	58	1.2	529	10.6	735	14.7
$n = C_4H_{10}$	1.0	58	0.6	550	5.5	766	7.7
	100.0		Average $M = 20.6$		Average $P_c = 668.2$ psia		Average $T_c = 398.3^\circ R$

[1] Table A.1 [2] $\left(\dfrac{80.2}{100}\right)(16) = 12.8$ [3] $\left(\dfrac{80.2}{100}\right)(673) = 539.0$ [4] $\left(\dfrac{80.2}{100}\right)(344) = 275.9$

$$P_r = \frac{1655 + 14.7}{688} = \frac{1670}{668} = 2.5$$

$$T_r = \frac{205 + 460}{398} = \frac{665}{398} = 1.67$$

$$z = 0.9 \text{ (Fig. A.9)}$$

$$\rho = 0.09337 \frac{zpM}{T}$$

$$\rho = \frac{(0.09337)(0.9)(1670)(20.6)}{665}$$

$$= 4.35 \text{ lb/cu ft}$$

Density of Liquids

Figures A.1, A.2, and A.3 and Table A.1 are useful in determining densities of liquids at a given temperature.   Additional data are contained in handbooks of chemistry.

Example A.3: Calculation for a Pure Hydrocarbon

Determine the density of liquid propane at 150°F. From Fig. A.2 the density is read from the portion of the curve above the "dot."

$$\rho = 0.425 \text{ gr/cc} \quad \text{or} \quad (0.425)(62.4)$$
$$= 26.55 \text{ lb/cu ft}$$

Calculation of Density from API Gravity

Determine the liquid density of a 30° API oil at 500°F.
  (a) From Fig. A.3, specific gravity at 60°F is 0.876.
  (b) From Fig. A.1, specific gravity at 500°F is 0.70 or density is (see pages 269 and 271):
       $\rho = (0.70)(62.4) = 39.0$ lb/cu ft

Example A.4: Calculation of Density for a Liquid Mixture

Determine the liquid density of a mixture of propane, *n*-butane, and pentane at 150°F and 200 psia.

	Mole Fraction
propane	0.10
n-butane	0.50
pentane	0.40

For ideal solutions, such as hydrocarbons at moderate conditions, the density of a mixture can be based on additive properties and estimated as follows:

From Fig. A.1 for specific gravity 0.60 at 60°F, specific gravity at 150°F = 0.542, or

$$\text{density} = (0.542)(62.4) = 38.8 \text{ lb/cu ft}$$

Note: For severe conditions and for greater accuracy see Hougen, Watson, and Ragatz, *Chemical Process Principles*, Vol. II 2nd ed., pp. 577, 578, and 861, John Wiley and Sons, New York, 1959.

VISCOSITY

From Figs. A.4 and A.5 viscosities in centipoises of various pure compounds as gases and liquids can be obtained.   Figure A.6 gives viscosities in centistokes of petroleum oils as a function of API gravity and characterization factor.   Centistokes are converted to centipoises by multiplying by the specific gravity at the temperature at which the viscosity is desired.

Viscosities of mixtures of hydrocarbons can be roughly approximated from pure component viscosities on a mole fraction basis.

$$\mu_m = x_1\mu_1 + x_2\mu_2 + x_3\mu_3 + \cdots$$

where μ_m = viscosity of mixture in centipoise.
$\mu_1, \mu_2, \mu_3, \cdots$ = viscosities of pure components in centipoise.
$x_1, x_2, x_3, \cdots$ = mole fractions of the respective pure components.

This method is rough but satisfactory for most fluid flow calculations.

Viscosities of mixtures of hydrocarbon vapors can also be roughly approximated from Figs. A.7 and A.8 as shown below.

Example A.5: Viscosity of the Gaseous Mixture of Example A.2 (page 263).

Average molecular weight, $M = 20.6$
μ at 1 atm and 205°F = 0.0125 cp (Fig. A.7)
μ at 1655 psig or 1670 psia:
   $P_r = 2.5$ and $T_r = 1.67$ (From Example A.2, page 263)

	Mole Fraction	Molecular Weight[1]	Weight lb	Sp. Gravity at 60°F/60°F[1]	Gravity Fraction
propane	0.10	44	4.4[2]	0.5077	0.036[3]
n-butane	0.50	58	29.0	0.5844	0.272
pentane	0.40	72	28.8	0.6312	0.292
	1.00		62.2	Avg. S.G. =	0.600

[1] Table A.1   [2] (44)(0.1) = 4.4   [3] (4.4/62.2)(0.5077) = 0.036

$$\frac{\mu_p}{\mu_{1\,atm}} = 1.30 \quad \text{(Fig. A.8)}$$

$$\mu_p = (0.014)(1.27) = 0.0163 \text{ cp}$$

For liquid or gaseous mixtures other than hydrocarbons and where greater accuracy is required, direct measurements or more accurate calculations are necessary. See Reid and Sherwood, *Properties of Gases and Liquids,* pp. 199–202 and 216–217, McGraw-Hill Book Co., New York, 1958.

TABLE A.1
PHYSICAL CONSTANTS OF HYDROCARBONS
Reproduced by permission: *Engineering Data Book*, Natural Gasoline
Supply Men's Association, Tulsa, Oklahoma, 1957.

No.	Compound	Formula	Molecular Weight	Boiling Point °F, 14.696 psi, abs	Vapor Pressure 100 F, psi, abs	Freezing Point °F, 14.696 psi, abs	Critical Constants Pressure, psi, abs	Temperature, °F	Volume, cu ft per lb	Liquid Density; 60 F, 14.696 psi, abs Specific Gravity 60 F/60 F [b]	Lb per Gal [c]	Gal/lb Mol	Temperature Coefficient of Density
1	Methane	CH_4	16.042	−258.68	...	−296.46[d]	673.1	−116.5	0.0993	...	...	6.4 (14)	...
2	Ethane	C_2H_6	30.068	−127.53	...	−297.89	708.3	+90.09	0.0787	...	...	9.67(14)	...
3	Propane	C_3H	44.094	−43.73	190	−305.84	617.4	206.26	0.0730	0.5077[h]	4.224[h]	10.44	0.00171[h]
4	n-Butane	C_4H_{10}	58.120	+31.10	51.6	−217.030	550.7	305.62	0.0704	0.5844[h]	4.863[h]	11.95	0.00111[h]
5	2-Methylpropane (isobutane)	C_4H_{10}	58.120	+10.89	72.2	−255.280	529.1	274.96	0.0725	0.5631[h]	4.685[h]	12.40	0.00123[h]
6	n-Pentane	C_5H_{12}	72.146	96.933	15.570	−201.498	489.5	385.92	0.0690	0.63116	5.2528	13.74	0.00086
7	2-Methylbutane (isopentane)	C_5H_{12}	72.146	82.134	20.44	−255.800	483.	370.0	0.0685	0.62476	5.1995	13.88	0.00091
8	2,2-Dimethylpropane (neopentane)	C_5H_{12}	72.146	49.105	...	+2.210	464.0	321.08	0.0671	0.601[h]	5.00[h]	14.43	0.00093[h]
9	n-Hexane	C_6H_{14}	86.172	155.736	4.956	−139.612	439.7	454.5	0.0685	0.66405	5.5271	15.59	0.00077
10	2-Methylpentane	C_6H_{14}	86.172	140.488	6.767	−244.622	440.1	436.8	0.0680	0.65790	5.4758	15.74	0.00079
11	3-Methylpentane	C_6H_{14}	86.172	145.908	6.098	−180.4	453.1	448.2	0.0680	0.66902	5.5686	15.48	0.00076
12	2,2-Dimethylbutane (neohexane)	C_6H_{14}	86.172	121.534	9.856	−147.825	450.5	421.2	0.0667	0.65399	5.4433	15.83	0.00081
13	2,3-Dimethylbutane	C_6H_{14}	86.172	136.378	7.404	−199.368	455.4	440.8	0.0667	0.66639	5.5466	15.54	0.00097
14	n-Heptane	C_7H_{16}	100.198	209.169	1.6199	−131.082	396.9	512.62	0.0682	0.68819	5.7284	17.49	0.00069
15	2-Methylhexane	C_7H_{16}	100.198	194.094	2.2712	−180.897	400.	496.2	0.0686	0.68299	5.6850	17.63	0.00069
16	3-Methylhexane	C_7H_{16}	100.198	197.330	2.1300	−182.9	413.	504.3	0.0667	0.69151	5.7561	17.41	0.00068
17	3-Ethylpentane	C_7H_{16}	100.198	200.251	2.0123	−181.487	420.	513.7	0.0667	0.70257	5.8483	17.13	0.00067
18	2,2-Dimethylpentane	C_7H_{16}	100.198	174.560	3.492	−190.858	417.	477.9	0.0645	0.67833	5.6462	17.75	0.00071
19	2,4-Dimethylpentane	C_7H_{16}	100.198	176.900	3.292	−182.636	403.	476.8	0.0671	0.67723	5.6370	17.78	0.00072
20	3,3-Dimethylpentane	C_7H_{16}	100.198	186.917	2.7727	−210.03[h]	440.	505.	(0.067)	0.69767	5.8075	17.25	0.00067
21	2,2,3-Trimethylbutane (triptane)	C_7H_{16}	100.198	177.584	3.374	−12.842	437.2	496.9	0.0629	0.69454	5.7813	17.33	0.00068
22	n-Octane	C_8H_{18}	114.224	258.197	0.537	−70.236	362.1	565.2	0.0682	0.70677	5.8833	19.41	0.00063
23	2,5-Dimethylhexane (diisobutyl)	C_8H_{18}	114.224	228.385	1.1012	−132.160	367.	534.	0.0671	0.69795	5.8098	19.66	0.00068
24	2.2.4-Trimethylpentane ("*isooctane*")	C_8H_{18}	114.224	210.628	1.7084	−161.257	374.7	520.7	0.0676	0.69625	5.7957	19.71	0.00067
25	n-Nonane	C_9H_{20}	128.250	303.436	0.179	−64.36	(345)	(613)	0.0673	0.72171	6.0079	21.35	0.00059
26	n-Decane	$C_{10}H_{22}$	142.276	345.2	0.073	−21.5	(320)	(655)	0.0671	0.73413	6.1140	23.27	0.00055
27	Cyclopentane	C_5H_{10}	70.130	120.672	9.915	−136.982	654.7	461.48	0.0593	0.75048	6.2478	11.23	0.000730
28	Methylcyclopentane	C_6H_{12}	84.156	161.262	4.503	−224.419	549.1	499.30	0.0607	0.75354	6.2733	13.42	0.000693
29	Cyclohexane	C_6H_{12}	84.156	177.328	3.265	43.797	561.4	538	0.0592	0.78344	6.5226	12.90	0.000662
30	Methylcyclohexane	C_7H_{14}	98.182	213.681	1.609	−195.860	504.4	570.43	0.0562	0.77398	6.4437	15.24	0.000623
31	Ethene (ethylene)	C_2H_4	28.052	−154.68	...	−272.47	742.1	49.82	0.0705	...	...	9.64(16)	...
32	Propene	C_3H_6	42.078	−53.86	226.4	−301.45	667	197.4	0.0689	0.5218	4.340	9.70	0.00177
33	1-Butene	C_4H_8	56.104	+20.73	63.05	−301.63	583	295.6	0.0690	0.6011	5.001	11.22	0.00113
34	cis-2-Butene	C_4H_8	56.104	38.70	45.54	−218.04	600	311	(0.0671)	0.6272	5.211	10.77	0.00105
35	trans-2-Butene	C_4H_8	56.104	33.58	49.80	−157.99	600	311	(0.0671)	0.6100	5.075	11.06	0.00105
36	2-Methylpropene (isobutene)	C_4H_8	56.104	19.58	63.40	−220.63	579.8	292.5	(0.0685)	0.6002	4.994	11.23	0.00113
37	1-Pentene	C_5H_{10}	70.130	85.95	19.14	−265.40	586	394	(0.0672)	0.6461	5.377	13.04	0.00087
38	1,2-Butadiene	C_4H_6	54.088	50.5	(20)	−213.3	653	339	(0.0649)	0.658	5.476	9.88	0.00114
39	1,3-Butadiene	C_4H_6	54.088	24.06	(60)	−164.05	628	306	(0.0654)	0.6272	5.220	10.36	0.00110
40	2-Methyl-1,3-butadiene (isoprene)	C_5H_8	68.114	93.34	(17)	−230.8	(558.4)	(412)	(0.0650)	0.6861	5.712	11.93	0.00085
41	Ethyne (acetylene)	C_2H_2	26.036	−119	...	−114	905	97.4	(0.0661)	0.615	5.118	5.09	...
42	Benzene	C_6H_6	78.108	176.185	3.224	+41.959	714	553.01	0.054	0.88458	7.3659	10.60	0.000686
43	Toluene	C_7H_8	92.134	231.121	1.032	−138.984	611	609.51	0.057	0.87190	7.2602	12.69	0.000592
44	Ethylbenzene	C_8H_{10}	106.160	277.137	0.3706	−138.955	560	654.8	0.056	0.87175	7.2589	14.63	0.000530
45	1,2-Dimethylbenzene (o-xylene)	C_8H_{10}	106.160	291.95	0.2636	−13.32	542	678.3	0.058	0.88482	7.3679	14.41	0.000533
46	1,3-Dimethylbenzene (m-xylene)	C_8H_{10}	106.160	282.38	0.3259	−54.17	526	654.9	0.058	0.86880	7.2344	14.67	0.000546
47	1,4-Dimethylbenzene (p-xylene)	C_8H_{10}	106.160	281.03	0.3421	+55.87	514	653.1	0.056	0.86576	7.2090	14.73	0.000561
48	Styrene (Phenyl Ethylene)	C_8H_8	104.144	293.4	(0.24)	−23.130	580	706.0	0.0541	0.91112	7.5872	13.73	0.000584
49	Isopropylbenzene (cumene)	C_9H_{12}	120.186	306.31	0.1880	−140.85	473	685	0.059	0.86642	7.2144	16.66	0.000550
50	Methyl Alcohol	CH_4O	32.042	148.1(2)	4.6(7)	−144.0(2)	1157(2)	464(2)	.0589(3)	.796(3)	6.637	4.83	...
51	Ethyl Alcohol	C_2H_6O	46.069	173.3(2)	2.3(3)	−179.1(2)	927(2)	469.6(2)	.0581(3)	.794(3)	6.619	6.96	...
52	Carbon Monoxide	CO	28.010	−313.6(2)	...	−340.6(2)	510(2)	−218(2)	.0515(3)	.801m(8)	...	...	...
53	Carbon Dioxide	CO_2	44.010	−109.3(2)	...	......	1073(2)	88.0(2)	.0348(3)	.8159(10)	6.802	6.47	...
54	Hydrogen Sulfide	H_2S	34.076	−76.5(4)	554.6(6)	−121.9(4)	1306(2)	212.7(2)	...	.79(6)	6.58	5.18	...
55	Sulfur Dioxide	SO_2	64.060	14.0(2)	84.5(7)	−98.9(2)	1142(2)	315.0(2)	.0308(3)	1.394(7)	11.62	5.51	...
56	Ammonia	NH_3	17.032	−28.1(2)	211.9(11)	−107.9(2)	1657(11)	271.4(11)	.0682(3)	.6173(11)	5.147	3.31	...
57	Air	N_2O_2	28.966	−317.7(2)	...	......	547(2)	−221.3(2)	.0517(3)	.856m(8)	...	...	...
58	Hydrogen	H_2	2.016	−422.9(2)	...	−434.4(2)	188(2)	−399.8(2)	.5168(3)	.07m(3)	...	...	...
59	Oxygen	O_2	32.000	−297.4(2)	...	−361.1(2)	730(2)	−181.8(2)	.0373(3)	1.14m(3)	...	...	...
60	Nitrogen	N_2	28.016	−320.4(2)	...	−345.6(2)	492(2)	−232.8(2)	.0515(3)	.808m(3)	...	...	...
61	Chlorine	Cl_2	70.914	30.3(2)	154(3)	−150.9(2)	1120(2)	291(2)	.0280(3)	1.423(10)	11.86	5.98	...
62	Water	H_2O	18.016	212.0	0.9492(9)	32.0	3206(12)	705.4(9)	.0400(3)	1.000	8.337	2.16	...
63	Hydrogen Chloride	HCl	36.465	−121(s)	900(t)	−173.6(u)	1199.2(s)	124.5(s)	0.0381(s)	0.8558(v) 7.1266 (h)	7.1266	5.117(q)	0.003346 (q)

TABLE A.1 (*Concluded*)

Gas Density, 60 F, 14.696 psi, abs, Ideal Gas — Specific Gravity Air = 1	Cu ft Gas per lb[b]	Cu ft Gas per gal Liquid	Specific Heat; Ideal Gas — C_p	C_v	Calorific Value Net — Btu per cu ft Vapor, 14.696 psi, abs	Gross — Btu per cu ft Vapor, 14.696 psi, abs	Gross — Btu per lb Liquid[e]	Gross — Btu per gal Liquid	Heat of Vaporization, 14.696 psi, abs at Boiling Point, Btu per lb	Refractive Index,[a] nD 68 F	Air Required for Combustion cu ft per cu ft	Flammability Lower	Flammability Higher	ASTM Octane Motor Method D-357	Research Method D-908	No.
0.555e	23.61e	...	0.5271	0.402	911	1012e	...	...	219.7	...	9.53	5.0	15.0	...	+6g	1
1.046e	12.52e	...	0.4097	0.343	1631	1783e	...	...	210.7	...	16.67	3.22	12.45	...	+6g	2
1.547e	8.471e	35.78e	0.3885	0.342	2353	2557e	21,554	91,044	183.5	...	23.82	2.37	9.50	...	+5.3g	3
2.071e	6.327e	30.77e	0.3970	0.363	3101	3369e	21,190	108,047	165.9	...	30.97	1.86	8.41	...	88g	4
2.067e	6.339e	29.70e	0.3872	0.352	3094	3354e	21,152	99,097	157.8	...	30.97	1.80	8.44	...	+0.2g	5
2.4906	5.2601	27.680	0.3972	0.370	3709	4009	20,965	110,125	153.8	1.35748	38.11	1.40	7.80	61.9	61.7	6
2.4906	5.2601	27.398	0.3880	0.361	3698	4001	20,929	108,820	145.9	1.35373	38.11	1.32	...	90.3	92.3	7
2.4906	5.2601	26.3	0.3914	0.364	3685	3987	20,878	104,390	135.8	1.342h	38.11	...	...	80.2	85.5	8
2.9749	4.4039	24.381	0.3984	(0.375)	4404	4756	20,819	115,069	144.2	1.37486	45.26	1.25	6.90	26.0	24.8	9
2.9749	4.4039	24.156	0.389	(0.366)	4396	4748	20,791	113,847	138.0	1.37145	45.26	...	...	73.5	73.4	10
2.9749	4.4039	24.564	0.397	(0.374)	4399	4751	20,804	115,849	140.3	1.37652	45.26	...	...	74.3	74.5	11
2.9749	4.4039	24.011	0.386	(0.363)	4383	4735	20,747	112,932	131.4	1.36876	45.26	...	...	93.4	91.8	12
2.9749	4.4039	24.467	0.391	(0.368)	4392	4744	20,779	115,253	136.3	1.37495	45.26	...	...	94.3	+0.3	13
3.4591	3.7875	21.731	0.3992	(0.379)	5101	5503	20,714	118,658	136.2	1.38765	52.41	1.0	6.0	0.0	0.0	14
3.4591	3.7875	21.567	(0.390)	(0.370)	5093	5495	20,692	117,628	131.8	1.38485	52.41	...	...	46.4	42.4	15
3.4591	3.7875	21.836	(0.378)	(0.358)	5096	5498	20,701	119,157	132.3	1.38864	52.41	...	...	55.0	52.0	16
3.4591	3.7875	22.185	(0.367)	(0.347)	5099	5501	20,712	121,128	133.0	1.39340	52.41	...	...	69.3	65.0	17
3.4591	3.7875	21.419	(0.359)	(0.339)	5380	5482	20,653	116,103	125.3	1.38215	52.41	...	...	95.6	92.8	18
3.4591	3.7875	21.385	(0.370)	(0.353)	5085	5487	20,669	116,511	126.8	1.38145	52.41	...	...	83.8	83.1	19
3.4591	3.7875	22.030	(0.347)	(0.327)	5085	5488	20,671	120,667	127.4	1.39092	52.41	...	...	86.6	80.8	20
3.4591	3.7875	21.932	(0.386)	(0.366)	5081	5484	20,660	119,442	124.4	1.38943	52.41	...	...	+0.1	+1.8	21
3.9432	3.3224	19.577	0.3998	(0.382)	5797	6250	20,638	121,420	131.9	1.39743	59.55	0.84	3.2	...	...	22
3.9432	3.3224	19.333	(0.373)	(0.355)	5781	6234	20,597	119,664	123.7	1.39246	59.55	...	...	55.7	55.5	23
3.9432	3.3224	19.286	(0.380)	(0.363)	5779	6232	20,602	119,403	116.8	1.39145	59.55	...	...	100	+100	24
4.4275	2.9590	17.805	0.400	(0.385)	6493	6996	20,577	123,625	126.9	1.40542	66.70	0.74	2.9	−20	...	25
4.9118	2.6673	16.325	0.401	(0.387)	7190	7743	20,683	126,456	120.2	1.41189	73.85	0.67	2.6	...	−53.0	26
2.4211	5.4113	33.86	0.2712	0.2429	3512	3763	20,216	126,310	167.34	1.40645	35.731	...	...	85.0	...	27
2.9053	4.5094	28.33	0.3010	0.2774	4199	4500	20,158	126,460	147.83	1.40970	42.878	1.33	8.35	80.0	91.3	28
2.9053	4.5094	29.45	0.2900	0.2664	4180	4482	20,066	130,880	153.7	1.42623	42.878	1.33	8.35	77.2	83.0	29
3.3896	3.8652	24.94	0.3170	0.2968	4864	5216	20,031	129,070	138.9	1.42312	50.024	...	...	71.1	74.8	30
0.9684	13.530	...	0.3622	0.2914	1499	1600	...	...	207.56	...	14.29	3.05	28.6	75.6	0.03	31
1.4526	9.020	39.23	0.3541	0.3069	2182	2333	20,943	90,891	188.19	...	21.44	2.00	11.1	84.9	0.24	32
1.9368	6.764	33.90	0.3703	0.3349	2880	3082	20,727	103,658	167.93	...	28.58	...	...	81.7	97.4	33
1.9368	6.764	35.37	0.3269	0.2915	2872	3074	20,655	107,633	178.91	...	28.58	...	...	...	...	34
1.9368	6.764	34.39	0.3654	0.3300	2867	3069	20,633	104,711	174.37	...	28.58	...	...	...	...	35
1.9368	6.764	33.85	0.3701	0.3347	2863	3065	20,618	102,964	169.48	...	28.58	...	...	88.1	0.26	36
2.4210	5.412	29.15	0.3817	0.3534	3574	3826	20,590	110,712	(149)	1.3714	35.73	...	...	77.1	90.9	37
1.8673	7.016	38.52	(0.3458)	(0.3091)	2792	2943	20,496	112,236	(181)	...	26.20	...	...	...	...	38
1.8673	7.016	36.69	(0.3412)	(0.3045)	2732	2883	20,095	104,898	(174)	...	26.20	...	...	...	...	39
2.3515	5.572	31.87	(0.357)	(0.328)	3510	3611	19,998	114,227	(153)	1.4216	33.35	...	...	...	...	40
0.8988	14.577	74.34	0.3966	0.3203	1422	1472	...	...	...	...	11.91	...	...	...	...	41
2.6965	4.8586	35.83	0.2404	0.2150	3591	3742	18,013	132,680	169.34	1.50112	35.731	1.41	6.75	+2.75	...	42
3.1808	4.1189	29.90	0.2599	0.2383	4273	4475	18,274	132,670	156.2	1.49693	42.878	1.27	6.75	+0.27	+5.82	43
3.6650	3.5747	25.98	0.2795	0.2608	4970	5222	18,517	134,410	145.7	1.49592	50.024	...	...	+97.9	+0.8	44
3.6650	3.5747	26.37	0.2914	0.2727	4958	5210	18,468	136,070	149.1	1.50543	50.024	1.00	6.00	100.	(>+6)	45
3.6650	3.5747	25.89	0.2782	0.2595	4956	5208	18,464	133,580	147.4	1.49721	50.024	...	...	<100.	(>+6)	46
3.6650	3.5747	25.80	0.2769	0.2582	4957	5209	18,468	133,140	146.1	1.49581	50.024	...	...	<100.	(>+6)	47
3.5954	3.6439	27.68	0.2711	0.2520	4830	5030	18,171	137,870	(151)	1.54682	47.641	1.10	6.10	+0.2	>+3	48
4.1492	3.1576	22.81	0.2917	0.2752	5661	5963	18,690	134,840	134.3	1.49146	57.170	...	...	99.3	+2.08	49
1.1062	14.479	96.097	0.27(13)	0.446	765	865(13)	9,750	64,643	473(8)	1.3288(13)	7.15(13)	6.72(5)	36.50	...	...	50
1.5905	20.818	137.794	0.307(13)	0.368	1446	1596(13)	12,755	84,311	368(8)	1.3614(13)	14.30(13)	3.28(5)	18.95	...	...	51
0.9670	13.550	...	.2484(13)	0.177	...	321(13)	...	...	91(8)	...	2.39	12.50(5)	74.20	...	...	52
1.5194	8.569	(58.3)(2)	.1991(13)	0.153	...	...	...	...	248n(3)	...	...	...	...	...	...	53
1.1764	11.050	(72.7)(2)	.254(13)	0.192	621(15)	672(13)	...	...	236(6)	...	(7.2)	4.30(5)	45.50	...	...	54
2.2116	5.824	(67.7)(3)	.147(13)	0.118	...	...	...	...	171(8)	...	...	...	...	...	...	55
0.5880	22.124	113.4(11)	.5232(3)	0.399	...	...	...	...	589.3(11)	...	...	15.50(5)	27.00	...	...	56
1.0000	13.089	...	.241(13)	0.171	...	...	...	...	92(3)	...	...	...	...	...	...	57
0.0696	188.679	...	3.408(13)	2.42	273(8)	324(13)	...	...	192(2)	...	2.39	4.00(5)	74.20	...	...	58
1.1047	11.848	...	0.2188(13)	0.156	...	...	...	...	92(3)	...	...	...	...	...	...	59
0.9672	13.532	...	0.2482(13)	0.177	...	...	...	...	86(3)	...	...	...	...	...	...	60
2.4482	(5.283)	(62.6)(3)	.1149(3)	0.0841	...	...	...	...	121(2)	...	...	...	...	...	...	61
0.6220		175.44	.4446(13)	0.332	...	...	...	...	970.3(9)	1.3330(13)	...	...	...	...	...	62
1.268	10.407 (q)	74.17(q)	0.1939(x)	0.1375(x)	...	...	...	...	190.4(y)	...	...	...	...	...	...	63

REFERENCE BIBLIOGRAPHY FOR TABLE A.1

REFERENCES

1. Values given for hydrocarbons 1 to 49 were selected or calculated from data in Special Technical Publication 109, "Physical Constants of Hydrocarbons Boiling Below 350 F," (1950), American Society for Testing Materials, 1916 Race St., Phila.
2. International Critical Tables.
3. Hodgmen, *Handbook of Chemistry and Physics,* 31st ed., 1949.
4. Giauque and Blue, *J. Am. Chem. Soc.,* **58,** 831 (1936).
5. Jones, *Chem. Rev.,* **22,** 1 (1938).
6. Baxter, et al., *J. Soc. Chem. Ind.,* **53,** 401 T (1934).
7. Perry, *Chemical Engineers Handbook,* 2nd ed., 1941.
8. Matteson and Hanna, *Oil and Gas Journal,* **41,** No. 2, 33 (1944).
9. Aston, *Thermodynamic Data on Hydrocarbons,* Standard Oil Development Company, 1944.
10. Lange, *Handbook of Chemistry,* Handbook Publishers, Inc., 1946.
11. NBS Circular No. 142, "Thermodynamic properties of Ammonia" (1923).
12. Keenan and Keyes, *Thermodynamic Properties of Steam,* 1936.
13. F. D. Rossini—API Project 44.
14. See table on page 137 of *Engineering Data Book,* NGSA.
15. Combustion-Industrial Gas Series, *AGA Reference Book,* 3rd ed., 1932.
16. J. B. Maxwell, *Data Book on Hydrocarbons,* Van Nostrand Co., 1950, page 3.
17. The Dow Chemical Co., Anhydrous HCl pamphlet.
18. *Encyclopedia of Chemical Technology,* vol. 7, p. 653, 1951.
19. J. W. Mellor, *Treatise on Inorganic and Theoretical Chemistry.* Vol. 2, 1927.
20. *API Toxicological Review,* HCl, 1948.

[a] Air saturated hydrocarbons.
[b] Absolute values from weights in vacuum.
[c] Apparent values from weights in air.
[d] At saturation pressure (triple point).
[e] Actual gas volumes corrected for deviation.
[f] The + sign and number following signify the octane number corresponding to that of 2, 2.4-trimethylpentane with the indicated number of ml. of TEL added.
[g] Equivalent octane number (gas).
[h] Saturation pressure.
[i] Critical solution temperatures.
[k] Minus octane numbers derived from blends with reference fuel.
[K] Values in parenthesis are estimated.
[m] Density of liquid gr/ml at normal boiling point.
[n] Heat of sublimation.
[p] Values reported at 15 C.
[q] Calculated from other properties.
[s] See References 16, 18 and 20.
[t] Reference 2.
[u] References 16 and 18.
[v] References 11 and h.
[w] References 19 and 19.
[x] References 18 and p.
[y] References 17 and 18.

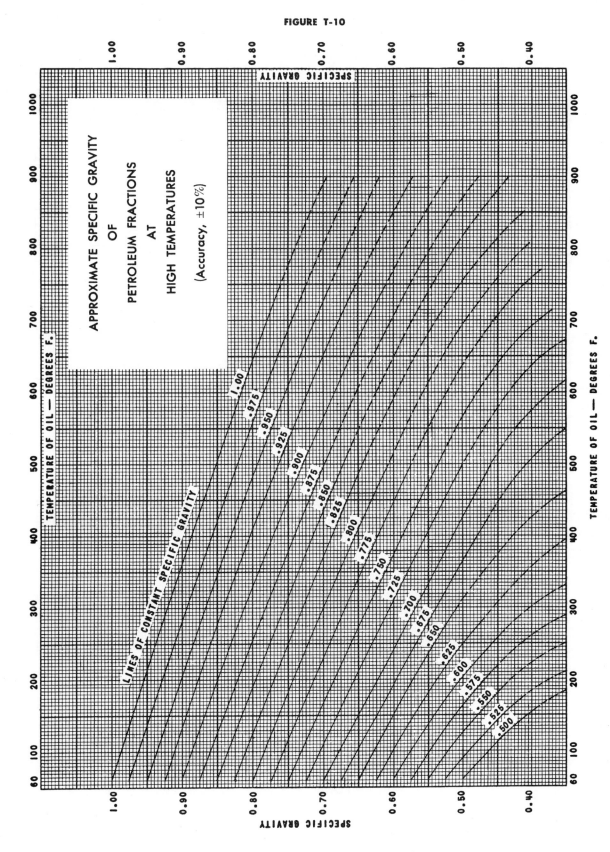

FIGURE T-10

APPROXIMATE SPECIFIC GRAVITY
OF
PETROLEUM FRACTIONS
AT
HIGH TEMPERATURES

(Accuracy, ±10%)

LINES OF CONSTANT SPECIFIC GRAVITY

TEMPERATURE OF OIL — DEGREES F.

SPECIFIC GRAVITY

FIG. A.1. Approximate specific gravity of petroleum fractions at high temperatures. (Reproduced by permission from *Standards of Tubular Exchanger Manufacturers Association*, 3rd ed., New York, 1952.)

269

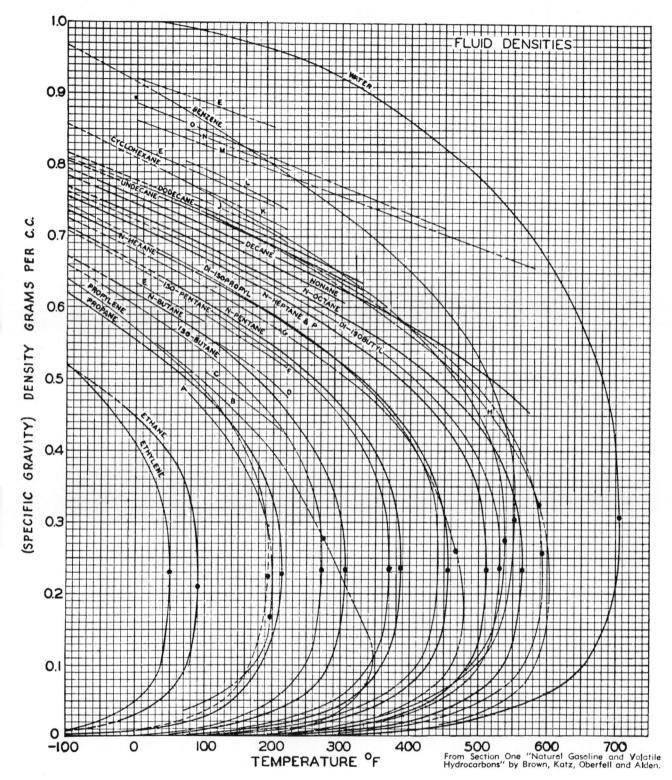

FIG. A.2. Hydrocarbon fluid densities. (Reproduced by permission: D. L. Katz and Associates, *Handbook of Natural Gas Engineering*, McGraw-Hill Book Co., New York, 1959.) [See notes on opposite page for identification of lines.]

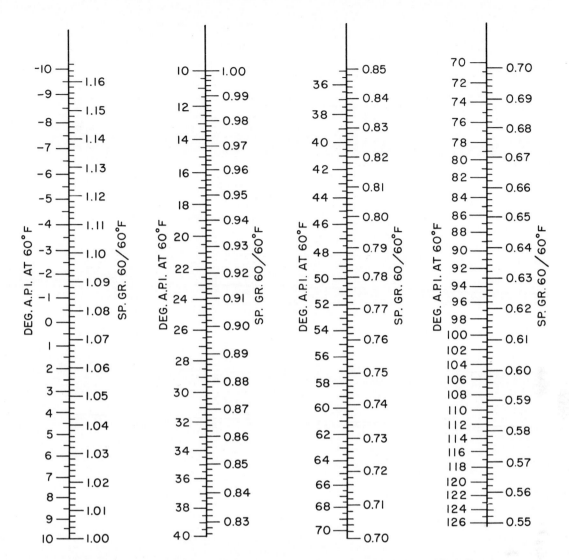

FIG. A.3. Gravity conversions. (Adapted from O. A. Hougen, K. M. Watson, and R. A. Ragatz, *Chemical Process Principles,* Vol. 1, 2nd ed., John Wiley and Sons, New York, 1954.)

Identification for Fig. A.2:

A 8 mol % CH_4—92 mol % C_3H_8 (370–739 lb)

B 50.25 wt. % C_2H_6—49.75 wt. % n-C_7H_{16}

C 19.2 wt. % CH_4—80.8 wt. % C_6H_{14} (2412–2506 lb)

D 7.15 wt. % CH_4—92.85 wt. % n-C_5H_{12} (854–1043 lb)

E National Standard Petroleum Oil Tables

F 7.15 wt. % CH_4—92.85 wt. % n-C_5H_{12} (at 3000 lb)

G 9.78 wt. % C_2H_6—90.22 wt. % n-C_7H_{16}

H Gasoline

I Naphtha

J Conroe Crude, sat. liq., changing composition (1642–2079 lb)

K 8 wt. % gas in Dominguez crude oil (2459–3096 lb)

L 5.11 wt. % gas in Santa Fe Springs crude oil (2118–2575 lb)

M Pennsylvania Spindle oil at atmospheric pressure

N 4. wt. % CH_4 in crystal oil (1945–2433 lb)

O Oklahoma lubricating oil at atmospheric pressure

P Kettleman Hills condensate (543–829 lb)

Note: The dots represent critical temperatures. The portion of each curve above the dot is for saturated liquid while that below is for saturated vapors.

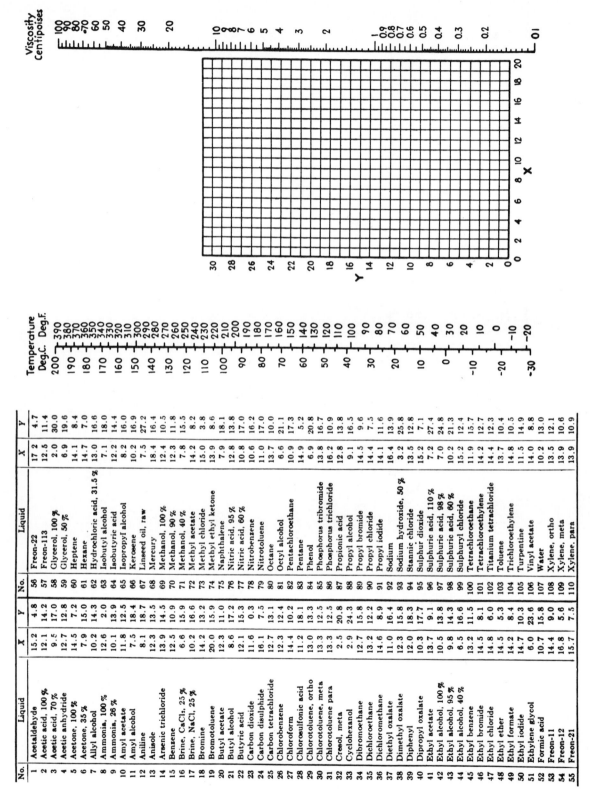

FIG. A.4. Viscosities of liquids at 1 atmosphere. (Reproduced by permission from R. H. Perry, *Chemical Engineers' Handbook*, 3rd ed., McGraw-Hill Book Co., New York, 1950.)

No.	Liquid	X	Y	No.	Liquid	X	Y
1	Acetaldehyde	15.2	4.8	56	Freon-22	17.2	4.7
2	Acetic acid, 100%	12.1	14.2	57	Freon-113	12.5	11.4
3	Acetic acid, 70%	9.5	17.0	58	Glycerol, 100%	2.0	30.0
4	Acetic anhydride	12.7	12.8	59	Glycerol, 50%	6.9	19.6
5	Acetone, 100%	14.5	7.2	60	Heptene	14.1	8.4
6	Acetone, 35%	7.9	15.0	61	Hexane	14.7	7.0
7	Allyl alcohol	10.2	14.3	62	Hydrochloric acid, 31.5%	13.0	16.6
8	Ammonia, 100%	12.6	2.0	63	Isobutyl alcohol	7.1	18.0
9	Ammonia, 26%	10.1	13.9	64	Isobutyric acid	12.2	14.4
10	Amyl acetate	11.8	12.5	65	Isopropyl alcohol	8.2	16.0
11	Amyl alcohol	7.5	18.4	66	Kerosene	10.2	16.9
12	Aniline	8.1	18.7	67	Linseed oil, raw	7.5	27.2
13	Anisole	12.3	13.5	68	Mercury	18.4	16.4
14	Arsenic trichloride	13.9	14.5	69	Methanol, 100%	12.4	10.5
15	Benzene	12.5	10.9	70	Methanol, 90%	12.3	11.8
16	Brine, CaCl₂, 25%	6.6	15.9	71	Methanol, 40%	7.8	15.5
17	Brine, NaCl, 25%	10.2	16.6	72	Methyl acetate	14.2	8.2
18	Bromine	14.2	13.2	73	Methyl chloride	15.0	3.8
19	Bromotoluene	20.0	15.9	74	Methyl ethyl ketone	13.9	8.6
20	Butyl acetate	12.3	11.0	75	Naphthalene	7.9	18.1
21	Butyl alcohol	8.6	17.2	76	Nitric acid, 95%	12.8	13.8
22	Butyric acid	12.1	15.3	77	Nitric acid, 60%	10.8	17.0
23	Carbon dioxide	11.6	0.3	78	Nitrobenzene	10.6	16.2
24	Carbon disulphide	16.1	7.5	79	Nitrotoluene	11.0	17.0
25	Carbon tetrachloride	12.7	13.1	80	Octane	13.7	10.0
26	Chlorobenzene	12.3	12.4	81	Octyl alcohol	6.6	21.1
27	Chloroform	14.4	10.2	82	Pentachloroethane	10.9	17.3
28	Chlorosulfonic acid	11.2	18.1	83	Pentane	14.9	5.2
29	Chlorotoluene, ortho	13.0	13.3	84	Phenol	6.9	20.8
30	Chlorotoluene, meta	13.3	12.5	85	Phosphorus tribromide	13.8	16.7
31	Chlorotoluene, para	13.3	12.5	86	Phosphorus trichloride	16.2	10.9
32	Cresol, meta	2.5	20.8	87	Propionic acid	12.8	13.8
33	Cyclohexanol	2.9	24.3	88	Propyl alcohol	9.1	16.5
34	Dibromoethane	12.7	15.8	89	Propyl bromide	14.5	9.6
35	Dichloroethane	13.2	12.2	90	Propyl chloride	14.4	7.5
36	Dichloromethane	14.6	8.9	91	Propyl iodide	14.1	11.6
37	Diethyl oxalate	11.0	16.4	92	Sodium	16.4	13.9
38	Dimethyl oxalate	12.3	15.8	93	Sodium hydroxide, 50%	3.2	25.8
39	Diphenyl	12.0	18.3	94	Stannic chloride	13.5	12.8
40	Dipropyl oxalate	10.3	17.7	95	Sulphur dioxide	15.2	7.1
41	Ethyl acetate	13.7	9.1	96	Sulphuric acid, 110%	7.2	27.4
42	Ethyl alcohol, 100%	10.5	13.8	97	Sulphuric acid, 98%	7.0	24.8
43	Ethyl alcohol, 95%	9.8	14.3	98	Sulphuric acid, 60%	10.2	21.3
44	Ethyl alcohol, 40%	6.5	16.6	99	Sulphuryl chloride	15.2	12.4
45	Ethyl benzene	13.2	11.5	100	Tetrachloroethane	11.9	15.7
46	Ethyl bromide	14.5	8.1	101	Tetrachloroethylene	14.2	12.7
47	Ethyl chloride	14.8	6.0	102	Titanium tetrachloride	14.4	12.3
48	Ethyl ether	14.5	5.3	103	Toluene	13.7	10.4
49	Ethyl formate	14.2	8.4	104	Trichloroethylene	14.8	10.5
50	Ethyl iodide	14.7	10.3	105	Turpentine	11.5	14.9
51	Ethylene glycol	6.0	23.6	106	Vinyl acetate	14.0	8.8
52	Formic acid	10.7	15.8	107	Water	10.2	13.0
53	Freon-11	14.4	9.0	108	Xylene, ortho	13.5	12.1
54	Freon-12	16.8	5.6	109	Xylene, meta	13.9	10.6
55	Freon-21	15.7	7.5	110	Xylene, para	13.9	10.9

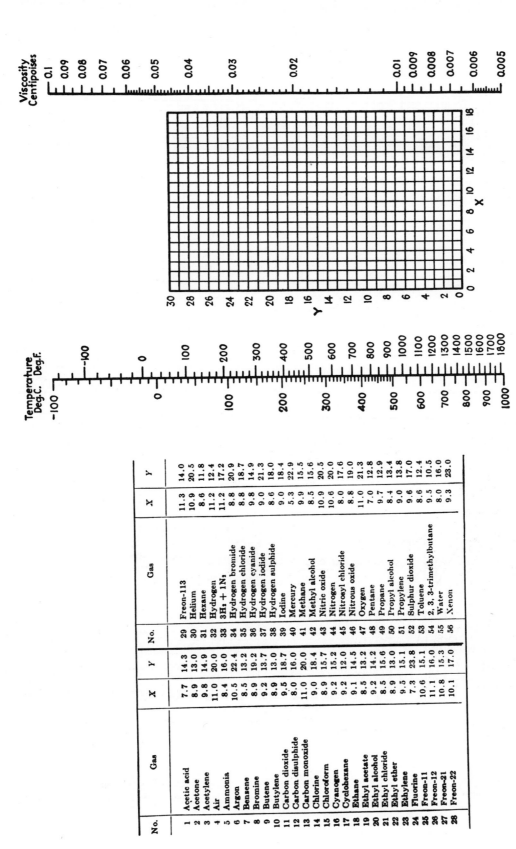

No.	Gas	X	Y	No.	Gas	X	Y
1	Acetic acid	7.7	14.3	29	Freon-113	11.3	14.0
2	Acetone	8.9	13.0	30	Helium	10.9	20.5
3	Acetylene	9.8	14.9	31	Hexane	8.6	11.8
4	Air	11.0	20.0	32	Hydrogen	11.2	12.4
5	Ammonia	8.4	16.0	33	3H₂ + 1N₂	11.2	17.2
6	Argon	10.5	22.4	34	Hydrogen bromide	8.8	20.9
7	Benzene	8.5	13.2	35	Hydrogen chloride	8.8	18.7
8	Bromine	8.9	19.2	36	Hydrogen cyanide	9.8	14.9
9	Butene	9.2	13.7	37	Hydrogen iodide	9.0	21.3
10	Butylene	8.9	13.0	38	Hydrogen sulphide	8.6	18.0
11	Carbon dioxide	9.5	18.7	39	Iodine	9.0	18.4
12	Carbon disulphide	8.0	16.0	40	Mercury	5.3	22.9
13	Carbon monoxide	11.0	20.0	41	Methane	9.9	15.5
14	Chlorine	9.0	18.4	42	Methyl alcohol	8.5	15.6
15	Chloroform	8.9	15.7	43	Nitric oxide	10.9	20.5
16	Cyanogen	9.2	15.2	44	Nitrogen	10.6	20.0
17	Cyclohexane	9.2	12.0	45	Nitrosyl chloride	8.0	17.6
18	Ethane	9.1	14.5	46	Nitrous oxide	8.8	19.0
19	Ethyl acetate	8.5	13.2	47	Oxygen	11.0	21.3
20	Ethyl alcohol	9.2	14.2	48	Pentane	7.0	12.8
21	Ethyl chloride	8.5	15.6	49	Propane	9.7	12.9
22	Ethyl ether	8.9	13.0	50	Propyl alcohol	8.4	13.4
23	Ethylene	9.5	15.1	51	Propylene	9.0	13.8
24	Fluorine	7.3	23.8	52	Sulphur dioxide	9.6	17.0
25	Freon-11	10.6	15.1	53	Toluene	8.6	12.4
26	Freon-12	11.1	16.0	54	2, 3, 3-trimethylbutane	9.5	10.5
27	Freon-21	10.8	15.3	55	Water	8.0	16.0
28	Freon-22	10.1	17.0	56	Xenon	9.3	23.0

FIG. A.5. Viscosities of vapors and gases at 1 atmosphere. (Reproduced by permission from R. H. Perry, *Chemical Engineers' Handbook*, 3rd ed., McGraw-Hill Book Co., New York, 1950.)

273

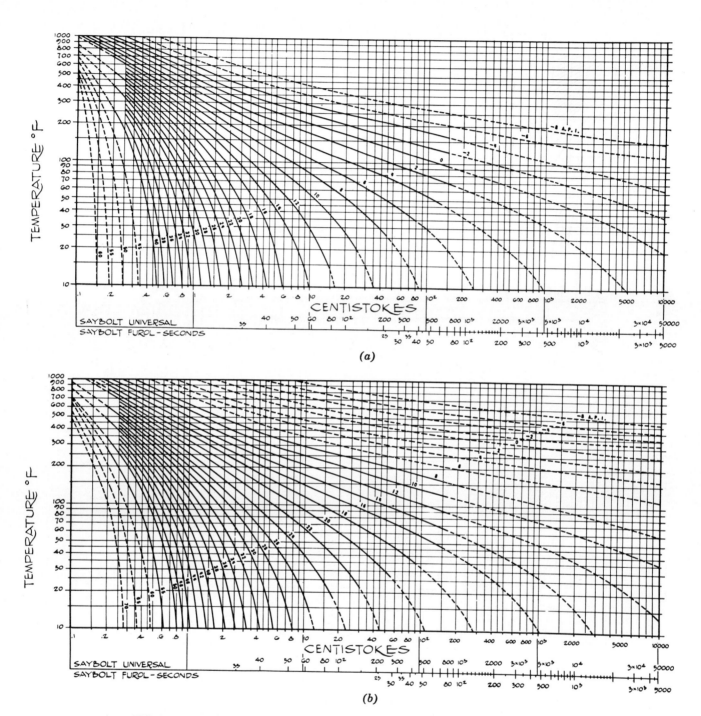

FIG. A.6.   Viscosity-temperature relationship for petroleum oils.   (Reproduced by permission from "Standards of Tubular Exchanger Manufacturers Association", 3rd ed., New York, 1952.)

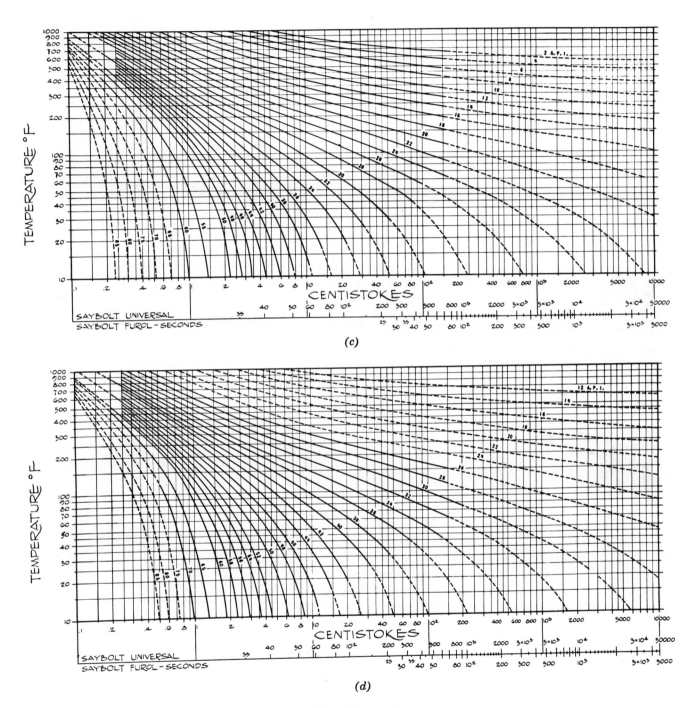

FIG. A.6 (*Concluded*).

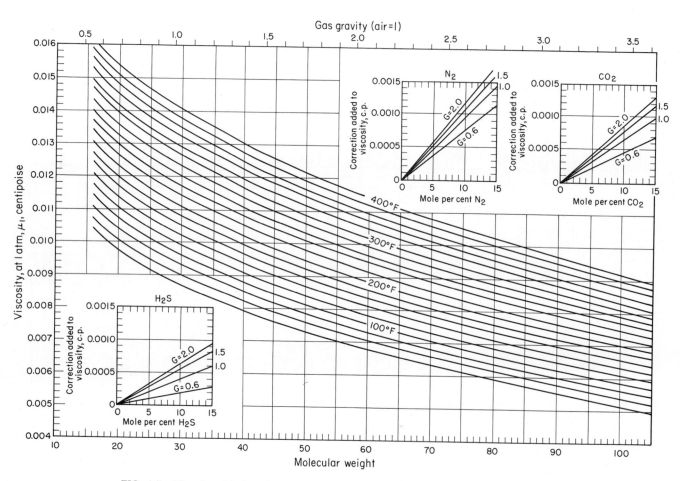

FIG. A.7.   Viscosity of hydrocarbon vapors at 1 atmosphere.   [Reproduced by permission from
N. L. Carr, R. Kobayashi, and O. B. Burrows, *Trans. AIME,* 201, 264 (1954).]

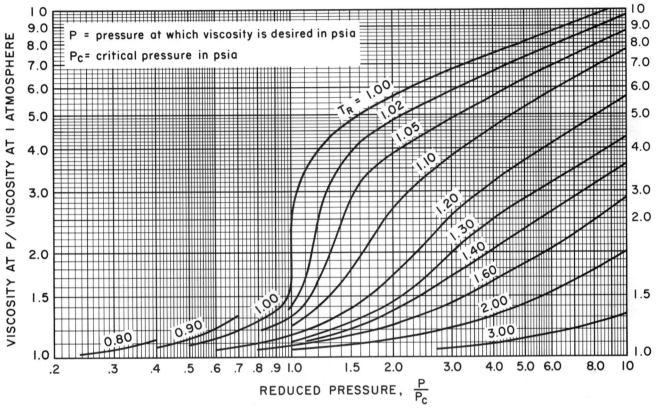

FIG. A.8. Viscosity of gases at high pressure. [Reproduced by permission from E. W. Commings and B. J. Mayland, *Chemical and Metallurgical Engineering,* **52,** No. 3, 115 (1945).]

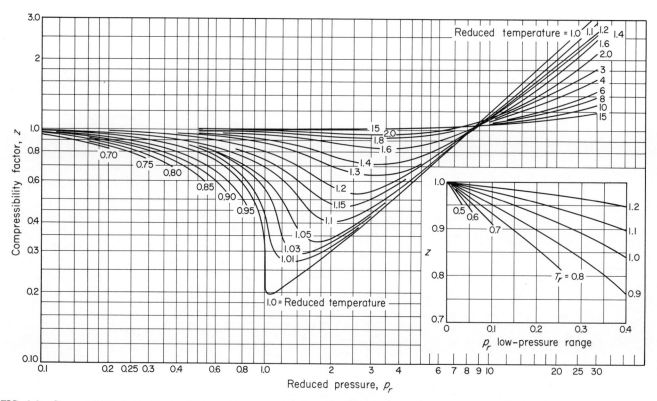

Appendix. Physical properties of fluids

FIG. A.9. Compressibility factor chart. (Reprinted by permission from O. A. Hougen and K. M. Watson, *Chemical Process Principles,* Vol. II, p. 489, John Wiley and Sons, New York, 1947.) These data are presented for ease of use. More accurate values, when required, can be obtained from the second edition of Hougen, Watson and Ragatz, or from L. C. Nelson and E. F. Obert, *Chemical Engineering,* pp. 203–208 (July 1954).

P_r = reduced pressure, P/P_c
T_r = reduced temperature, T/T_c
P = pressure in psia

P_c = critical pressure from Table A.1 in psia
T = temperature in °R = °F + 460
T_c = critical temperature in °R = °F + 460 (See Table A.1)

NOTE: Other consistent units of temperature and pressure may be used. Units shown correspond to those given in Table A.1.

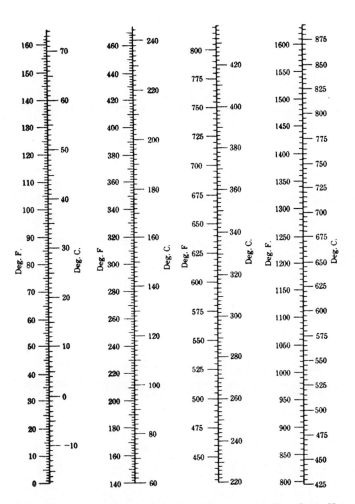

FIG. A.10. Temperature conversions. (Reproduced by permission from O. A. Hougen, K. M. Watson, and R. A. Ragatz, *Chemical Process Principles,* Vol. 1, 2nd ed., John Wiley and Sons, New York, 1954.)

TABLE A.2
CONVERSION FACTORS
(Reproduced by permission from *Piping Design and Engineering*, Grinnell Co., Inc., Providence, R.I., 1951.)

Multiply	by	To Obtain
Absolute viscosity (poise)	1	Gram/second centimeter
Absolute viscosity (centipoise)	0.01	Poise
Acceleration due to gravity (g)	32.174	Feet/second^2
	980.6	Centimeters/second^2
Ares	0.4047	Hectares
	10	Square Chains
	43,560	Square Feet
	4047	Square Meters
	0.001562	Square Miles
	4840	Square Yards
	160	Square Rods
Acre-feet	43,560	Cubic Feet
	325,851	Gallons (US)
	1233.49	Cubic Meters
	1,233,490	Liters
Acre-feet/hour	726	Cubic feet/Minute
	5430.86	Gallons/Minute
Angstroms	10^{-10}	Meters
Ares	0.01	Hectares
	1076.39	Square Feet
	0.02471	Acres
Atmospheres	76.0	Cms of Hg at 32° F
	29.921	Inches of Hg at 32° F
	33.94	Feet of Water at 62° F
	10,333	Kgs/Square meter
	14.6963	Pounds/Square inch
	1.058	Tons/Square foot
	1013.15	Millibars
	235.1408	Ounces/Square inch
Bags of cement	94	Pounds of cement
Barrels of oil	42	Gallons of oil (US)
Barrels of cement	376	Pounds of cement
Barrels (not legal) or	31	Gallons (US)
	31.5	Gallons (US)
Board feet	144 × 1 in.*	Cubic inches
Boiler horse power	33,479	BTU/hour
	9.803	Kilowatts
	34.5	Pounds of water evaporated/hour at 212° F
BTU	252.016	Calories (gm)
	0.252	Calories (Kg)
	777.54	Foot pounds
	0.0003927	Horse power hours
	1054.2	Joules
	107.5	Kilogram meters
	0.0002928	Kilowatt hours
BTU/Cu foot	8.89	Calories (Kg)/Cu meter at 32° F
BTU/Hr/ft^2/°F/ft	0.00413	Cal (gm)/Sec/cm^2/°C/cm
	1.49	Cal (Kg)/Hr/M^2/°C/Meter
BTU/minute	12.96	Foot pounds/second
	0.02356	Horse power
	0.01757	Kilowatts

* For thickness less than 1 in. use actual thickness in decimals of an inch.

Multiply	by	To Obtain
BTU/minute	17.57	Watts
BTU/pound	0.556	Calories (Kg)/Kilogram
Bushels	2150.4	Cubic inches
	35.24	Liters
	4	Pecks
	32	Quarts (dry)
Cables	120	Fathoms
Calories (gm)	0.003968	BTU
	0.001	Calories (Kg)
	3.088	Foot pounds
	1.558×10^{-6}	Horse power hours
	4.185	Joules
	0.4265	Kilogram meters
	1.1628×10^{-6}	Kilowatt hours
	0.0011628	Watt hours
Cal (gm)/sec/cm^2/°C/cm	242.13	BTU/Hr/ft^2/°F/ft
Calories (Kg)	3.968	BTU
	1000	Calories (gm)
	3088	Foot pounds
	0.001558	Horse power hours
	4185	Joules
	426.5	Kilogram meters
	0.0011628	Kilowatt hours
	1.1628	Watt hours
Calories (Kg)/Cu meter	0.1124	BTU/Cu foot at 0° C
Cal (Kg)/Hr/M^2/°C/M	0.671	BTU/Hr/ft^2/°F/foot
Calories (Kg)/Kg	1.8	BTU/pound
Calories (Kg)/minute	51.43	Foot pounds/second
	0.09351	Horse power
	0.06972	Kilowatts
Carats (diamond)	200	Milligram
Centares (Centiares)	1	Square meters
Centigram	0.01	Grams
Centiliters	0.01	Liters
Centimeters	0.3937	Inches
	0.032808	Feet
	0.01	Meters
	10	Millimeters
Centimeters of Hg at 32°F	0.01316	Atmospheres
	0.4461	Feet of water at 62° F
	136	Kgs/Square meter
	27.85	Pounds/Square foot
	0.1934	Pounds/Square inch
Centimeters/second	1.969	Feet/minute
	0.03281	Feet/second
	0.036	Kilometers/hour
	0.6	Meters/minute
	0.02237	Miles/hour
	0.0003728	Miles/minute
Centimeters/second^2	0.03281	Feet/second^2
Centipoise	0.000672	Pounds/sec foot
	2.42	Pounds/hour foot
	0.01	Poise
Chains (Gunter's)	4	Rods
	66	Feet
	100	Links

TABLE A.2 *(Continued)*

CONVERSION FACTORS

Multiply	by	To Obtain	Multiply	by	To Obtain
Cheval-vapeur	1	Metric horse power	Cubit	18	Inches
	75	Kilogram meters/second	Days (mean)	1440	Minutes
	0.98632	Horse power		24	Hours
Circular inches	10^6	Circular mils		86,400	Seconds
	0.7854	Square inches	Days (sidereal)	86,164.1	Solar seconds
	785,400	Square mils	Decigrams	0.1	Grams
Circular mils	0.7854	Square mils	Deciliters	0.1	Liters
	10^{-6}	Circular inches	Decimeters	0.1	Meters
	7.854×10^{-5}	Square inches	Degrees (angle)	60	Minutes
Cubic centimeters	3.531×10^{-5}	Cubic feet		0.01745	Radians
	0.06102	Cubic inches		3600	Seconds
	10^{-6}	Cubic meters	Degrees F [less 32]	0.5556	Degrees C
	1.308×10^{-6}	Cubic yards	Degrees F	1 [plus 460]	Degrees F above absolute 0
	0.0002642	Gallons (US)			
	0.001	Liters	Degrees C	1.8 [plus 32]	Degrees F
	0.002113	Pints (liq. US)		1 [plus 273]	Degrees C above absolute 0
	0.001057	Quarts (liq. US)			
	0.0391	Ounces (fluid)	Degrees/second	0.01745	Radians/second
Cubic feet	28,320	Cubic centimeters		0.1667	Revolutions/minute
	1728	Cubic inches		0.002778	Revolutions/second
	0.02832	Cubic meters	Dekagrams	10	Grams
	0.03704	Cubic yards	Dekaliters	10	Liters
	7.48052	Gallons (US)	Dekameters	10	Meters
	28.32	Liters	Diameter (circle)	3.14159265359	Circumference
	59.84	Pints (liq. US)	(approx)	3.1416	
	29.92	Quarts (liq. US)	(approx)	3.14	
	2.296×10^{-5}	Acre feet	(approx)	$\frac{22}{7}$	
	0.803564	Bushels	Diameter (circle)	0.88623	Side of equal square
Cubic feet of water	62.4266	Pounds at 39.2° F		0.7071	Side of inscribed **square**
	62.3554	Pounds at 62° F	Diameter³ (sphere)	0.5236	Volume (sphere)
Cubic feet/minute	472	Cubic centimeters/sec	Diam (major) × diam (minor)	0.7854	Area of ellipse
	0.1247	Gallons (US)/second			
	0.472	Liters/second	Diameter² (circle)	0.7854	Area (circle)
	62.36	Pounds water/min at 62°F	Diameter² (sphere)	3.1416	Surface (sphere)
	7.4805	Gallons (US)/minute	Diam (inches) × RPM	0.262	Belt speed ft/minute
	10,772	Gallons/24 hours	Digits	0.75	Inches
	0.033058	Acre feet/24 hours	Drams (avoirdupois)	27.34375	Grains
Cubic feet/second	646,317	Gallons (US)/24 hours		0.0625	Ounces (avoir.)
	448.831	Gallons/minute		1.771845	Grams
	1.98347	Acre feet/24 hours	Fathoms		Feet
Cubic inches	16.387	Cubic centimeters	Feet	30.48	Centimeters
	0.0005787	Cubic feet		12	Inches
	1.639×10^{-5}	Cubic meters		0.3048	Meters
	2.143×10^{-5}	Cubic yards		$\frac{1}{3}$	Yards
	0.004329	Gallons (US)		0.06061	Rods
	0.01639	Liters	Feet of water at 62	0.029465	Atmospheres
	0.03463	Pints (liq. US)		0.88162	Inches of Hg at 32° F
	0.01732	Quarts (liq. US)		62.3554	Pounds/square foot
Cubic meters	10^6	Cubic centimeters		0.43302	Pounds/square inch
	35.31	Cubic feet		304.44	Kilogram/sq meter
	61,023	Cubic inches			
	1.308	Cubic yards			
	264.2	Gallons (US)			
	1000	Liters			
	2113	Pints (liq. US)			
	1057	Quarts (liq. US)			
Cubic yards	764,600	Cubic centimeters			
	27	Cubic feet			
	46,656	Cubic inches			
	0.7646	Cubic meters			
	202	Gallons (US)			
	764.6	Liters			
	1616	Pints (liq. US)			
	807.9	Quarts (liq. US)			
Cubic yards/minute	0.45	Cubic feet/second	Feet/minute	0.5080	Centimeters/second
	3.367	Gallons (US)/second		0.01667	Feet/second
	12.74	Liters/second		0.01829	Kilometers/hour

TABLE A.2 (*Continued*)

CONVERSION FACTORS

Multiply	by	To Obtain	Multiply	by	To Obtain
Feet/minute	0.3048	Meters/minute	Grains/gallon (US)	17.118	Parts/million
	0.01136	Miles/hour		142.86	Pounds/million gallons (US)
Feet/second	30.48	Centimeters/second	Grams	980.7	Dynes
	1.097	Kilometers/hour		15.43	Grains
	0.5921	Knots		0.001	Kilograms
	18.29	Meters/minute		1000	Milligrams
	0.6818	Miles/hour		0.03527	Ounces (avoir.)
	0.01136	Miles/minute		0.03215	Ounces (troy)
Feet/second²	30.48	Centimeters/second²		0.002205	Pounds
	0.3048	Meters/second²	Grams/centimeter	0.0056	Pounds/inch
Flat of a hexagon	1.155	Distance across corners	Grams/cubic centimeter	62.43	Pounds/cubic foot
Flat of a square	1.414	Distance across corners		0.03613	Pounds/cubic inch
Foot pounds	0.0012861	BTU		4.37	Grains/100 cubic ft
	0.32412	Calories (gm)	Grams/liter	58.417	Grains/gallon (US)
	0.0003241	Calories (Kg)		8.345	Pounds/100 gallons (US)
	5.05 × 10⁻⁷	Horse power hours		0.062427	Pounds/cubic foot
	1.3558	Joules		1000	Parts/million
	0.13826	Kilogram meters	Gravity (g)	32.174	Feet/second²
	3.766 × 10⁻⁷	Kilowatt hours		980.6	Centimeters/second²
	0.0003766	Watt hours	Hand	4	Inches
Foot pounds/minute	0.001286	BTU/minute		10.16	Centimeters
	0.01667	Foot pounds/second	Hectares	2.471	Acres
	3.03 × 10⁻⁵	Horse power		107,639	Square feet
	0.0003241	Calories (Kg)/minute		100	Ares
	2.26 × 10⁻⁵	Kilowatts	Hectograms	100	Grams
Foot pounds/second	0.07717	BTU/minute	Hectoliters	100	Liters
	0.001818	Horse power	Hectometers	100	Meters
	0.01945	Calories (Kg)/minute	Hectowatts	100	Watts
	0.001356	Kilowatts	Hogshead	63	Gallons (US)
Furlong	40	Rods		238.4759	Liters
	220	Yards	Horse power	42.44	BTU/minute
	660	Feet		33,000	Foot pounds/minute
	0.125	Miles		550	Foot pounds/second
	0.2042	Kilometers		1.014	Metric horse power (Cheval vapeur)
Gallons (Imperial)	277.42	Cubic inches		10.7	Calories (Kg)/min
	4.543	Liters		0.7457	Kilowatts
	1.20095	Gallons (US)		745.7	Watts
Gallons (US)	3785	Cubic centimeters	Horse power (boiler)	33,479	BTU/hour
	0.13368	Cubic feet		9.803	Kilowatts
	231	Cubic inches		34.5	Pounds of water evaporated/hour at 212° F
	0.003785	Cubic meters			
	0.004951	Cubic yards	Horse power hours	2546.5	BTU
	3.785	Liters		641,700	Calories (gm)
	8	Pints (liq. US)		641.7	Calories (Kg)
	4	Quarts (liq. US)		1,980,000	Foot pounds
	0.83267	Gallons (Imperial)		2,684,500	Joules
	3.069 × 10⁻⁶	Acre feet		273,740	Kilogram meters
Gallons (US) of water at 62° F	8.3357	Pounds of water		0.7455	Kilowatt hours
				745.5	Watt hours
Gallons (US) of water/minute	6.0086	Tons of water/24 hours	Inches	2.54	Centimeters
				0.08333	Feet
Gallons (US)/minute	0.002228	Cubic feet/second		1000	Mils
	0.13368	Cubic feet/minute		12	Lines
	8.0208	Cubic feet/hour		72	Points
	0.06309	Liters/second	Inches of Hg at 32° F	0.03342	Atmospheres
	3.78533	Liters/minute		345.3	Kilograms/square meter
	0.0044192	Acre feet/24 hours		70.73	Pounds/square foot
Grains	1	Grains (avoirdupois)		0.49117	Pounds/square inch
	1	Grains (apothecary)		1.1343	Feet of water at 62° F
	1	Grains (troy)			
	0.0648	Grams			
	0.0020833	Ounces (troy)			
	0.0022857	Ounces (avoir.)			

TABLE A.2 (*Continued*)
CONVERSION FACTORS

Multiply	by	To Obtain	Multiply	by	To Obtain
Inches of Hg at 32° F	13.6114	Inches of water at 62° F	Kilowatt hours	860,500	Calories (gm)
	7.85872	Ounces/square inch		860.5	Calories (Kg)
				2,655,200	Foot pounds
Inches of water at 62° F	0.002455	Atmospheres		1.341	Horse power hours
	25.37	Kilograms/square meter		3,600,000	Joules
	0.5771	Ounces/square inch		367,100	Kilogram meters
	5.1963	Pounds/square foot		1000	Watt hours
	0.03609	Pounds/square inch	Knots	1	Nautical miles/hour
	0.07347	Inches of Hg at 32° F		1.1516	Miles/hour
Joules	0.00094869	BTU		1.8532	Kilometers/hour
	0.239	Calories (gm)			
	0.000239	Calories (Kg)	Leagues	3	Miles
	0.73756	Foot pounds	Lines	0.08333	Inches
	3.72×10^{-7}	Horse power hours	Links	7.92	Inches
	0.10197	Kilogram meters	Liters	1000	Cubic centimeters
	2.778×10^{-7}	Kilowatt hours		0.03531	Cubic feet
	0.0002778	Watt hours		61.02	Cubic inches
	1	Watt second		0.001	Cubic meters
Kilograms	980,665	Dynes		0.001308	Cubic yards
	2.205	Pounds		0.2642	Gallons (US)
	0.001102	Tons (short)		0.22	Gallons (Imp)
	1000	Grams		2.113	Pints (liq. US)
	35.274	Ounces (avoir.)		1.057	Quarts (liq. US)
	32.1507	Ounces (troy)		8.107×10^{-7}	Acre Feet
Kilogram meters	0.009302	BTU		2.2018	Pounds of water at 62° F
	2.344	Calories (gm)	Liters/minute	0.0005886	Cubic feet/second
	0.002344	Calories (Kg)		0.004403	Gallons (US)/second
	7.233	Foot pounds		0.26418	Gallons (US)/minute
	3.653×10^{-6}	Horse power hours	Meters	100	Centimeters
	9.806	Joules		3.281	Feet
	2.724×10^{-6}	Kilowatt hours		39.37	Inches
	0.002724	Watt hours		1.094	Yards
Kilograms/cubic meter	0.06243	Pounds/cubic foot		0.001	Kilometers
Kilograms/meter	0.6720	Pounds/foot		1000	Millimeters
Kilograms/sq centimeter	14.223	Pounds/sq inch	Meters/minute	1.667	Centimeters/second
	1	Metric atmosphere		3.281	Feet/minute
Kilogram/sq meter	9.678×10^{-5}	Atmospheres		0.05468	Feet/second
	0.003285	Feet of water at 62° F		0.06	Kilometers/hour
	0.002896	Inches of Hg at 32° F		0.03728	Miles/hour
	0.2048	Pounds/square foot	Meters/second	196.8	Feet/minute
	0.001422	Pounds/square inch		3.281	Feet/second
	0.007356	Centimeters of Hg at 32° F		3.6	Kilometers/hour
				0.06	Kilometers/minute
Kiloliters	1000	Liters		2.237	Miles/hour
Kilometers	100,000	Centimeters		0.03728	Miles/minute
	1000	Meters	Microns	10^{-6}	Meters
	3281	Feet		0.001	Millimeters
	0.6214	Miles		0.03937	Mils
	1094	Yards	Mils	0.001	Inches
Kilometers/hour	27.78	Centimeters/second		0.0254	Millimeters
	54.68	Feet/minute		25.4	Microns
	0.9113	Feet/second	Miles	160,934	Centimeters
	16.67	Meters/minute		5280	Feet
	0.6214	Miles/hour		63,360	Inches
	0.5396	Knots		1.609	Kilometers
Kilometers/hr/sec	27.78	Centimeters/sec/sec		1760	Yards
	0.9113	Feet/sec/sec		80	Chains
	0.2778	Meters/sec/sec		320	Rods
Kilowatts	56.92	BTU/minute		0.8684	Nautical miles
	44,250	Foot pounds/minute	Miles/hour	44.70	Centimeters/second
	737.6	Foot pounds/second		88	Feet/minute
	1.341	Horse power		1.467	Feet/second
	14.34	Calories (Kg)/min		1.609	Kilometers/hour
	1000	Watts		0.8684	Knots
Kilowatt hours	3413	BTU		26.82	Meters/minute

TABLE A.2 (*Continued*)
CONVERSION FACTORS

Multiply	by	To Obtain	Multiply	by	To Obtain
Miles/minute	2682	Centimeters/second	Poncelots	100	Kilogram meters/second
	88	Feet/second		1.315	Horse power
	1.609	Kilometers/minute	Pounds (avoirdupois)	16	Ounces (avoir.)
	60	Miles/hour		256	Drams (avoir.)
Millibars	0.000987	Atmosphere		7000	Grains
Milliers	1000	Kilograms		0.0005	Tons (short)
Milligrams	0.001	Grams		453.5924	Grams
	0.01543	Grains		1.21528	Pounds (troy)
				14.5833	Ounces (troy)
Milligrams/liter	1	Parts/million	Pounds (troy)	5760	Grains
Milliliters	0.001	Liters		240	Pennyweights (troy)
Million gals/24 hours	1.54723	Cubic feet/second		12	Ounces (troy)
Millimeters	0.1	Centimeters		373.24177	Grams
	0.03937	Inches		0.822857	Pounds (avoir.)
	39.37	Mils		13.1657	Ounces (avoir.)
	1000	Microns		0.00036735	Tons (long)
				0.00041143	Tons (short)
Miner's inches	1.5	Cubic feet/minute		0.00037324	Tons (metric)
Minutes (angle)	0.0002909	Radians	Pounds of water at 62° F	0.01604	Cubic feet
Nautical miles	6080.2	Feet		27.72	Cubic inches
	1.1516	Miles		0.120	Gallons (US)
Ounces (avoirdupois)	16	Drams (avoir.)	Pounds of water/min at 62° F	0.0002673	Cubic feet/second
	437.5	Grains			
	0.0625	Pounds (avoir.)	Pounds/cubic foot	0.01602	Grams/cubic centimeter
	28.349527	Grams		16.02	Kilograms/cubic meter
	0.9115	Ounces (troy)		0.0005787	Pounds/cubic inch
Ounces (fluid)	1.805	Cubic inches	Pounds/cubic inch	27.68	Grams/cubic centimeter
	0.02957	Liters		27,680	Kilograms/cubic meter
	29.57	Cubic centimeters		1728	Pounds/cubic foot
	0.25	Gills	Pounds/foot	1.488	Kilograms/meter
Ounces (troy)	480	Grains	Pounds/inch	178.6	Grams/centimeter
	20	Pennyweights (troy)	Pounds/hour foot	0.4132	Centipoise
	0.08333	Pounds (troy)		0.004132	Poise grams/sec cm
	31.103481	Grams	Pounds/sec foot	14.881	Poise grams/sec cm
	1.09714	Ounces (avoir.)		1488.1	Centipoise
Ounces/square inch	0.0625	Pounds/square inch	Pounds/square foot	0.016037	Feet of water at 62° F
	1.732	Inches of water at 62° F		4.882	Kilograms/square meter
	4.39	Centimeters of water at 62° F		0.006944	Pounds/square inch
	0.12725	Inches of Hg at 32° F		0.014139	Inches of Hg at 32° F
	0.004253	Atmospheres		0.0004725	Atmospheres
Palms	3	Inches	Pounds/square inch	0.068044	Atmospheres
Parts/million	0.0584	Grains/gallon (US)		2.30934	Feet of water at 62° F
	0.07016	Grains/gallon (Imp)		2.0360	Inches of Hg at 32° F
	8.345	Pounds/million gal (US)		703.067	Kilograms/square meter
				27.912	Inches of water at 62° F
Pennyweights (troy)	24	Grains	Quadrants (angular)	90	Degrees
	1.55517	Grams		5400	Minutes
	0.05	Ounces (troy)		324,000	Seconds
	0.0041667	Pounds (troy)		1.751	Radians
Pints (liq. US)	4	Gills	Quarts (dry)	67.20	Cubic inches
	16	Ounces (fluid)	Quarts (liq. US)	2	Pints (liq. US)
	0.5	Quarts (liq. US)		0.9463	Liters
	28.875	Cubic inches		32	Ounces (fluid)
	473.1	Cubic centimeters		57.75	Cubic inches
Pipe	126	Gallons (US)		946.3	Cubic centimeters
Points	0.01389	Inches	Quintal, Argentine	101.28	Pounds
			Brazil	129.54	Pounds
Poise	0.0672	Pounds/sec foot	Castile, Peru	101.43	Pounds
	242	Pounds/hour foot	Chile	101.41	Pounds
	100	Centipoise	Metric	220.46	Pounds
			Mexico	101.47	Pounds

TABLE A.2 (*Concluded*)

CONVERSION FACTORS

Multiply	by	To Obtain	Multiply	by	To Obtain
Quires	25	Sheets	Square miles	27,878,400	Square feet
				2.590	Square kilometers
Radians	57.30	Degrees		259	Hectares
	3438	Minutes		3,097,600	Square yards
	206,625	Seconds		102,400	Square rods
	0.637	Quadrants		1	Sections
Radians/second	57.30	Degrees/second	Square millimeters	0.01	Square centimeters
	0.1592	Revolutions/second		0.00155	Square inches
	9.549	Revolutions/minute		1550	Square mils
				1973	Circular mils
Radians/second²	573.0	Revolutions/minute²			
	0.1592	Revolutions/second²	Square mils	1.27324	Circular mils
				0.0006452	Square millimeters
Reams	500	Sheets		10^{-6}	Square inches
Revolutions	360	Degrees	Square yards	0.0002066	Acres
	4	Quadrants		9	Square feet
	6.283	Radians		0.8361	Square meters
				3.228×10^{-7}	Square miles
Revolutions/minute	6	Degrees/second			
	0.1047	Radians/second	Stere	1	Cubic meters
	0.01667	Revolutions/second			
			Stone	14	Pounds
Revolutions/minute²	0.001745	Radians/second²		6.35029	Kilograms
	0.0002778	Revolutions/second²			
			Tons (long)	1016	Kilograms
Revolutions/second	360	Degrees/second		2240	Pounds
	6.283	Radians/second		1.12	Tons (short)
	60	Revolutions/minute			
			Tons (metric)	1000	Kilograms
Revolutions/second²	6.283	Radians/second²		2205	Pounds
	3600	Revolutions/minute²		1.1023	Tons (short)
Rods	16.5	Feet	Tons (short)	2000	Pounds
	5.5	Yards		32,000	Ounces
				907.185	Kilograms
Seconds (angle)	4.848×10^{-6}	Radians		0.90718	Tons (metric)
				0.89286	Tons (long)
Sections	1	Square miles			
			Tons of refrigeration	12,000	BTU/hour
Side of a square	1.4142	Diameter of inscribed circle		288,000	BTU/24 hours
	1.1284	Diameter of circle with equal area	Tons of water/24 hours at 62° F	83.33	Pounds of water/hour
				0.16510	Gallons (US)/minute
Span	9	Inches		1.3263	Cubic feet/hour
Square centimeters	0.001076	Square feet	Watts	0.05692	BTU/minute
	0.1550	Square inches		44.26	Foot pounds/minute
	0.0001	Square meters		0.7376	Foot pounds/second
	100	Square millimeters		0.001341	Horse power
				0.01434	Calories (Kg)/minute
Square feet	2.296×10^{-5}	Acres		0.001	Kilowatts
	929.0	Square centimeters		1	Joule/second
	144	Square inches			
	0.0929	Square meters	Watt hours	3.413	BTU
	3.587×10^{-8}	Square miles		860.5	Calories (gm)
	0.1111	Square yards		0.8605	Calories (Kg)
				2655	Foot pounds
Square inches	6.452	Square centimeters		0.001341	Horse power hours
	0.006944	Square feet		3600	Joules
	645.2	Square millimeters		367.1	Kilogram meters
	1.27324	Circular inches		0.001	Kilowatt hours
	1,273,239	Circular mils			
	1,000,000	Square mils	Watts/square inch	8.2	BTU/square foot/minute
Square kilometers	247.1	Acres		6373	Foot pounds/sq ft/minute
	10,760,000	Square feet		0.1931	Horse power/square foot
	1,000,000	Square meters			
	0.3861	Square miles	Yards	91.44	Centimeters
	1,196,000	Square yards		3	Feet
				36	Inches
Square meters	0.0002471	Acres		0.9144	Meters
	10.764	Square feet		0.1818	Rods
	1.196	Square yards			
	1	Centares	Year (365 days)	8760	Hours
Square miles	640	Acres			

TABLE A.3
DECIMAL EQUIVALENTS OF WIRE AND SHEET METAL GAUGE NUMBERS
(Reproduced by permission from *Piping Design and Engineering,* Grinnell Co., Inc., Providence, R.I., 1951.)

Gauge No.	American wire gauge, or Brown and Sharpe (for copper wire)	Steel wire gauge, or Washburn and Moen or Roebling (for steel wire)	Birmingham wire gauge (B.W.G.) (for steel wire or sheets)	Stubs steel wire gauge	U.S. standard gauge for sheet metal (iron and steel) 480 lb per cu ft	AISI inch equivalent for U.S. steel sheet thickness
0000000		0.4900			0.500	
000000		0.4615			0.469	
00000		0.4305			0.438	
0000	0.460	0.3938	0.454		0.406	
000	0.410	0.3625	0.425		0.375	
00	0.365	0.3310	0.380		0.344	
0	0.325	0.3065	0.340		0.312	
1	0.289	0.2830	0.300	0.227	0.281	
2	0.258	0.2625	0.284	0.219	0.266	
3	0.229	0.2437	0.259	0.212	0.250	0.2391
4	0.204	0.2253	0.238	0.207	0.234	0.2242
5	0.182	0.2070	0.220	0.204	0.219	0.2092
6	0.162	0.1920	0.203	0.201	0.203	0.1943
7	0.144	0.1770	0.180	0.199	0.188	0.1793
8	0.128	0.1620	0.165	0.197	0.172	0.1644
9	0.114	0.1483	0.148	0.194	0.156	0.1495
10	0.102	0.1350	0.134	0.191	0.141	0.1345
11	0.091	0.1205	0.120	0.188	0.125	0.1196
12	0.081	0.1055	0.109	0.185	0.109	0.1046
13	0.072	0.0915	0.095	0.182	0.094	0.0897
14	0.064	0.0800	0.083	0.180	0.078	0.0747
15	0.057	0.0720	0.072	0.178	0.070	0.0673
16	0.051	0.0625	0.065	0.175	0.062	0.0598
17	0.045	0.0540	0.058	0.172	0.056	0.0538
18	0.040	0.0475	0.049	0.168	0.050	0.0478
19	0.036	0.0410	0.042	0.164	0.0438	0.0418
20	0.032	0.0348	0.035	0.161	0.0375	0.0359
21	0.0285	0.0317	0.032	0.157	0.0344	0.0329
22	0.0253	0.0286	0.028	0.155	0.0312	0.0299
23	0.0226	0.0258	0.025	0.153	0.0281	0.0269
24	0.0201	0.0230	0.022	0.151	0.0250	0.0239
25	0.0179	0.0204	0.020	0.148	0.0219	0.0209
26	0.0159	0.0181	0.018	0.146	0.0188	0.0179
27	0.0142	0.0173	0.016	0.143	0.0172	0.0164
28	0.0126	0.0162	0.014	0.139	0.0156	0.0149
29	0.0113	0.0150	0.013	0.134	0.0141	0.0135
30	0.0100	0.0140	0.012	0.127	0.0125	0.0120
31	0.0089	0.0132	0.010	0.120	0.0109	0.0105
32	0.0080	0.0128	0.009	0.115	0.0102	0.0097
33	0.0071	0.0118	0.008	0.112	0.0094	0.0090
34	0.0063	0.0104	0.007	0.110	0.0086	0.0082
35	0.0056	0.0095	0.005	0.108	0.0078	0.0075
36	0.0050	0.0090	0.004	0.106	0.0070	0.0067
37	0.0045	0.0085		0.103	0.0066	0.0064
38	0.0040	0.0080		0.101	0.0062	0.0060
39	0.0035	0.0075		0.099		
40	0.0031	0.0070		0.097		
41		0.0066		0.095		
42		0.0062		0.092		
43		0.0060		0.088		
44		0.0058		0.085		
45		0.0055		0.081		
46		0.0052		0.079		
47		0.0050		0.077		
48		0.0048		0.075		
49		0.0046		0.072		
50		0.0044		0.069		

TABLE A.4

STANDARD DIMENSIONS OF PIPE, VALVES, AND FITTINGS

(Reproduced by permission of C. F. Braun & Co., Engineers and Contractors.)

PIPE SIZE NOM	PIPE SIZE OD	WALL SCHEDULE 40	WALL SCHEDULE 80	FLANGES WELD NECK LENGTH (including raised face) 150	WELD NECK LENGTH 300	ALL TYPES THICK (including raised face) 150	THICK 300	OD 150	OD 300	NO. HOLES AND BOLT DIA. 150	NO. HOLES AND BOLT DIA. 300
1/2	.840	.109	.147	1 7/8	2 1/16	7/16	9/16	3 1/2	3 3/4	4 – 1/2	4 – 1/2
3/4	1.050	.113	.154	2 1/16	2 1/4	1/2	5/8	3 7/8	4 5/8	4 – 1/2	4 – 5/8
1	1 5/16	.133	.179	2 3/16	2 7/16	9/16	11/16	4 1/4	4 7/8	4 – 1/2	4 – 5/8
1 1/2	1 7/8	.145	.200	2 7/16	2 11/16	11/16	13/16	5	6 1/8	4 – 1/2	4 – 3/4
2	2 3/8	.154	.218	2 1/2	2 3/4	3/4	7/8	6	6 1/2	4 – 5/8	8 – 5/8
2 1/2	2 7/8	.203	.276	2 3/4	3	7/8	1	7	7 1/2	4 – 5/8	8 – 3/4
3	3 1/2	.216	.300	2 3/4	3 1/8	15/16	1 1/8	7 1/2	8 1/4	4 – 5/8	8 – 3/4
4	4 1/2	.237	.337	3	3 3/8	15/16	1 1/4	9	10	8 – 5/8	8 – 3/4
6	6 5/8	.280	.432	3 1/2	3 7/8	1	1 7/16	11	12 1/2	8 – 3/4	12 – 3/4
8	8 5/8	.322	.500	4	4 3/8	1 1/8	1 5/8	13 1/2	15	8 – 3/4	12 – 7/8
10	10 3/4	.365	.593	4	4 5/8	1 3/16	1 7/8	16	17 1/2	12 – 7/8	16 – 1
12	12 3/4	.406	.687	4 1/2	5 1/8	1 1/4	2	19	20 1/2	12 – 7/8	16 – 1 1/8
14	14	.437	.750	5	5 5/8	1 3/8	2 1/8	21	23	12 – 1	20 – 1 1/8
16	16	.500	.843	5	5 3/4	1 7/16	2 1/4	23 1/2	25 1/2	16 – 1	20 – 1 1/4
18	18	.562	.937	5 1/2	6 1/4	1 9/16	2 3/8	25	28	16 – 1 1/8	24 – 1 1/4
20	20	.593	1.031	5 11/16	6 3/8	1 11/16	2 1/2	27 1/2	30 1/2	20 – 1 1/8	24 – 1 1/4
24	24	.687	1.218	6	6 5/8	1 7/8	2 3/4	32	36	20 – 1 1/4	24 – 1 1/2

NOM	FLANGED VALVES GATE * 150	GATE * 300	GATE 150 L OPEN	GATE 150 D	GATE 300 L OPEN	GATE 300 D	GLOBE * 150	GLOBE * 300	GLOBE 150 L OPEN	GLOBE 150 D	GLOBE 300 L OPEN	GLOBE 300 D	SWING CHECK * 150	SWING CHECK * 300	CONTROL VALVES (AIR TO CLOSE) 150	CONTROL VALVES 300	CONTROL VALVES L **	CONTROL VALVES D
1/2		■6 1/2			■6 3/8	■3 1/2		•6 1/2			•7 3/8	•3 1/2		+6 1/2		▲7 1/2	▲20	▲11
3/4		■7 1/2			■7 13/16	■4		•7 1/2			•9	•4		+7 1/2	7 1/4	7 5/8	20	11
1		■8 1/2			■9 3/16	■5		•8 1/2			•9 3/4	•5		+8 1/2	7 1/4	7 3/4	20	11
1 1/2	••6 1/2	■9 1/2	••14 7/8	••7	■12	■6		•9 1/2			•12 3/16	•7		+9 1/2	8 3/4	9 1/4	20 3/4	11
2	7	8 1/2	16 1/2	8	18 1/2	8	8	10 1/2	13 3/4	8	17 3/4	9	8	10 1/2	10	10 1/2	23 1/2	13
2 1/2	7 1/2	9 1/2	18	8	20 1/4	8	8 1/2	11 1/2	14 1/2	8	19	10	8 1/2	11 1/2	10 7/8	11 1/2	24	13
3	8	11 1/8	20 3/4	9	25	9	9 1/2	12 1/2	16 1/2	9	20 1/2	10	9 1/2	12 1/2	11 3/4	12 1/2	29	15
4	9	12	25 3/4	10	31	10	11 1/2	14	19 3/4	10	24 3/4	14	11 1/2	14	13 7/8	14 1/2	29 1/4	15
6	10 1/2	15 7/8	35 1/4	14	38 1/2	14	16	17 1/2	24 1/2	12	29 3/4	18	14	17 1/2	17 3/4	18 5/8	35	17 1/2
8	11 1/2	16 1/2	43	14	47	16	19 1/2	22	26	16	35 1/2	24	‡19 1/2	21	21 3/8	22 3/8	39 3/4	20 3/4
10	13	18	52	18	56 1/2	20							‡24 1/2	24 1/2	24 5/8	27 7/8	41 1/4	20 3/4
12	14	19 3/4	60 1/2	18	64 1/4	20							‡27 1/2	28	28 3/4	30 1/4		
14	15	30	70 1/4	22	74 3/4	27									▲30 1/2	▲32		
16	16	33	79 3/4	24	80 1/2	27									▲36	▲37 5/8		
18	17	36	89	27	91	30												
20	18	39	97 1/4	30	100 1/2	36												
24	20	45	112 3/4	30	120 1/2	36												

NOM	WELD FITT. CENTER TO END 45°	CENTER TO END T	RED. END TO END	SCREWED FITTINGS SWAGES CONCENTRIC AND ECCENTRIC BY	SWAGES L	THREAD PENETRATION TIGHT JOINT	TEE ELL ℄ TO END 300 MI	TEE ELL ℄ TO END 3000 FS	UNION END TO END 300 MI	UNION END TO END FS	NOMINAL PIPE SIZE
1/2				1/8 THRU 3/8	2 3/4	1/2	1 1/4	1 5/16	1 15/16	1 15/16	1/2
3/4	9/16	1 1/8		1/8 THRU 1/2	3	9/16	1 7/16	1 1/2	2 1/4	2 1/4	3/4
1	7/8	1 1/2	2	1/8 THRU 3/4	3 1/2	11/16	1 5/8	1 3/4	2 7/16	2 7/16	1
1 1/2	1 1/8	2 1/4	2 1/2	1/4 THRU 1 1/4	4 1/2	11/16	2 1/8	2 3/8	3	3	1 1/2
2	1 3/8	2 1/2	3	1/4 THRU 1 1/2	6 1/2	3/4	2 1/2	2 1/2	3 3/8	3 3/8	2
2 1/2	1 3/4	3	3 1/2	1/4 THRU 2	7	15/16	2 15/16	3 3/8	4	4	2 1/2
3	2	3 3/8	3 1/2	1/4 THRU 2 1/2	8	1	3 3/8	3 3/4	4 5/16	4 5/16	3
4	2 1/2	4 1/8	4	1/4 THRU 3 1/2	9	1 1/8	4 1/2	4 1/2			4
6	3 3/4	5 5/8	5 1/2	1/2 THRU 5	12	1 5/16	6 1/4				6
8	5	7	6	2 THRU 7	13	1 7/16					8
10	6 1/4	8 1/2	7	2 THRU 8	15	1 5/8					10
12	7 1/2	10	8	2 THRU 10	16	1 3/4					12
14	▲	▲	13	2 THRU 12	17						14
16	▲	▲	14	2 THRU 14	18						16
18	▲	▲	15	2 THRU 16	20						18
20	▲		20	2 THRU 18	22						20
24	▲		20	6 THRU 20	24						24

* FOR 150 LB & 300 LB RTJ ADD DEPTH OF GROOVES
\+ 600 LB CRANE #3686X
‡ 150 LB CRANE # 147
** FOR AIRFINS ADD ABOUT 5" TO L DIM.

■ 600 LB CRANE #3615
• 600 LB CRANE #3656
▲ USE SUPPLIERS CATALOG DIMENSIONS
•• PACIFIC VALVE #150

DIMENSIONS INSIDE DOUBLE LINE ARE ASA STANDARD

Index